FOREST MEASUREMENTS

McGraw-Hill Series in Forest Resources

Avery and Burkhart: Forest Measurements
Dana and Fairfax: Forest and Range Policy
Daniel, Helms, and Baker: Principles of Silviculture
Davis and Johnson: Forest Management
Duerr: Introduction to Forest Resource Economics
Ellefson: Forest Resources Policy: Process, Participants, and Programs
Harlow, Harrar, Hardin, and White: Textbook of Dendrology
Knight and Heikkenen: Principles of Forest Entomology
Laarman and Sedjo: Global Forests: Issues for Six Billion People
Panshin and De Zeeuw: Textbook of Wood Technology
Sharpe, Hendee, and Sharpe: Introduction to Forestry
Sinclair: Forest Products Marketing
Stoddart, Smith, and Box: Range Management

Walter Mulford was Consulting Editor of this series from its inception in 1931 until January 1, 1952.

Henry J. Vaux was Consulting Editor of this series from January 1, 1952, until July 1, 1976.

Paul V. Ellefson, University of Minnesota, is currently our Consulting Editor.

Consolidated Papers tree farm, Wisconsin Rapids, WI. (Photograph by Dick Durrance, II.)

FOREST MEASUREMENTS

FOURTH EDITION

Thomas Eugene Avery

Independent Resource Consultant

Harold E. Burkhart

Virginia Polytechnic Institute and State University

Boston, Massachusetts   Burr Ridge, Illinois
Dubuque, Iowa   Madison, Wisconsin   New York, New York
San Francisco, California   St. Louis, Missouri

This book was set in Times Roman by The Clarinda Company.
The editors were Anne C. Duffy and John M. Morriss;
the production supervisor was Friederich W. Schulte.
The cover was designed by Rafael Hernandez.
Cover photo by Dick Durrance, II.
Project supervision was done by The Total Book.
New drawings were done by Hadel Studio.

McGraw-Hill

A Division of The McGraw·Hill Companies

Forest Measurements

7 8 9 10 11 12 13 14 BKMBKM 9 9 8 7

ISBN-0-07-002556-8

Library of Congress Cataloging-in-Publication Data

Avery, Thomas Eugene.
    Forest measurements / Thomas Eugene Avery, Harold E. Burkhart.—
4th ed.
        p.   cm.—(McGraw-Hill series in forest resources)
    Includes bibliographical references and index.
    ISBN 0-07-002556-8
    1. Forests and forestry—Mensuration.   I. Burkhart, Harold E.
II. Title   III. Series.
SD555.A93   1994
634.9′285—dc20   93-3783

ABOUT THE AUTHORS

THOMAS EUGENE AVERY (B.S. University of Georgia, M.F. Duke University, Ph.D. University of Minnesota) is the author of three university textbooks in current use and has published more than 60 technical articles in professional journals. He has worked for the private forest industry, the U.S. Forest Service, and several major universities, and was formerly Head of the Department of Forestry at the University of Illinois. Dr. Avery has had professional forestry experience in West Germany, New Zealand, and Australia; he has held national office in the Society of American Foresters and the American Society of Photogrammetry. In 1987 he received the Alan Gordon Memorial Award from the American Society of Photogrammetry, and he has been listed in *American Men of Science* and *Who's Who in America*. At present, Dr. Avery is a semiretired consultant and writer and spends most of his time in Texas and Florida.

HAROLD E. BURKHART holds a B.S. degree in forestry from Oklahoma State University and M.S. and Ph.D. degrees in forest biometrics from the University of Georgia. He has been a faculty member in the Department of Forestry, Virginia Polytechnic Institute and State University since 1969. From 1976–1977, Dr. Burkhart was Senior Research Fellow at the Forest Research Institute in Rotorua, New Zealand. He has published extensively in professional journals on the subjects of forest growth and yield prediction, and on forest inventory and sampling. His contributions to forestry education have been recognized through awards from several organizations, including the International Union of Forestry Research Organizations Scientific Achievement Award, the Virginia Academy of Science J. Shelton Horsley Research Award, the Virginia Tech Alumni Award for Research Excellence, the State Council of Higher Education for Virginia Outstanding Faculty Award, and the Society of American Foresters Barrington Moore Memorial Award. A former editor of the journal *Forest Science*, he is a Fellow in the American Association for the Advancement of Science and the Society of American Foresters.

CONTENTS IN BRIEF

CONTENTS

PREFACE

The fourth edition of *Forest Measurements*, prepared mainly by the coauthor, is intended for introductory courses in forest measurements. Emphasis is on the measurement of timber, with detailed coverage on measuring products cut from tree boles, measuring attributes of standing trees, inventorying volumes of forest stands, and predicting growth of individual trees and stands of trees. Background information on statistical methods, sampling designs, land measurements, and use of aerial photographs is also provided. An introduction to assessing range, wildlife, water, and recreation resources associated with forested lands comprises the last chapter. The measurement principles and techniques discussed apply to any inventory that includes assessment of the tree overstory, regardless of whether the inventory is conducted for timber, range, wildlife, watershed, recreation, or other management objectives.

With an introductory text of this nature, the final arbiters of what should and should not be included are the course instructors who adopt and use the book for their classes. Thus, the contents of this fourth edition were determined largely through a detailed questionnaire sent to forestry instructors in the United States and Canada. Most of the respondents were pleased with the third edition format, but several requested expansion or addition of specific topics. We have accommodated requests for expanded coverage where possible.

New timber-measurement material includes log grading, tarif tables, and regeneration surveys. In addition, we divided the third edition chapter on sampling and estimation into two chapters, one on statistical methods (Chap. 2) and another on sampling designs (Chap. 8), while greatly expanding the scope and coverage of sampling designs. The material on site, stocking, and stand density has also been greatly expanded, and a chapter on assessing range, wildlife, water, and recreation resources was added. The appendix has been enhanced with material on basic mathematical operations, thus enabling students to review this subject matter readily.

As with previous editions, we have attempted to present the text in a straightforward fashion that is easily grasped by students. It is presumed that readers will have some background in algebra and plane trigonometry; a prior knowledge of basic statistics and sampling methods will also be helpful, although basic concepts are presented herein. A knowledge of calculus, while not essential, will be useful for some of the material. Explanations which assume a background in calculus are placed in separate sections which can be omitted without loss of continuity.

English units of measurement are used, although metric equivalents or examples are also given in some instances. It is virtually impossible to give *both* systems equal treatment, because many basic tree measurements are not directly comparable in English and metric units.

Where masculine pronouns are used for succinctness, they are intended to refer to both males and females.

We would like to extend our thanks to those instructors in the United States and Canada who responded to our questionnaire; their suggestions ultimately determined the contents of this book. We are also grateful to the Literary Executor of the late Sir Ronald A. Fisher, F.R.S., to Dr. Frank Yates, F.R.S., and to Longman Group Ltd., London, for permission to reprint Appendix Table 6 (The Distribution of *t*) from their book *Statistical Tables for Biological, Agricultural, and Medical Research* (6th edition, 1974).

Chapter 17, "Assessing Rangeland, Wildlife, Water, and Recreational Resources," is a condensed and updated treatment of selected chapters from Part Three of the second edition (published in 1975 under the title *Natural Resources Measurements*). The contributions of C. Roger Hungerford, Robert L. Beschta, Peter F. Ffolliott, David A. King, and A. Jay Schultz to the topics included from the second edition are gratefully acknowledged.

The fourth edition manuscript was reviewed by Loukas G. Arvanitis, University of Florida; John Bell, Oregon State University; Vincent S. Cegelka, Oregon State Community College; Craig J. Davis, Syracuse College of Environmental Science and Forestry; Paul V. Ellefson, University of Minnesota; George Gertner, University of Illinois; W. Groman, Northern Arizona University; Valerie Le May, University of B.C., Canada; Charles C. Myers, Southern Illinois University; and Charles T. Stiff, University of Idaho. Their suggestions resulted in improvements to the final manuscript.

Colleagues at Virginia Polytechnic Institute and State University were very helpful during the preparation of the fourth edition. Special thanks are extended to Timothy G. Gregoire and Richard G. Oderwald for their evaluation of the third edition and their assistance with preparation of the fourth edition manuscript. James L. Smith provided valuable advice on the sections dealing with land measurements, aerial photography, and geographic information systems. In addition, Dean F. Stauffer reviewed the section on measuring wildlife resources, W. Michael Aust evaluated the material on measuring water resources, and Gregory J. Buhyoff and Daniel R. Williams commented on the treatment of measuring recreational resources. Finally, for expert assistance with typing and other details of manuscript preparation, we thank Gerry Kwiatkowski.

Thomas Eugene Avery

Harold E. Burkhart

INTRODUCTION

1-1 Purpose of Book This book is intended for introductory courses in forest measurements. Although a "how-to-do-it" approach is employed, there are still many measurement problems for which no satisfactory solutions exist. Furthermore, there is room for considerable improvement in currently employed techniques and instruments. During recent years, new ideas have been responsible for such practices as weight scaling of wood, 3P sampling, and adaptation of electronic computers to mensurational problems. To a large degree, however, we are still measuring timber volumes, tree form, growth, cull factors, and mortality much as foresters have done for decades. The continued need for personnel with imagination and inventiveness is clearly apparent.

1-2 Need for Measurements Management of forested land requires knowledge of the location and current volume of timber resources. Because forests are dynamic, biological systems, estimates of growth for various management strategies are also required. Forest measurements can be considered a part of forest management; the role of measurements is to supply the numerical data required to make prudent management decisions.

The field of forest measurements is concerned with direct measurements, sampling, and prediction. Direct measurements require appropriate use of instruments to obtain the desired data. Examples of direct measurements are measuring tree diameter at breast height using calipers and measuring tree height using hypsometers. Because forested lands are typically extensive in area and the property of interest is likely to contain tens of thousands of trees, foresters com-

1

monly measure only a sample of the trees and then expand the sample values appropriately to obtain estimates for the population of interest. Prediction also is an important aspect of forest measurements. For instance, the weight of a standing tree cannot be measured directly, but it can be predicted using easily measured tree attributes such as diameter at breast height and total tree height. The science of statistics, in particular sampling techniques and regression analysis (quantifying associations between variables), plays a central role in forest measurements. Mathematics is fundamental to statistics and forest measurements, and with the vast amounts of data and complexity of the analyses involved, computer science has become an integral component.

This book is concerned with the measurement of the tree overstory on forested land. The coverage includes measuring products cut from tree boles, measuring attributes (diameter, height, form, age) of standing trees, quantifying stand characteristics (volumes, weights, etc.), and measuring past growth and predicting future growth of individual trees and stands of trees. Regardless of the land management objectives—timber, wildlife, recreation, watershed, or a combination of these resources—the timber overstory must be quantified for informed decision making. Forest cover is an important part of wildlife habitat, and the understory component can often be successfully related (using regression analysis) to the overstory characteristics. The recreation potential of wildland is a function of many variables, including the timber overstory. Water yields are related to the composition and density of the tree canopy. The sampling methods and measurement principles discussed in this book are applicable to all natural resource management situations that require quantitative information about the tree component of the land base. While an in-depth treatment of specialized techniques for sampling and measuring wildlife populations, recreation resources, water yields, and other resources associated with forested land is outside the scope of a text primarily concerned with tree measurements, an introduction to these topics is provided (Chap. 17). Additional information on measurement of nontimber resources associated with forests can be obtained from the *Forestry Handbook* (Wenger, 1984), *The Wildlife Management Techniques Manual* (Schemnitz, 1980), and textbooks and references on subjects such as wildlife, recreation, range, and watershed management.

1-3   Measurement Cost Considerations   In almost all resource inventories, cost factors are of primary importance; the forester must continually seek out more efficient methods for counting, measuring, and appraisal. The basic objective of most inventories is to obtain an estimate of acceptable statistical precision for the lowest possible expenditure. To achieve this objective requires a sound knowledge of sampling methods, because once the specific needs of management have been determined, the resource inventory becomes essentially a sampling problem.

The measurement of various resource parameters adds no real value to the materials or benefits being assessed; therefore, such measurements are regarded as service functions rather than control functions. Measuring techniques must be subordinate to the productive or beneficial phases of an operation, for the operation itself cannot be modified just to accommodate an inventory requirement. For example, every visitor to a crowded public campground cannot be delayed and required to complete a detailed questionnaire on recreational preferences— nor can a sawmill be shut down in order to measure or weigh a recent delivery of logs. Instead, an appropriate sampling scheme must be designed and employed to obtain the essential resource measurements without disrupting normal activities.

It is an obvious though commonly overlooked fact that the amount expended for a given inventory task should be geared to the value of the products or services being measured. Also, the nearer one approaches the finished product or ultimate benefit, the greater can be the allowable cost of measurement. Thus the measurement of high-quality black walnut trees, which may be worth several thousand dollars each, justifies a much greater unit expense than the assessment of small pine trees for pulpwood. Similarly, the value of finished lumber warrants a greater inventory cost than the scaling of logs. Forest managers who become "cost conscious" early in their careers have an attribute that will be highly respected by their employers.

1-4 Scales of Measurement In the broadest sense, measurement is the assignment of numerals to objects or events according to rules. The fact that numerals can be assigned under different rules leads to different kinds of scales and different kinds of measurement (Stevens, 1946).

Four scales of measurement—nominal, ordinal, interval, and ratio—are recognized. The *nominal scale* is used to number objects for identification. For instance, one might develop numerical codes for species identification in a forest inventory. Because the numbers are assigned only for identification purposes, no meaningful analyses (except perhaps frequency of occurrence) can be performed on the numerical data.

The *ordinal scale* is used to express rank or position in a series. Numerical codes of 1, 2, 3, and 4 could be used to designate the tree-crown classes dominant, codominant, intermediate, and suppressed, respectively. When applying ordinal scales, the successive intervals on the scale are not necessarily equal, nor can one necessarily infer that an equal difference on the scale (e.g., a difference of one unit) means the same thing at all positions along the scale. However, the rank ordering does have meaning when an ordinal scale is used to quantify phenomena such as tree-crown classes, lumber grades, or site-quality classes.

The *interval scale* involves a series of graduations marked off at uniform intervals from an arbitrary origin. The zero point on an interval scale is a matter of convention or convenience. An example of an interval scale is temperature as measured on the Celsius or Fahrenheit scale. Equal intervals of temperature are scaled off from an arbitrary zero agreed upon for each scale.

Ratio scales are the ones most commonly applied in forest measurements. For ratio scales, as with interval scales, there is equality of intervals between successive points on the scale; however, an absolute zero is always implied. Quantities such as the height of trees, the volume per unit area of stands, and the weight of truckloads of logs are measured on a ratio scale.

Foresters routinely take measurements with different scales. For instance, on a forest ·inventory plot a forester might record the timber type using a nominal scale, the site productivity using an ordinal scale, the date of stand origin using an interval scale, and the height of a tree using a ratio scale. One cannot legitimately perform all mathematical operations on measurements from the various scales. The analysis and interpretation of data must take into account the measurement scale. From the standpoint of arithmetic operations, only counting is appropriate for nominal- or ordinal-scale measurements. Counting, addition, subtraction, multiplication, and division are all appropriate for data obtained on interval and ratio scales. Percentage changes are also permissible for ratio-scale measurement data, because there is a true zero point.

1-5 English versus Metric Systems In spite of its obvious complexities and disadvantages, the English system persists as the primary basis for measurements in the United States. The more logical metric system, devised and adopted in France around 1790, has gained limited acceptance in scientific research, but foresters are still surveying by feet and acres rather than meters and hectares. Bills requiring universal adoption of the metric system have been introduced several times in the Congress of the United States, but none have yet been enacted into law.

Admittedly, an abrupt changeover to the metric system would result in considerable confusion for an extended period of time. For example, the conversion of real estate records alone would require years of revising deeds and property descriptions; highway markers and automobile speedometers would require changes from miles to kilometers, and so on. The myriad of problems that would be generated seems to assure that adoption of the metric system in this country will proceed gradually.

Although the English system appears to be grounded in concepts of human anatomy, the metric system was formulated from geodetic measurements. The fundamental metric unit, the meter, was originally defined as being equal to one ten-millionth of the meridional distance from the equator to the earth's poles. In terms of English units, the meter is approximately 39.37 in. in length or slightly longer than 1 yd. Following are several common equivalents for converting English to metric units and vice versa:

Converting English units to metric system	
1 in. or 1000 mils	= 2.5400 cm
1 ft or 12 in.	= 30.4800 cm
1 yd or 3 ft	= 0.9144 m
1 U.S. statute mile or 5280 ft	= 1.6093 km
1 acre or 43,560 sq ft	= 0.4047 ha
1 cu ft or 1728 cu in.	= 0.0283 m^3

Converting metric units to English system	
1 cm or 10 mm	= 0.3937 in.
1 dm or 10 cm	= 3.9370 in.
1 m or 10 dm	= 39.3700 in.
1 km or 1000 m	= 0.6214 U.S. statute mile
1 ha or 10,000 m^2	= 2.4710 acres
1 m^3 or 1,000,000 cm^3	= 35.3147 cu ft

1-6 Abbreviations and Symbols In many scientific disciplines, there are periodic attempts to standardize the nomenclature, symbols, and abbreviations associated with various quantities. Unfortunately, symbols adopted for one discipline may have entirely different connotations in another scientific field. In this book, we have attempted to employ abbreviations and symbols commonly found in forestry literature in the United States.

In accordance with the publisher's standards, abbreviations and symbols for measurement units are used without periods, except when they spell a word (e.g., in. for inch). With this provision in mind, the more common symbols and abbreviations used herein are as follows:

Abbreviation or symbol	Meaning
B, b	cross-sectional areas of logs or bolts
BA, ba	basal area
BAF, baf	basal-area factor (point sampling)
bd ft	board feet
cd	cord
CFI	continuous forest inventory
cu ft	cubic feet
D, d	tree or log diameter (at any specified point)
dbh	diameter breast height
dib	diameter inside bark
dob	diameter outside bark
f	frequency (statistical notation)
H, h	height

Abbreviation or symbol	Meaning
L, l	log or bolt length
M	thousand
MBF	thousand board feet
N, n	number of (statistical notation)
RF	representative fraction
sp gr	specific gravity
V, v	volume

1-7   Preparation of Graphs   The presence of a meaningful relationship between two variables can be quickly and clearly depicted by plotting paired values. Ordinary numerical tabulations can be immediately visualized and interpreted, and trends can be established. Furthermore, errors and abnormal values are easily detected, and minor irregularities in a relationship may often be eliminated by establishing a curve through a series of plotted points.

In the plotting of graphical data, independent variables are placed on the horizontal, or *X*, axis, and dependent variables are plotted along the vertical, or *Y*, axis. Measurements along the *X* axis are known as *abscissas*; those on the *Y* axis are termed *ordinates*. Graduations for the *X* and *Y* axes need not be identical; instead, each scale may be expanded to the maximum degree in keeping with the

FIGURE 1-1
Graph of tree volume-dbh relationship for 32 red pines in Chippewa County, Michigan.

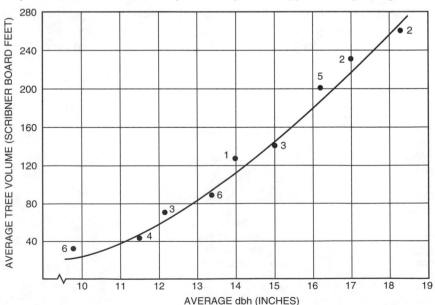

ranges of data that must be accommodated. Though not absolutely essential, it is often desirable to arrange each scale to show the graph origin, i.e., the *zero-zero* coordinate point. Other general rules of graphical presentation are as follows:

1 Scale units and complete identifications of variables should be clearly lettered on each axis. All labels should be oriented for easy reading, i.e., as illustrated by the graph in Fig. 1-1.

2 Plotted points should be denoted by dots, small circles, or other appropriate symbols, and weights (frequency) should be indicated for each point.

3 When the graphs are hand-drafted, freehand curves may be sketched in as guides, but all final curves should be drawn with the ruling edge of a flexible spline or French curve.

4 Each graph should carry a figure number and a complete descriptive title.

1-8 Preparation of Technical Reports As a professional group, foresters are sometimes inclined to minimize the value of neat, concise, and well-written technical reports. Yet in many instances, such reports may provide the only concrete evidence of work accomplished; thus they may constitute the prime basis for judgment of field proficiency by supervisors.

It goes without saying that one must be more than an accomplished grammarian; no amount of flowing penmanship can compensate for deficiencies in fieldwork and data collection. Nevertheless, the importance of producing technically accurate and grammatically correct reports can hardly be overemphasized.

Whenever feasible, all but the most routine reports should be typewritten. Figures and tables are preferably placed on separate pages and numbered consecutively. Line drawings, graphs, and charts should be drawn in black ink and presented on an appropriate drafting medium. Although a single format cannot be expected to meet the requirements for all reports, the following outline may prove useful for student term papers or technical reports on assigned experiments:

Title page Title in centered caps, followed by author's name. Lower part of page should show location of study (e.g., Cripple Creek National Forest) and the date (month and year completed).

Table of contents Listing of chapter headings and major subdivisions, along with corresponding page numbers.

Introduction Statement of the problem, justification and importance of the study, specific objectives, and practical considerations.

Review of previous work A concise, critical review of published literature bearing on the problem, including a statement on the relationship of the present study to previous research.

The study area Location of the study and a description of the area involved (e.g., physiography, forest types, site conditions, climatic factors, legal description, size of area, ownership, management or silvicultural history).

Collection of field data or laboratory procedure   For some studies "Design of the experiment" may be a more appropriate heading. List all data collected, arrangement of field samples, special instruments or techniques employed, illustration of field forms, size of crews, time or expense involved, and special problems encountered.

Analysis of results   Compilation of field data, statistical procedures, and presentation and discussion of results.

Summary and conclusions   A brief synopsis of the study undertaken, results obtained, and implications of the findings. For some types of reports, a brief summary or abstract may be required at the beginning of the discussion, i.e., preceding the introduction.

Literature cited   Arranged in standard form according to an acceptable style manual or in conformance with requirements of a specific technical publication. (See references at end of chapter.)

Appendix   Copies of field forms and/or original raw data are often included here. Detailed statistical formulas or computations may also be shown. The various sections of the appendix should be designated by alphabetical divisions or by use of Roman numerals.

1-9   Reviews of Technical Literature   Among principal sources of forest inventory literature are research papers issued by state and federal experiment stations, the *Journal of Forestry, Forest Science, Forestry Chronicle*, and the *Canadian Journal of Forest Research*. The forester who expects to comprehend and evaluate such articles must adopt a disciplined attitude to the reading of scientific literature. If an abstract precedes the main report, this should be read first, followed by a rapid scanning of the entire article. Then, if the study appears to be of special interest or utility, the article should be carefully reread.

Although an abstract may obviate the necessity of making notes on each article, it is well to look for salient points. After noting the locale of the study and the author's affiliation, the reader should ask, What were the real objectives of this study? Next, it may be appropriate to note the laboratory procedure or statistical design employed, along with the number and type of samples measured. Finally, any tables or graphical presentations should be studied to see whether they fully substantiate the author's principal findings or conclusions. Only by taking such an analytical approach can the reader expect to gain any real benefit from reports of specialized research.

PROBLEMS

1-1 Convert these measurements as specified:
  a 51.3 miles to kilometers
  b 50 m^3 per ha to cubic feet per acre
  c 1500 cu ft per acre to cubic meters per hectare
  d 500 trees per ha to trees per acre

1-2 How many acres are included within the boundaries of a football (or soccer) field, including end zones? Convert this value to hectares.

1-3 Be able to write equations for:

    a Determining the radius for a circular plot when the area is known

    b Determining the length of the hypotenuse of a right triangle

    c Determining the length of one side of a square plot when the area is known

1-4 Without using instruments or scales, explain how you might determine the height of a tree from its shadow length.

1-5 For each of the following, indicate if the measurement would be made on a nominal, ordinal, interval, or ratio scale:

    a Number of trees per acre

    b Cubic volume in a log

    c Log grade

    d Designation for forest types on a map

1-6 Refer to a recent issue of the *Journal of Forestry* or the *Forestry Chronicle*. What is the preferred style for preparing references to literature citations?

1-7 Prepare a typewritten abstract of not more than 250 words for a published technical article dealing with some phase of forest inventory.

REFERENCES

Avery, T. E. 1978. *Student's guide to thesis research*. Burgess Publishing Company, Minneapolis, Minn. 108 pp.

CBE Style Manual Committee. 1983. *CBE style manual: A guide for authors, editors and publishers in the biological sciences*. 5th ed. Council of Biology Editors, Bethesda, Md. 324 pp.

Cleveland, W. S. 1985. *The elements of graphing data*. Wadsworth Advanced Book Program, Monterey, Calif. 323 pp.

Ffolliott, P. F., Robinson, D. W., and Space, J. C. 1982. Proposed metric units in forestry. *J. Forestry* **80:**108–109.

Rains, M. T., and Larson, D. E. 1978. Graphics in forestry: A guide to effective display of data. U.S. Forest Service, Atlanta, Ga. 12 pp.

Schemnitz, S. D. (ed.) 1980. *The wildlife management techniques manual*. The Wildlife Society, Washington, D.C. 686 pp.

Schuster, E. G., and Zuuring, H. R. 1986. Quantifying the unquantifiable. *J. Forestry* **84:**25–30.

Stevens, S. S. 1946. On the theory of scales of measurement. *Science* **103:**677–680.

————.1968. Measurement, statistics, and the schemapiric view. *Science* **161:**849–856.

Struck, W., Jr., and White, E. B. 1972. *The elements of style*, 2d ed. MacMillan Company, New York. 78 pp.

Tufte, E. R. 1983. *The visual display of quantitative information*. Graphics Press, Cheshire, Conn. 197 pp.

U.S. Department of Commerce. 1972. *The International System of Units (SI)*. National Bureau of Standards, Special pub. 330, Government Printing Office, Washington, D.C. 42 pp.

Wenger, K. F. (ed.). 1984. *Forestry handbook*, 2d ed. John Wiley & Sons, New York. 1335 pp.

STATISTICAL METHODS

2-1 Introduction To the practicing forester, an understanding of statistical techniques and sampling methods has become as important as a knowledge of dendrology or type mapping. Whether designing a timber inventory or reading a scientific article, a background in statistics is essential. Because forestry students usually complete one or more statistics courses prior to work in forest mensuration, this chapter is intended only as a brief review of applied techniques. Emphasis is placed on how to handle routine computations and (to a lesser degree) how to interpret the meaning of certain statistical quantities. Derivations and theory are purposely avoided, because they are best treated in textbooks devoted strictly to these subjects.

The reader is reminded that the procedures discussed in this chapter were not designed specifically for the solution of forestry problems. Rather, they are standard statistical methods which have been found useful in forest-oriented situations.

2-2 Rounding Off Numbers To minimize personal bias and assure a degree of consistency in computations, it is desirable to adopt a systematic technique for rounding off numbers. The necessity for such a method arises when a calculated value apparently falls exactly halfway between the units being used, i.e., when the number 5 immediately follows the digit positions to be retained.

As an example, suppose the values of 27.65 and 104.15 are to be rounded off to 1 decimal place. A commonly used rule is to ignore the 5 when the digit preceding it is an even number; thus 27.65 becomes 27.6. Conversely, if the digit preceding the 5 is an odd number its value is raised by one unit. Therefore, in the example here, 104.15 would be recorded as 104.2.

Rounding off should be done after all intermediate calculations have been completed. Intermediate calculations should be carried at least two places beyond that of the final rounded figures.

2-3 Bias, Accuracy, and Precision   Although most persons have a general idea of the distinction among these three terms, it appears appropriate to define the terms from the statistical viewpoint. *Bias* is a systematic distortion arising from such sources as a flaw in measurement or an incorrect method of sampling. Measurements of 100-ft units with a tape only 99 ft long will be biased; similarly, biases may occur when timber cruisers consistently underestimate tree heights or arbitrarily shift field plot locations to obtain what they regard as more typical samples.

Accuracy refers to the success of estimating the true value of a quantity, and *precision* refers to the clustering of sample values about their own average. A biased estimate may be precise, but it cannot be accurate; thus it is evident that accuracy and precision are not synonymous or interchangeable terms. As an example, a forester might make a series of careful measurements of a single tree with an instrument that is improperly calibrated or out of adjustment. If the measurements closely cluster about their average value, they are precise. However, as the instrument is out of adjustment, the measured values will be biased and considerably off the true value; thus the estimate is not accurate. The failure to attain an accurate result may be due to the presence of bias, the lack of precision, or both.

2-4 Calculating Probabilities   For purposes of discussion, probability may be defined as the relative frequency with which an event takes place "in the long run." If an observed event A occurs x times in n trials, the probability or relative frequency is

$$P(\text{A}) = \frac{x}{n}$$

For example, if a balanced coin is tossed in an unbiased fashion, one would expect to obtain *heads* about 50 percent of the time; i.e., the probability is 0.50. If the same coin is tossed 100 times and heads occur only 41 times, the *observed* relative frequency of heads is $^{41}/_{100}$, or 0.41. Still, the likelihood of getting *heads* on any given toss is 0.50, and "over the long run" (thousands of unbiased tosses) one would expect the observed relative frequency to closely approximate 0.50.

Coin flipping is an example of *independent* events; i.e., the occurrence of heads or tails on one toss has no predictable effect on the outcomes of subsequent tosses. As the expected probability of obtaining heads on a single toss is

$1/2$, the probability of obtaining two heads (or two tails) in a row is $1/2 \times 1/2 = 1/4$, or one chance in four. Thus for two *independent* events, the probability that both will occur is the *product* of their individual probabilities.

As another example of events that are apparently independent, assume that the probability of owning a bicycle is 0.17, the probability of having red hair is 0.04, and the probability of being a college student is 0.21. If the assumption of independence is correct, the probability that a randomly selected individual will be a red-headed college student with a bicycle is $0.17 \times 0.04 \times 0.21$, or 0.001428 (roughly 14 chances in 10,000). These events have been referred to as *apparently* independent, because truly independent happenings are difficult to establish, except by statistical design and randomization.

If the occurrence of one event A precludes the occurrence of some other event B and vice versa, A and B are said to be *mutually exclusive*. In a single appearance at bat, a baseball player may walk or hit safely but cannot do both. If the probability of drawing a walk is 0.104 and the probability of a safe hit is 0.310, the probability that the player will *either* draw a walk *or* hit safely is $P(0.310) + P(0.104) = 0.414$ (roughly 41 chances in 100). Thus for mutually exclusive events, the probability that at least one *or* the other will occur is the *sum* of their individual probabilities.

Probabilities are always positive numbers, and they range between 0 and 1. The probability that the earth will continue to revolve on its axis for another year is unknown but assumed to be 1. If this is true, the probability that it will not do so is 0. Nevertheless, there are few events that can be described in such absolute terms. When the probability of an event happening is 0.75 (three times in four), the probability that it will *not* occur is $1 - 0.75$, or 0.25 (one chance in four).

Foresters who employ statistical procedures must learn to accept the fact that they are dealing with probabilities and not with certainties. Even when we say that we are 95 percent certain that the confidence interval for a timber stand includes the true mean, there are still 5 chances in 100 that it does not!

2-5   Factorial Notation, Permutations, and Combinations   When very few events are involved, various outcomes can be simply counted; in other instances, special mathematical formulas are helpful. Assume, for example, that a four-volume set of books is placed upright on a shelf in a completely random order. The number of possible *arrangements* (permutations) of n things is n factorial (designated as $n!$) or $(n)(n-1)(n-2) \ldots (2)(1)$. For our four books, this is $4! = (4)(3)(2)(1) = 24$. Because the books can be shelved in 24 possible ways, the probability of their being put in correct order is 1 out of 24.

A useful formula for calculating the number of possible *different events* (combinations) involving a things, b things, $\ldots$, z things, is $n = (a)(b)(c) \ldots (z)$. As an illustration, suppose a forest cover-type map is to be prepared to depict six species composition classes, five tree-height classes, three stand-density classes, and three soil-site conditions. The total number of possible cover types is $n = (6)(5)(3)(3) = 270$ combinations.

As described in foregoing instances, the term *combination* implies that two combinations are composed of different items; the sets *ABC, ABD,* and *ACD* are all different combinations. By contrast, the term *permutation* denotes the arrangement of a set of items. The sets *ABC, ACB,* and *BAC* are all the same combination, but they are different permutations. It is therefore obvious that a given population will have many more permutations than combinations. To illustrate this point, it may be presumed that one wishes to determine how many slates of officers (permutations) and how many different committees (combinations) of 4 individuals each can be selected from a group of 10 persons.

The number of *permutations* or arrangements of r items that can be formed from a total of n items is computed as $n!/(n - r)!$. For our example, this is $10!/6! = 5040$ slates of officers. In most inventory situations, the forester is concerned more with combinations of sampling units than with permutations.

The number of different *combinations* of r items that can be formed from a total of n items is computed as $n!/r!(n - r)!$. In this example, $r = 4$ and $n = 10$; therefore,

$$\text{Combinations} = \frac{10!}{4!6!} = \frac{10 \cdot 9 \cdot 8 \cdot 7 \cdot 6 \cdot 5 \cdot 4 \cdot 3 \cdot 2 \cdot 1}{(4 \cdot 3 \cdot 2 \cdot 1)(6 \cdot 5 \cdot 4 \cdot 3 \cdot 2 \cdot 1)}$$
$$= 210 \text{ committees}$$

STATISTICAL CONCEPTS

2-6   Analysis of Data   Statistical data can be obtained by means of a sample survey (Chap. 8) or an experiment. The raw data usually consist of an unorganized set of numerical values. Before these data can be used as a basis for inference about a population or phenomenon of interest, or as a basis for decision, they must be summarized and the pertinent information must be extracted.

Tabular and graphical forms (Sec. 1-7) of presentation may be used to summarize and describe quantitative data. While these techniques are valuable for showing important features of the data, statistical methods for the most part require concise numerical descriptions. Measures of central tendency or location and measures of dispersion or scatter, as well as measures of association between variables, are important tools for analyzing and interpreting forestry data. These statistical concepts, along with the methods of computation, are discussed in the sections that follow.

2-7   Populations, Parameters, and Variables   A *population* may be defined as the aggregate of all arbitrarily defined, nonoverlapping *sample units.* If individual trees are the sample unit, then all trees on a given area of land could be considered a population.

Constants that describe the population as a whole are termed *parameters.* A *statistic* is a quantitative characteristic that describes a sample obtained from a population. Statistics are used to estimate population parameters. The *sample* itself is merely the aggregate of sample units from which measurements or observations are taken.

Populations are generally classed as being *finite* or *infinite.* A finite population is one for which the total number of sample units can be expressed as a finite number. The number of trees in a tract of land and the number of sawmills in a geographic region are both examples of finite populations.

Infinite populations are those in which the number of sample units is not finite. Also, populations from which samples are selected and replaced after each drawing may be regarded as equivalent to infinite populations. From a practical viewpoint, all the gray squirrels in North America or all the ponderosa pines in southwestern United States may be treated as infinite populations. As described in later sections of this chapter, the distinction between these two classes of populations becomes important when a relatively large number of sample units is drawn from a finite population. In statistical notation, finite population size is denoted by N, and the number of sample units observed is indicated by n.

Without variation in forest characteristics such as tree volumes, there would be few sampling problems. Any characteristic that may vary from one sample unit to another is referred to as a *variable.* Variables that may occupy any position along a scale are termed *continuous* variables. Tree weights and heights are conceptually continuous variables, as are air temperature, wind velocity, and atmospheric pressure.

Qualitative variables and variables that are commonly described by simple counts (integers) are termed *discrete* variables. Most of the statistical procedures described in this chapter are applicable to continuous, rather than discrete, variables.

2-8 Frequency Distributions The frequency distribution defines the relative frequency with which different values of a variable occur in a population. Each population has its own distinct type of distribution. If the form of the distribution is known, it is possible to predict the proportion of individuals that are within any specified limits.

The most common distribution forms are the normal, binomial, and Poisson. The normal distribution is associated with continuous variables, and it is the form most used by foresters. The arithmetic techniques for handling data from normally distributed populations are relatively simple in comparison with methods developed for other distributions. Regardless of the distribution followed by a given variable, the means of large samples from the distribution are expected to have a distribution that approaches normality. Consequently, estimates and inferences may be based on this assumption.

STATISTICAL COMPUTATIONS

2-9   Mode, Median, and Mean   These values are referred to as measures of central tendency. The *mode* is defined as the most frequently appearing value or class of values in a set of observations. The *median* is the middle value of the series of observations when they have been arranged in order of magnitude, and the arithmetic *mean* is simply the arithmetic average of the set of observations. For a majority of statistical analyses, the mean is the most useful value of the three. In populations that are truly normally distributed, values for the mode, median, and mean are identical.

Following are observations of diameters (in inches) taken on a sample of 26 trees. These values are listed haphazardly (as tallied) at the left and arranged in a frequency table at the right. In the frequency table, the indicated diameter at breast height (dbh) is the midpoint of the diameter class.

			Frequency table	
Haphazard listing, dbh			dbh	No. of trees
8	9	10	5	3
8	9	9	6	0
5	7	7	7	6
10	5	8	8	9
9	8	9	9	5
10	7	8	10	3
8	7	7		26
5	8	8		
7	8			
$n = 26$				

For this set of observations, 8 in. is the modal diameter class. This class is easily detected in the frequency table but is less discernible in the unorganized listing. If there had been nine trees in any other class as well as nine in the 8-in. class, the distribution would have been termed *bimodal*. When three or more values have the same frequency or when each value appears only once, no apparent mode can be specified.

The *median* position is found by adding 1 to the number in the sample and dividing by 2, that is $(n + 1)/2$. With an odd number of observations, the median is merely picked out as the middle ranking value. Thus in a sample of seven observations ranked as 2, 4, 9, 12, 17, 24, and 50, the 12 is the median value. Had there been only six observations (eliminating the 50), the median *position* would have fallen between the 9 and 12. Its *value* would be recorded as the arithmetic average of these two numbers, or $(9 + 12)/2 = 10.5$. For the 26 tree diameters previously noted, the median position is $(26 + 1)/2$, or 13.5. As both the thirteenth and fourteenth values fall within the 8-in.-dbh class, the median value is recorded as 8 in.

It will be noted that both median and mode are unaffected by extreme values. Thus as measures of central tendency, the median and mode may be more informative than the arithmetic mean when a few extreme values are observed.

The sample *mean* or arithmetic average, commonly designated as $\bar{x}$, is computed from

$$\bar{x} = \frac{\Sigma x}{n}$$

where Σ = sum of (over entire sample)
  x = value of an individual observation
  n = number of observations in sample

For the 26 tree diameters under consideration, the *sample mean* is $204 \div 26 = 7.85$ in.

2-10   The Range and Average Deviation   In a series of sample values, the *range* is merely the difference between the highest and the lowest value recorded. For the 26 tree diameters listed previously, the range is therefore $10 - 5$, or 5 in. Although based solely on extreme values, the range is a useful indicator of the dispersion or variability of a set of observations, and it also has some utility in providing estimates of the standard deviation. A rough estimate of the standard deviation (Sec. 2-11) can be computed as

$$s = \frac{R}{4}$$

where s = standard deviation
  R = range (largest minus smallest value)

The *average deviation* (AD), though largely supplanted by the standard deviation, provides an easily computed measure of the dispersion of individual variables about their sample mean. It is computed as the arithmetic average of deviations from the mean (ignoring algebraic signs). Using the same symbols as in the previous section, the formula is

$$AD = \frac{\Sigma |x - \bar{x}|}{n}$$

Again referring to the 26 tree diameters, the average deviation is calculated as

$$AD = \frac{27.20}{26} = 1.05 \text{ in.}$$

The calculated value indicates that the average deviation of the individual dbh measurements from their mean of 7.85 in. is 1.05 in. Although this measure of dispersion about the arithmetic mean is not widely used for making statistical inferences, it is sometimes used as a measure of the reliability of prediction equations or tables (e.g., tree volume tables).

2-11 Variance and Standard Deviation   The variance and the standard deviation are measures of the dispersion of individual observations about their arithmetic mean. In a normally distributed population, approximately two-thirds (68 percent) of the observations will be within ± 1 standard deviation of the mean. About 95 percent will be within 1.96 standard deviations and roughly 99 percent within 2.58 standard deviations. In succeeding sections, the standard deviation will be used to evaluate the reliability of sample estimates.

The standard deviation of a population is a parameter, and it is commonly denoted by the Greek letter sigma (σ). The sample standard deviation is a statistic that is an estimate of the population parameter σ, and it is symbolized by s. For sample data, the variance, which is defined as the sum of squared deviations from the mean divided by $n - 1$, is denoted by s^2 and is computed first; then the standard deviation is derived by taking the square root of the variance. Again employing the symbols previously identified, the estimated standard deviation is calculated from

$$s = \sqrt{\frac{\Sigma x^2 - (\Sigma x)^2 / n}{n - 1}}$$

This is equivalent to the formula

$$s = \sqrt{\frac{\Sigma (x - \bar{x})^2}{n - 1}}$$

where $\bar{x}$ is the arithmetic mean and $(x - \bar{x})^2$ is the squared deviation of an individual observation from the arithmetic mean.

The first formula is a shortcut version of the second and is easier to use for calculations. For the 26 measurements of tree diameters, the standard deviation is

$$s = \sqrt{\frac{1650 - (204)^2 / 26}{25}} = \sqrt{1.98} = 1.41 \text{ in.}$$

If the population sampled is normally distributed, it is expected that about two-thirds of the individual tree diameters will fall within ± 1.41 in. of the population mean.

2-12   Coefficient of Variation   The ratio of the standard deviation to the mean is known as the *coefficient of variation* (CV). It is usually expressed as a percentage value. Because populations with large means tend to have larger standard deviations than those with small means, the coefficient of variation permits a comparison of relative variability about different-sized means. The magnitude of the variance and standard deviation also depends on the measurement units used, but the coefficient of variation will be the same for a given set of observations regardless of the unit of measure. A standard deviation of 5 for a population with a mean of 15 indicates the same relative variability as a standard deviation of 30 with a mean of 90. The coefficient of variation in each instance would be 0.33, or 33 percent.

For the 26 tree diameters, the mean is 7.85 and the standard deviation is 1.41. The coefficient of variation CV from the sample is

$$CV = \frac{s}{\bar{x}}(100) = \frac{1.41}{7.85}(100) = 18\%$$

2-13   Standard Error of the Mean   The standard deviation is a measure of the variation of individual sample observations about their mean. Inasmuch as individuals vary, there will also be variation among means computed from different samples of these individuals. A measure of the variation among sample means is the standard error of the mean. It may be regarded as a standard deviation among the means of samples of a fixed size n. As described in succeeding sections, the standard error of the mean can be used to compute confidence limits for a population mean or to determine the sample size required to achieve a specified sampling precision.

Calculation of the standard error of the mean depends on the manner in which the sample was selected. For simple random sampling from an infinite population, the formula for the estimated standard error of the mean $s_{\bar{x}}$ is

$$s_{\bar{x}} = \sqrt{\frac{s^2}{n}}$$

When sampling without replacement from a finite population, the formula becomes

$$s_{\bar{x}} = \sqrt{\frac{s^2}{n}\left(\frac{N-n}{N}\right)}$$

The term $(N - n)/N$ is referred to as the finite population correction; in this term, N denotes the population size, and n is the actual sample size. The finite population correction serves to reduce the standard error of the mean when relatively large samples are drawn without replacement from finite populations.

If it is assumed that the 26 tree diameters were drawn without replacement from a population of only 200 stems, the standard error of the mean would be computed as

$$s_{\bar{x}} = \sqrt{\frac{1.98}{26}\left(\frac{200 - 26}{200}\right)} = 0.26 \text{ in.}$$

This value indicates that if several samples of 26 units each were randomly drawn from the same population, the standard deviation among the sample means might be expected to be approximately 0.26 in. The value of the finite population correction is always less than unity, but it approaches unity when the sampling intensity is very low. If less than 5 percent of the population appears in the sample, the finite population correction is sometimes omitted.

If the 26 tree diameters had been drawn from an infinite population or from one that was quite large in relation to the sample size, the standard error of the mean would have been computed simply as

$$s_{\bar{x}} = \sqrt{\frac{1.98}{26}} = 0.28 \text{ in.}$$

2-14 Confidence Limits It is recognized that sample means vary about the true mean of the population. Thus an estimate of the mean, by itself, is not very informative. To make an estimate more meaningful, confidence limits can be computed to establish an interval which, at some specified probability level, would be expected to include the true mean. The standard error of the mean and a table of t values (see Appendix) are used for establishing confidence limits. For simple random samples from normally distributed populations, the confidence limits for the population mean are computed by

$$\text{Mean} \pm t \text{ (standard error)} \qquad \text{or} \qquad \bar{x} - ts_{\bar{x}} \text{ to } \bar{x} + ts_{\bar{x}}$$

Under ordinary circumstances, one does not know the underlying distribution of the population being sampled. However, the distribution of *means* from reasonably large samples, almost regardless of the distribution of the underlying parent population, will approach normality and the confidence interval, as shown, is appropriate.

In using the Appendix table for the distribution of t, the column labeled df refers to *degrees of freedom*, which in the case of a simple random sample will be equal to one less than the sample size (that is, $n - 1$). The columns labeled *probability* refer to the level of odds demanded. If one wishes to state that the confidence interval will include the true mean unless a 1-in-20 chance occurs (95 percent probability level), the t values in the 0.05 column are used. If one wishes to establish confidence limits at the 99 percent probability level, the 0.01 column in the t table is used, and so forth.

For the sample tree diameters previously listed, the estimated mean was 7.85 in. and the standard error of the mean (for sampling without replacement) was ±0.26 in. Because only 26 observations were taken, there are 26 − 1, or 25, *df.* The 95 and 99 percent *t* values are read from the Appendix table as 2.060 and 2.787, respectively. Confidence limits for these probability levels are

$$P = 0.95; 7.85 - (2.060)(0.26) \text{ to } 7.85 + (2.060)(0.26) = 7.31 \text{ to } 8.39 \text{ in.}$$
$$P = 0.99; 7.85 - (2.787)(0.26) \text{ to } 7.85 + (2.787)(0.26) = 7.13 \text{ to } 8.57 \text{ in.}$$

Therefore, if the 26 units were randomly selected from a normally distributed population, the interval between 7.31 and 8.39 in. includes the true population mean, unless a 1-in-20 chance has occurred in sampling. In other words, the population mean will be included in the interval unless this random sample is one of those which, by chance, yields a sample mean and standard error such that the interval constructed from it will not include the mean. Such would happen, on the average, once in every 20 samples. Similarly, unless a 1-in-100 chance has occurred, the true mean is included in the interval of 7.13 to 8.57 in. It can be seen from these examples that the higher the probability level, the wider the confidence limits must be expanded.

The forester must remember that confidence limits and accompanying statements of probability account for *sampling variation only.* It is assumed that sampling procedures are unbiased, field measurements are without error, and no computational mistakes are included. If these assumptions are incorrect, confidence statements may be misleading.

2-15   Covariance   The covariance is a measure of how two variables vary in relation to each other. If there is little or no association between two variables, the covariance will be close to zero. In cases where large values of one variable tend to be associated with small values of another variable, the covariance will be negative. When large values of one variable tend to be associated with large values of another variable, the covariance will be positive.

The sample covariance of two variables x and y is symbolized by s_{xy} and is defined as

$$s_{xy} = \frac{\Sigma(x - \bar{x})(y - \bar{y})}{n - 1}$$

The computing formula is

$$s_{xy} = \frac{\Sigma xy - \dfrac{(\Sigma x)(\Sigma y)}{n}}{n - 1}$$

2-16 Simple Correlation Coefficient   The magnitude of the covariance is related to the units of measure used for x and y. A measure of the degree of *linear* association between two variables that is independent of the units of measure is the simple correlation coefficient *(r)*

$$r = \frac{s_{xy}}{\sqrt{s_x^2 \, s_y^2}}$$

where s_x^2 and s_y^2 represent the sample variances of x and y, respectively.

The correlation coefficient can range from -1 to $+1$ with values near -1 or $+1$ indicating very strong association. Negative correlation coefficients indicate that large values of one variable are associated with small values of the other variable. When the correlation coefficient is positive, large values of one variable tend to be associated with large values of the other variable. If two variables are independent, the correlation coefficient will tend to be near zero. The converse is not necessarily true, however. A correlation near zero does *not* mean that two variables are independent, but that there is no apparent *linear* association between the variables (they could be strongly related in a curvilinear manner).

2-17 Expansion of Means and Standard Errors   In most instances, estimates of means per sample plot are multiplied by a constant to scale the estimates to a more useful basis. If a forest inventory utilizes $1/4$-acre plots, for example, the mean volume per plot is multiplied by 4 to put the estimated mean on a per acre basis. Or for a tract of 500 acres, the mean volume per plot would be multiplied by 2000 (the number of possible $1/4$ acres in the tract) to estimate the total volume.

The rule to remember is that expansion of sample means must be accompanied by a similar expansion of standard errors. Thus if the mean volume per $1/4$-acre plot is 1500 bd ft with a standard error of 60, the mean volume per acre is $4(1500) \pm 4(60)$, or 6000 ± 240 bd ft. Variances are expanded by multiplying by the constant squared. In this example the variance of the estimate of the mean volume per $1/4$ acre is $60^2 = 3600$, and the variance on an acre basis would be $(4^2)(3600) = 57{,}600$. Taking the square root of 57,600 gives 240 bd ft for the estimate of the standard error of the mean on an acre basis.

When area is known, estimates on a tract basis are expanded similarly—that is, the mean is multiplied by the area, the standard error by the area, and the variance by the area squared. Continuing with the preceding example, for a tract of 500 acres, the total volume would be expressed as $2000(1500) \pm 2000(60)$, or $3{,}000{,}000 \pm 120{,}000$ bd ft.

The foregoing examples presume the use of expansion factors having no error. However, sample-based estimates of area are (or should be) also accompanied by standard errors. Thus the expansion of per acre volume to total tract volume becomes one of deriving the product of volume times area and computing a standard error applicable to this product (Meyer, 1963). The computation may be illustrated by assuming *independent* inventories that produced the following estimates and standard errors for volume $\overline{V}$ and area A:

Mean volume: 18 ± 2 cd per acre

Tract acreage: 52 ± 1.5 acres

If these two estimates are independent (and only if they are), their product and its standard error may be computed by

$$\overline{V}A \pm \sqrt{\overline{V}^2(s_A)^2 + A^2(s_{\overline{V}})^2}$$

where $\overline{V} =$ estimated mean volume per acre

$A =$ estimated number of acres

$s_{\overline{V}} =$ standard error of mean volume per acre

$s_A =$ standard error of area estimate

Substituting the sample problem data yields the total volume and its standard error:

$$(18)(52) \pm \sqrt{(18)^2(1.5)^2 + (52)^2(2)^2} = 936 \pm 107.45 \text{ cd (tract total)}$$

2-18   Mean and Variance of Linear Functions   Quite often the variable of interest is a linear function of two or more other variables. In general terms, if the variable of interest Z is a linear function of k variables, that is,

$$Z = c_1 x_1 + c_2 x_2 + \cdots + c_k x_k$$

where $c_1, c_2 \ldots c_k$ are constants, then the mean of Z would be estimated as

$$\overline{Z} = c_1 \overline{x}_1 + c_2 \overline{x}_2 + \cdots + c_k \overline{x}_k$$

or $\qquad\qquad = \Sigma c_i \overline{x}_i$

When variables are *independent,* the variance of the sum is the sum of the variances. Thus the variance of Z would be

$$s_Z^2 = c_1^2 s_1^2 + c_2^2 s_2^2 + \cdots + c_k^2 s_k^2$$

or $\qquad\qquad = \Sigma c_i^2 s_i^2$

where s_1^2 denotes the variance of x_1, etc. If the x's are not independent, the variance of Z is

$$s_Z^2 = c_1^2 s_1^2 + c_2^2 s_2^2 + \cdots + c_k^2 s_k^2$$
$$+ 2c_1 c_2 s_{12} + \cdots + 2c_{k-1} c_k s_{k-1,k}$$

or

$$= \Sigma c_i^2 s_i^2 + 2 \underset{i \neq j}{\Sigma \Sigma} c_i c_j s_{ij}$$

where s_{12} denotes the covariance of x_1 and x_2, etc.

These general relationships are very important and are commonly employed in forestry. If, for example, the mean volume per acre on a timbered tract was estimated to be $\bar{x}_1$ by a simple random sample and 5 years later another completely independent simple random sample was taken and the mean volume per acre was $\bar{x}_2$, growth per acre *(G)* could be estimated as

$$G = \bar{x}_2 - \bar{x}_1$$

If the relationships just specified for linear functions of variables are applied, the variance of G would be

$$s_G^2 = (1)^2 s_{\bar{x}_2}^2 + (-1)^2 s_{\bar{x}_1}^2$$
$$= s_{\bar{x}_2}^2 + s_{\bar{x}_1}^2$$

because $\bar{x}_1$ and $\bar{x}_2$ are independent. If, on the other hand, the second inventory was conducted by returning to the *same plot centers* as those used on the first occasion, the two estimates $\bar{x}_1$ and $\bar{x}_2$ would not be independent but would be expected to have a positive covariance term (s_{12}). The mean growth per acre would be estimated as with independent samples, but the variance for the estimate of growth would now be

$$s_G^2 = (1)^2 s_{\bar{x}_2}^2 + (-1)^2 s_{\bar{x}_1}^2 + (2)(1)(-1)s_{12}$$
$$= s_{\bar{x}_2}^2 + s_{\bar{x}_1}^2 - 2s_{12}$$

Since s_{12} is expected to be a relatively large positive quantity, the variance of estimated growth is expected to be much smaller with repeated measurements on the same plots than with independent sample plots. This relationship is the basis for the use of permanent sample plots when the primary objective is to estimate growth (Chap. 10).

SIMPLE LINEAR REGRESSION

2-19 Definitions   In analyzing various resource measurements, it may be important to quantify the degree of association between two or more variables. Such associations can often be examined by regression analysis. The simplest type of relationship that can exist between two quantities is one that can be represented by a straight line. Thus a *simple linear regression* describes a straight-line relationship that exists between two quantities: one *dependent variable Y* and one *independent variable X.*

The quantity that is being estimated by the regression line is termed the *dependent variable,* and the quantity measured in order to predict the associated value is called the *independent variable.* When these paired quantities are shown graphically, it is conventional to plot Y values along the vertical axis *(ordinate)* of the graph and X values along the horizontal axis *(abscissa).* This would be termed a *relationship* of Y on X.

When each Y value is graphically plotted against its corresponding X value, the resulting representation is termed a *scatter diagram.* Since the purpose of such a diagram is to determine whether or not a relationship exists between the two variables, this should be the first analytical step following data collection. Careful inspection of the scatter diagram will also provide an indication of the strength of the relationship and its probable form, i.e., whether or not the association can be logically represented by a straight line (Fig. 2-1).

For those rare situations where all the plotted points fall exactly on a line, a perfect linear relationship exists. Such will rarely (if ever) be the case with biological data, but the smaller the deviations from a line, the stronger the linear relationship between the two variables. Where Y values increase with X values this is termed *positive,* or *direct, correlation;* if Y values decrease as X values become larger, a *negative,* or *inverse, correlation* is said to exist. If the plotted points are not indicative of a straight-line relationship, it is sometimes possible to make a simple *transformation* of one or both of the variables so that the relationship becomes linear in form. (See Appendix for some common transformations.) Squaring a variable or expressing it in logarithmic terms are examples of transformations.

2-20   A Linear Equation   When a linear trend exists between two variables, a regression equation of the form $\hat{Y} = b_0 + b_1X$ may be fitted to the plotted

FIGURE 2-1
Scatter diagrams that suggest a linear relationship (left), a curvilinear relationship (middle), and no relationship (right).

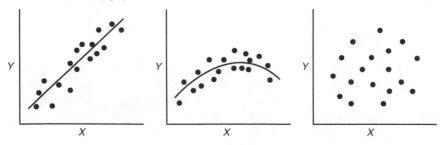

points. In this equation, $\hat{Y}$ refers to the estimated value of the dependent variable, X is the value of the independent variable, and b_0 and b_1 are regression coefficients established from analysis of the data. After these coefficients have been determined, b_0 denotes the value of the Y intercept, which is the value of Y when X equals zero; the coefficient b_1 is the value that establishes the slope of the straight line. Therefore, a line represented by the equation $\hat{Y} = -3 + 2X$ would intercept the Y axis at an ordinal value of -3, i.e., three units below the origin of the graph. The slope coefficient of 2 means that the line would rise two units vertically along the Y axis for each unit horizontally along the X axis.

When the regression line is fitted to the plotted points by the method of "least squares," the line will pass through the point defined by the means of X and Y. The principle of least squares is that the sum of the squared deviations of the observed values of Y from the regression line will be a minimum. The least squares method gives the best unbiased estimates of the slope and intercept if the following assumptions are satisfied:

1 The X values are measured without error.

2 The variance of Y is the same at all levels of X (i.e., a homogeneous variance is assumed).

3 For each value of X, Y is normally and independently distributed.

2-21   A Sample Problem   Suppose we observe that pine trees with large crowns appear to grow faster than trees with small crowns. Since we would like to be able to predict the growth of trees from their relative crown sizes, we decide to use regression analysis to determine whether a strong relationship exists between the two variables. A sample from the area of interest results in 62 paired measurements of tree-crown volumes X and basal-area growth Y. In Table 2-1, crown volume is in 100 cu ft, and basal-area growth is in square feet.

First, the paired measurements are plotted on graph paper to determine whether there is any visual evidence of a relationship between the two variables. The resulting scatter diagram (Fig. 2-2) does indicate a general linear trend of direct correlation, and so it was decided to fit an equation of the form $\hat{Y} = b_0 + b_1X$ to the plotted points by the method of least squares (Freese, 1967).

After selecting the model to be fitted, the next step is to calculate the corrected sums of squares *(SS)* and cross products *(SP)* by applying the following equations.

The corrected sum of squares for Y:

$$SS_y = \sum^{n} Y^2 - \frac{\left(\sum^{n} Y\right)^2}{n}$$

$$= (0.36^2 + 0.09^2 + \cdots + 0.42^2) - \frac{26.62^2}{62}$$

$$= 2.7826$$

TABLE 2-1
PAIRED VALUES OF TREE-CROWN VOLUME AND BASAL-AREA GROWTH

Crown volume X	Basal-area growth Y	Crown volume X	Basal-area growth Y	Crown volume X	Basal-area growth Y
22	0.36	7	0.25	81	0.66
6	0.09	2	0.06	93	0.69
93	0.67	53	0.47	99	0.71
62	0.44	70	0.55	14	0.14
84	0.72	5	0.07	51	0.41
14	0.24	90	0.69	75	0.66
52	0.33	46	0.42	6	0.18
69	0.61	36	0.39	20	0.21
104	0.66	14	0.09	36	0.29
100	0.80	60	0.54	50	0.56
41	0.47	103	0.74	9	0.13
85	0.60	43	0.64	2	0.10
90	0.51	22	0.50	21	0.18
27	0.14	75	0.39	17	0.17
18	0.32	29	0.30	87	0.63
48	0.21	76	0.61	97	0.66
37	0.54	20	0.29	33	0.18
67	0.70	29	0.38	20	0.06
56	0.67	50	0.53	96	0.58
31	0.42	59	0.58	<u>61</u>	<u>0.42</u>
17	0.39	70	0.62		
			Total	3050	26.62
			Mean ($n = 62$)	49.1935	0.42935

The corrected sum of squares for X:

$$SS_x = \sum_{}^{n} X^2 - \frac{\left(\sum_{}^{n} X\right)^2}{n}$$

$$= (22^2 + 6^2 + \cdots + 61^2) - \frac{3050^2}{62}$$

$$= 59,397.6775$$

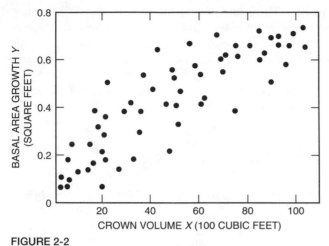

FIGURE 2-2
Scatter diagram of basal-area growth Y on crown volume X for
62 trees. *(From Freese, 1967.)*

The corrected sum of cross products:

$$SP_{xy} = \sum^n (XY) - \frac{\left(\sum^n X\right)\left(\sum^n Y\right)}{n}$$

$$= [(22)(0.36) + (6)(0.09) + \cdots + (61)(0.42)] - \frac{(3050)(26.62)}{62}$$

$$= 354.1477$$

According to the principle of least squares, the best estimates of the regression coefficients (b_0 and b_1) are obtained as follows:

$$b_1 = \frac{SP_{xy}}{SS_x} = \frac{354.1477}{59,397.6775} = 0.005962$$

$$b_0 = \overline{Y} - b_1\overline{X} = 0.42935 - (0.005962)(49.1935) = 0.13606$$

Substituting these estimates in the general equation gives

$$\hat{Y} = 0.13606 + 0.005962X$$

where $\hat{Y}$ is used to indicate that we are dealing with an estimated value of Y.

With this equation, we can estimate the annual basal-area growth $\hat{Y}$ from measurements of crown volume X.

2-22   Indicators of Fit   There are several methods of determining how well the regression line fits the sample data. One method is to compute the proportion of the total variation in Y that is associated with the regression on X. This ratio is sometimes called the *coefficient of determination,* or the *squared correlation coefficient (r^2).*

First, we must calculate the sum of squares due to the regression (also called the *reduction sum of squares*). Referring to the determination of b_1, we have

$$\text{Reduction } SS = \frac{\left(SP_{xy}\right)^2}{SS_x} = \frac{(354.1477)^2}{59,397.6775} = 2.1115$$

Next, the total variation in Y is estimated by $SS_y = 2.7826$ (as previously calculated), and

$$\text{Coefficient of determination } (r^2) = \frac{\text{reduction } SS}{\text{total } SS} = \frac{2.1115}{2.7826} = 0.76$$

A common means of interpreting the r^2 value is that "76 percent of the variation in Y is associated with X." Or, in this particular example, 76 percent of the variation in basal-area growth can be "explained" by measurements of crown volume.

In addition to the coefficient of determination, the standard error of estimate $S_{y \cdot x}$ is customarily reported as an indicator of fit to the data. The standard error of estimate is defined as

$$S_{y \cdot x} = \sqrt{\frac{\sum\limits^{n}(Y - \hat{Y})^2}{n - 2}}$$

and is easily computed as

$$S_{y \cdot x} = \sqrt{\frac{SS_y - (SP_{xy})^2/SS_x}{n - 2}}$$

For the present example,

$$S_{y \cdot x} = \sqrt{\frac{2.7826 - (354.1477)^2/59,397.6775}{60}}$$
$$= 0.106 \text{ sq ft}$$

2-23   Regression through the Origin   In many situations with biological data, theory calls for a straight line that passes through the origin. That is, when the independent variable X is zero, the value of the dependent variable Y must also be zero. Such a straight-line relationship can be written as

$$\hat{Y} = b_1 X$$

When no intercept term is estimated (i.e., the equation is conditioned to pass through the origin), the estimate of the slope coefficient is computed as

$$b_1 = \frac{\sum\limits^{n} XY}{\sum\limits^{n} X^2}$$

where $\sum\limits^{n} XY$ is the *uncorrected* sum of cross products and $\sum\limits^{n} X^2$ is the *uncorrected* sum of squares for X.

2-24   Hazards of Interpretation   A strong correlation between two variables (e.g., an r^2 of 0.90 or greater) implies only that the variables are closely associated. Such correlations are *not* evidence of a cause-and-effect relationship; in many instances, both quantities may be directly affected by a third element that has not been taken into consideration. For example, if prices for both pork and eggs rise at similar rates, one might find a high correlation between these two values over a period of time. However, instead of one price *causing* the other to rise, both prices are probably being pushed upward by a third factor, such as general increases in the cost of producing farm products.

Since many completely unrelated variables can be associated by "nonsense correlations," the necessity for rational thinking in data collection is of paramount importance. Before attempting to employ regression analysis, the forest manager should have sound biological reasons for associating changes in one quantity with those in another quantity. Unless reliable and representative data are collected through use of an unbiased sampling plan, regression analysis may prove to be a futile exercise.

2-25   Multiple Regression   Finally, it should be again stressed that the preceding discussion has dealt only with simple linear regression, i.e., the treatment of one dependent and one independent variable. Frequently, however, the dependent variable is related to more than one independent variable. If this relationship can be estimated by using *multiple regression* analysis, it may allow more precise predictions of the dependent variable than is possible by a simple regression. The general model for a multiple regression is

$$\hat{Y} = b_0 + b_1 X_1 + b_2 X_2 + \cdots + b_k X_k$$

where $b_0, b_1, b_2, \ldots, b_k$ are regression coefficients that are estimated from analysis of the data. Details for handling multiple regression analysis are not treated in this chapter but may be found in most textbooks on statistical methods.

PROBLEMS

2-1 For the sample data presented below

1	7	17	16
5	14	7	7
10	15	18	

    a Compute the following measures of central tendency:
      1 Mean
      2 Median
      3 Mode
    b Compute the following measures of dispersion:
      1 Variance
      2 Standard deviation
      3 Range
      4 Average deviation
    c Calculate the coefficient of variation.
    d Place 95 percent confidence limits on the mean.

2-2 Draw a simple random sample of at least 30 observations from a population in your field of interest. Then, (a) place the 95 percent confidence limits on the sample mean, and (b) compute the total number of sample units that would be required to estimate the population mean within ±5 percent at a confidence probability of 90 percent. (*Note:* Remember to apply the finite population correction, if applicable.)

2-3 Given the following pairs of values

X	Y	X	Y
2	3	4	3
4	2	7	7
6	7	6	5
8	5	11	9
10	9		

    a Find the regression equation, $\hat{Y} = b_0 + b_1 X$
    b Compute the following indicators of fit:
      1 Coefficient of determination (r^2)
      2 Standard error of estimate ($s_{y \cdot x}$)
    c What is the predicted value of Y, given $X = 7$?

2-4 By simple random sampling, obtain paired measurements of two variables that you believe to be linearly correlated. If a scatter diagram indicates that a straight-line relationship exists, then (a) fit a simple linear regression to the plotted points by the method of least squares, and (b) compute the coefficient of determination for the association. (*Note:* If a linear relationship is not indicated by the scatter diagram, attempt to transform one or both variables so that the trend of plotted points becomes a linear one.)

2-5 Given the following sample data on X, Y pairs

X	Y	X	Y
0	0	−1	1
1	1	3	9
−3	9	−4	16
4	16	2	4
−2	4		

a Compute the simple correlation coefficient (r) between X and Y.
b Plot the values of Y versus X. Is there a relationship between X and Y? Interpret the computed value of r in view of the relationship exhibited in the data plot.

2-6 Theory indicates that the relationship between two variables of interest, X and Y, should be linear and should pass through the origin. The following sample data on X, Y pairs were collected:

X	Y	X	Y
20	24	1	1
15	17	27	28
6	7	12	12
11	10	3	5

a Plot Y versus X. Does the assumption of a straight-line relationship that passes through the origin seem reasonable?
b Fit the regression line

$$\hat{Y} = b_1 X$$

2-7 The following are the average heights of dominant trees, age in years, and measures of available water for 16 stands of a commercially important tree species.

Height, ft	Age	Available water	Height, ft	Age	Available water
33	13	0.90	20	9	0.98
34	12	0.92	23	10	1.65
21	9	0.89	38	17	0.88
30	13	0.59	47	18	0.77
35	12	1.24	21	8	0.78
25	10	0.57	39	15	0.94
21	8	0.77	38	14	0.70
48	16	0.91	40	15	1.00

a Find the equation for the regression of height on age.
b Find the equation for the regression of height on available water.
c Which variable, age or available water, is the better predictor of height?

REFERENCES

Draper, N. R., and Smith, H. 1981. *Applied regression analysis.* 2d ed. John Wiley & Sons, New York. 709 pp.

Freese, F. 1962. Elementary forest sampling, *U.S. Dept. Agr. Handbook* 232, Government Printing Office, Washington, D.C. 91 pp.

————. 1967. Elementary statistical methods for foresters. *U.S. Dept. Agr. Handbook* 317, Government Printing Office, Washington, D.C. 87 pp.

Huntsberger, D. V., and Billingley, P. 1973. *Elements of statistical inference.* 3d ed. Allyn and Bacon, Inc., Boston. 349 pp.

Meyer, W. H. 1963. Some comments on the error of the total volume estimate. *J. Forestry* **61:**503–507.

Myers, R. H. 1990. *Classical and modern regression with applications,* 2d ed. *PWS-Kent* Publishing Co., Boston. 488 pp.

LAND MEASUREMENTS

3-1 Applications of Surveying A knowledge of the elements of land surveying is essential to the inventory forester. Although foresters may rarely be responsible for original property surveys, they are often called upon to retrace old lines, locate property boundaries, and measure land areas. To adequately perform these tasks, foresters should be adept in pacing, chaining, running compass traverses, and various methods of area estimating. They should also be familiar with the principal systems of land subdivision found in a particular region of the country.

Surveying is the art of making field measurements that are used to determine the lengths and directions of lines on the earth's surface. If a survey covers such a small area that the earth's curvature may be disregarded, it is termed *plane surveying*. For larger regions, where the curvature of the earth must be considered, *geodetic surveys* are required. Under most circumstances, the forester is concerned with plane surveying, viz., the measurement of distances and angles, the location of boundaries, and the estimation of areas.

The fundamental unit of horizontal measurement employed by foresters is the surveyor's, or Gunter's, chain of 66 ft. The chain is divided into 100 equal parts that are known as links; each link is thus 0.66 ft, or 7.92 in., in length. Distances on all U.S. Government Land Surveys are measured in chains and links. The simple conversion from chained dimensions to acres is one reason for the continued popularity of this measurement standard. Areas expressed in square chains can be immediately converted to acres by dividing by 10. Thus a tract 1 mile square (80 chains on a side) contains 6400 square chains, or 640 acres.

3-2   Pacing Horizontal Distances   Pacing is perhaps the most rudimentary of all techniques for determining distances in the field; nonetheless, accurate pacing is an obvious asset to the timber cruiser or land appraiser who must determine distances without the aid of an assistant. With practice and frequently measured checks, an experienced pacer can expect to attain an accuracy of 1 part in 80 when traversing fairly level terrain.

The pace is commonly defined as the average length of two natural steps; i.e., a count is made each time the same foot touches the ground.[1] A natural walking gait is recommended, because this pace can be most easily maintained under difficult terrain conditions. One should never attempt to use an artificial pace based on a fixed step length such as exactly 3 ft. Experienced pacers have demonstrated that the natural step is much more reliable.

In learning to pace, a horizontal distance of 10 to 40 chains should be staked on level or typical terrain. This course should be paced over and over until a consistent gait has been established. The average number of paces required, divided by the measured course distance, gives the number of paces per chain. Most foresters of average height and stride have a natural pace of 12 to 13 paces (double steps) per chain. It is also helpful to compute the exact number of feet per pace, for pacing is often relied upon in locating boundaries of sample plots during a timber cruise.

Uniform pacing is difficult in mountainous terrain, because measurement of horizontal rather than slope distance is the prime objective. Steps are necessarily shortened in walking up and down steep hillsides, and special problems are created when obstructions such as deep stream channels are encountered. Thus some individual technique must be devised to compensate for such difficulties.

The inevitable shortening of the pace on sloping ground can be handled by repeating the count at certain intervals, as 1, *2, 2,* 3, 4, 5, *6, 6,* and so on. Or if pace lengths are cut in half, counts may be restricted to every other pace. For obstructions that cannot be traversed at all (such as streams and rivers), the distance to some well-defined point ahead can be ocularly estimated; then a non-paced detour can be made around the obstacle.

Paced distances should always be field-recorded as horizontal distances in chains or feet—not in terms of actual paces. When accurate mental counts become tedious, a reliable pacing record can be kept by a written tally or by using a hand-tally meter. The importance of regular pacing practice cannot be overstressed; without periodic checks, neither accuracy nor consistency can be expected.

3-3   Chaining Horizontal Distances   Two persons, a head chainman and a rear chainman, are needed for accurate measurement with a steel tape. On level terrain, the chain can be stretched directly on the ground. If 11 chaining pins are

---

[1] Some foresters prefer to count *every step* as a pace. Advantages claimed for this technique are (1) less chance of losing count, (2) fewer problems with fractional paces, and (3) easier adjustments for slope.

used, one is placed at the point of origin, and the head chainman moves ahead with 10 pins and the "zero end" of the chain. If the head chainman carries the compass, he must keep himself and the chain on the correct bearing line at all times; otherwise, it is the rear chainman's duty to keep his partner on a straight and proper course.

A good head chainman paces the length of the tape so that he can anticipate when he has moved to the approximate chaining interval. When the end of the tape approaches the rear chainman, he calls, "Chain!" The head chainman pulls the tape taut until the rear chainman yells, "Stick!" Upon sticking the pin, the head chainman replies, "Stuck!" The rear chainman picks up the first pin, and the procedure is repeated until the desired length has been measured. When the head chainman "sticks" his last pin, 10 chain intervals will have been covered. The rear chainman then passes the 10 pins he has collected to the head chainman for continuing the measurement. A distance of 12 chains and 82 links is recorded as 12.82 chains.

In rough terrain, where the chain is held high off the ground, plumb bobs may be used at each end to aid in proper pin placement and accurate measurement of each interval. On steep slopes, it may be necessary to "break chain," i.e., to use only short sections of the tape for holding a level line. In mountainous country where horizontal distance cannot be chained directly, slope distances can be measured and converted to horizontal measurements by use of a trailer tape. Experienced foresters can expect an accuracy of 1 part in 1000 to 1 in 2500 by careful chaining of horizontal distances.

The principal sources of error in chaining are (1) allowing the chain to sag instead of keeping it taut at the moment of measurement, (2) aligning incorrectly, i.e., not keeping on the proper compass bearing, (3) making mistakes in counting pins, and (4) reading or recording the wrong numbers.

Precautions must be observed in chaining to avoid loops or tangles that will result in a broken or permanently "kinked" tape. The ends of the tape should be equipped with leather thongs, and the loss of chaining pins can be minimized by tying colored plastic flagging to each. Chains should be lightly oiled occasionally to prevent rust and properly coiled or "thrown" when not in use.

3-4 Methods of Tape Graduation Steel tapes used in forest surveying are commonly 1 or 2 chains in length. They are usually graduated in link intervals, with the first (and often the last) link graduated in tenths to measure fractional distances. Two commonly used methods of graduating the fractional length at the zero end are shown in Fig. 3-1. Before using an unfamiliar tape, one should carefully observe where zero is located and the method used to graduate the fractional link.

Tapes with an additional link beyond the zero end, graduated from zero to 1 link in tenths, are called *adding tapes*. With the rear chainman holding a full-link graduation, the head chainman reads the additional length beyond the zero mark and *adds* this reading to the full-link reading to obtain the distance between the

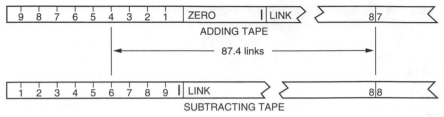

FIGURE 3-1
Reading fractional distances with adding and subtracting tapes.

two points. For example, if the rear chainman is holding 87 links and the head chainman 0.4, the distance between the two points is 87.4 links, or 0.874 chain.

The other type of tape commonly found in practice is graduated from zero to the last link by full-link intervals with the first link (and often the last) further graduated in tenths; these are called *subtracting tapes.* With a full-link graduation held by the rear chainman, the head chainman reads the tenths from the graduated link at the zero end and *subtracts* the reading from the full-link value to obtain the distance between the two points. Thus if the rear chainman were holding 88 links and the head chainman were holding 0.6 link, the distance between two points would be 88.0 − 0.6, or 87.4 links. The adding tape is generally preferred, since errors in measurement and recording are less likely to occur.

3-5 Nomenclature of the Compass In elemental form, a compass consists of a magnetized needle on a pivot point, enclosed in a circular housing that has been graduated in degrees. Because the earth acts as a huge magnet, compass needles in the northern hemisphere point in the direction of the horizontal component of the magnetic field, commonly termed *magnetic north.* If a sighting base is attached to the compass housing, it is then possible to measure the angle between the line of sight and the position of the needle. Such angles are referred to as magnetic *bearings,* or *azimuths.*

Bearings are horizontal angles that are referenced to one of the quadrants of the compass, viz., NE, SE, SW, or NW. Azimuths are comparable angles measured clockwise from due north, thus reading from 0 to 360°. Relationships between bearings and azimuths are illustrated in Fig. 3-2. It will be seen that a bearing of N60°E corresponds to an azimuth of 60°, while a bearing of S60°W is the same as an azimuth of 240°. The angle formed between magnetic north and true north is called *magnetic declination,* and allowance must be made for this factor in converting magnetic bearings and azimuths to true angular readings.

3-6 Magnetic Declination In North America, corrections may be required for either *east* or *west* declination, the former when the north magnetic pole is east of true north and the latter when it is west of true north. Isogonic charts illustrating magnetic declination are issued periodically by government

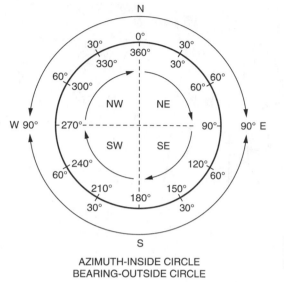

FIGURE 3-2
Relationship of compass bearings and azimuths.

AZIMUTH-INSIDE CIRCLE
BEARING-OUTSIDE CIRCLE

agencies (Fig. 3-3). On such maps, points having equal declination are connected by lines known as *isogons*. The line of zero declination (no corrections required) passing through the eastern section of the country is called the *agonic* line. It will be noted that areas east of the agonic line have west declination, while areas west of the agonic line have east declination.

The agonic line has been shifting westward at a rate of approximately 1 min per year. In some parts of the conterminous United States, however, the change in declination is as high as 4 to 5 min annually. Because the position of the north magnetic pole is constantly shifting, it is important that current declination values be used in correcting magnetic bearings. Where reliable data cannot be obtained from isogonic charts, the amount of declination can be determined by establishing a true north-south line through observations on the sun or Polaris. The magnetic bearing of this true line provides the declination for that locality. As an alternative to this approach, any existing survey line whose true bearing is known can be substituted.

3-7 Allowance for Declination In establishing or retracing property lines, angles should preferably be recorded as *true* bearings or azimuths. The simplest and most reliable technique for handling declination is to set the allowance directly on the compass itself. Thus the graduated degree circle must be rotated until the north end of the compass needle reads true north when the line of sight points in that direction. For most compasses, this requires that the graduated degree circle be turned counterclockwise for east declination and clockwise for west declination.

When there is no provision for setting the declination directly on the compass, the proper allowance can be made mentally in the field, or magnetic bear-

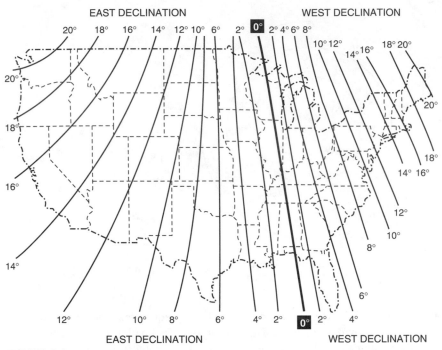

FIGURE 3-3
Lines of equal magnetic declination in the conterminous United States, 1990. *(U.S. Geological Survey, Department of the Interior.)*

ings may be recorded and corrected later in the office. For changing magnetic azimuths to true readings, east declinations are added, and west declinations are subtracted. Thus if a magnetic azimuth of 105° is recorded where the declination is 15° east, the true azimuth would be 120°.

Changing magnetic bearings to true bearings is slightly more confusing than handling azimuths, because declinations must be added in two quadrants and subtracted in the other two. The proper algebraic signs to be used in making such additions or subtractions are illustrated in Fig. 3-4.

Accordingly, if a magnetic bearing of S40°E is recorded where the declination is 5° west, the true bearing, obtained by addition, would be S45°E. In those occasional situations where true bearings and azimuths must be converted back to magnetic readings, all algebraic signs in Fig. 3-4 should be reversed.

3-8 Use of the Compass Whether hand or staff compasses are used, care must be exercised to avoid local magnetic attractions such as wire fences, overhead cables, and iron deposits. In running a traverse, "backsights" of 180° should be taken to check all compass bearings. When such backsights fail to

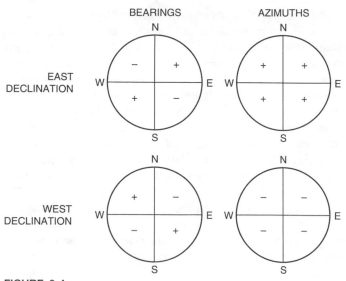

FIGURE 3-4
Algebraic signs for changing magnetic bearings and azimuths to true angles.

agree with foresights and no instrument errors can be detected, it is likely that some form of local attraction is present. Here it may be necessary to shorten or prolong the bearing line in question so that a new turning point outside the attraction area can be used for the compass setup. Compasses having needles immersed in liquid are generally less susceptible to local attractions than nondampened types.

Most good compasses are provided with a means of clamping the needle in a fixed position while the instrument is being transported. After each bearing is read, the needle should be tightened before moving to a new compass position; adherence to this practice will save considerable wear on the sensitive needle pivot point. To ensure accurate compass readings, novices must be cautious to see that (1) the compass is perfectly level, (2) the sights are properly aligned, (3) the needle swings freely before settling, and (4) all readings are taken from the *north end* of the needle. Hand compass shots should not normally exceed 5 chains, and staff compass sights should be limited to about 10 chains per setup.

AREA DETERMINATION

3-9 Simple Closed Traverse   For purposes of this discussion, it is assumed that the primary objectives of a field survey are to locate the approximate boundaries of a tract and determine the area enclosed. Where there are no ownership disputes involved, a simple closed traverse made with staff compass and chain will often suffice for the purposes stated. For most surveys, three persons constitute a minimum crew, and a fourth may be used to advantage. The party chief serves as a compassman and noteman, two individuals chain horizontal distances, and the fourth member handles a range pole at each compass station.

The most reliable property corner available is selected as a starting point; the traverse may be run clockwise or counterclockwise around the tract from this origin. Backsights and front sights should be taken on each line and numbered stakes driven at all compass stations. Immediately upon completion of the traverse, *interior angles* should be computed. If bearings have been properly read and recorded, the sum of all the interior angles should be equal to $(n - 2)$ 180°, where n is the number of sides in the traverse.

After interior angles have been checked, the traverse should be plotted at a convenient scale (for example, 10 chains per in.). If horizontal distances between stations have been correctly chained, the plotted traverse should appear to "close." At this point, the tract area included may be accurately computed by the double meridian distance (DMD) method, or the area can be closely approximated by other techniques. DMD procedures are detailed in all surveying textbooks; hence this approach will not be presented here. Graphical techniques, dot grids, planimeters, and transects are often used by foresters in estimating areas.

3-10   Graphical Area Determination   It may be presumed that a closed compass traverse is plotted at a scale of 10 chains per in. on cross-section paper having 100 subdivisions per sq in. All 1-in. squares thus represent 10 acres, and small squares are 0.1 acre. The total acreage can be quickly determined by counting all small squares enclosed. Where less than one-half of a square is inside the tract boundary, it is ignored; squares bisected by an exterior line are alternately counted and disregarded. The method is fast and reasonably accurate when traverses are correctly plotted and when finely subdivided cross-section paper is employed.

3-11   Dot Grids   If a piece of clear tracing material were placed over a sheet of cross-section paper and pin holes punched at all grid intersections, the result would be a dot grid. Thus dot grid and graphical methods of area determination are based on the same principle; dots *representing* squares or rectangular areas are merely counted in lieu of the squares themselves. The principal gain enjoyed is that fractional squares along tract boundaries are less troublesome, for the dot determines whether or not the square is to be tallied. If an area is mapped at a scale of 10 chains per in. as in the previous example, and a grid having 25 dots per sq in. is used, each dot will represent $^{10}/_{25}$, or 0.4, acre.

Dot grids are commonly used to approximate areas on aerial photographs as well as maps. If the terrain is essentially level and photo scales can be accurately determined, this technique provides a quick and easy method of area estimation

(Fig. 3-5). In regions of rough topography where photographic scales fluctuate widely, area measurements should be made on maps of controlled scale rather than directly on contact prints. Additional uses of aerial photographs are given in Chap. 13.

The optimum number of dots to be counted per square inch depends on the map scale employed, size of area involved, and precision desired. Grids commonly used by foresters may have from 4 to more than 100 dots per sq in. Denser grids are used when the size of the region is small or when more precision is needed. Grids with low dot densities are used for larger tracts or when less precision is needed. In all cases, however, it is recommended that an *average* dot count be obtained by several random orientations of a grid over the same area.

FIGURE 3-5
Dot grid positioned over part of an aerial photograph. Photo scale is 566 ft/in., or about 7.35 acres/sq in. As there are 40 dots per sq in., each dot has a conversion value of approximately 0.18 acre. *(Photograph courtesy Tobin Surveys, Inc.)*

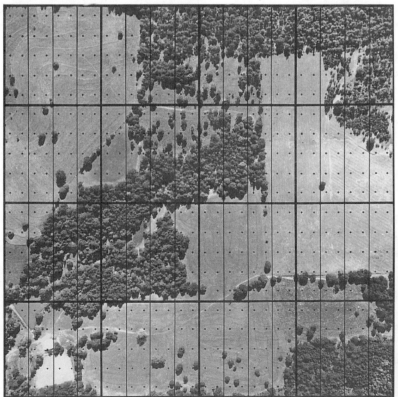

In summary, dot grids are a relatively simple and inexpensive tool for estimating areas on maps or photographs. However, care must be taken because dot grids are considered to be a less precise method for area determination than planimeters.

3-12   Planimeters   A planimeter is a device designed to estimate map or photo area; it is composed of three basic parts: a weighted polar arm of fixed length, a tracer arm hinged on the unweighted end of the polar arm, and a rolling wheel that rests on the map and to which is attached a vernier scale.

There are two basic types of planimeter, polar compensating and linear rolling. A polar compensating planimeter rotates around a fixed point and is limited in the sizes of the areas it can measure. A linear rolling planimeter has no fixed point and thus is not limited by a fixed point. Both types of planimeter are available in electronic form, where the results can be read directly from a display on the device.

In use, the pointer of the instrument is run around the boundaries of an area in a *clockwise* direction; usually the perimeter is traced two or three times for an average reading. From the vernier scale, the area in *square inches* (or other units) is read directly and converted to desired area units on the basis of the map or photo scale. Prolonged use of the planimeter is somewhat tedious, and a steady hand is essential for tracing irregular tract boundaries.

It is often useful to check planimeter estimates of area by use of dot grids, and vice versa. Relative accuracy of the two methods can be approximated by measuring a few tracts of known area. Since individual preferences vary, it may also be informative to compare the *time* required for each estimation technique.

3-13   Transects   The transect method is basically a technique for proportioning a known area among various types of land classifications, such as forests, cultivated fields, and urban uses. An engineer's scale is aligned on a photograph or map so as to cross topography and drainage at right angles. The length of each type along the scale is typically recorded to the nearest 0.1 in. Proportions are developed by relating the total measure of a given classification to the total linear distance. For example, if 10 equally spaced, parallel lines 15 in. long are established on a given map, the total transect length is 150 in. If forest land is intercepted for a total measure of 30 in., this particular classification would be assigned an acreage equivalent to $^{30}/_{150}$, or 20 percent, of the total area. The transect method is simple and requires no special equipment when lines are established with an engineer's scale. Common area conversions for the English and metric systems are given in Table 3-1.

3-14   Topographic Maps   Topographic quadrangle maps (Fig. 3-6) have been prepared for sizable areas of the United States by various governmental agencies. Persons concerned with land surveying often find such maps useful in

TABLE 3-1
CONVERSIONS FOR SEVERAL UNITS OF AREA MEASUREMENT

Square feet	Square chains	Acres	Square miles	Square meters	Hectares	Square kilometers
4,356	1	0.1	0.000156	404.687	0.040469	0.000405
43,560	10	1	0.0015625	4,046.87	0.404687	0.004047
27,878,400	6400	640	1	2,589,998	258.9998	2.589998
107,638.7	24.7104	2.47104	0.003861	10,000	1	0.01
10,763,867	2471.04	247.104	0.386101	1,000,000	100	1

FIGURE 3-6
Portion of a U.S. Geological Survey 7^1/$_2$-min topographic quadrangle map showing the Hayters Gap, Virginia, area. *(Photograph by Rick Griffiths.)*

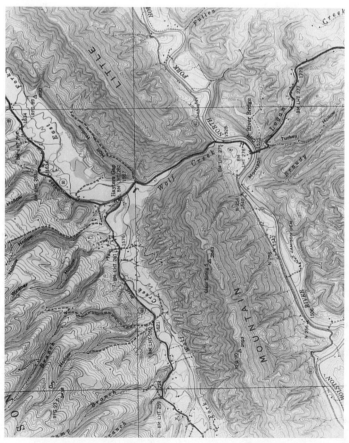

retracing ownership lines, and estimating areas. Topographic maps contain information important to foresters. Contours, or lines connecting points of equal elevation, help forest managers visualize the terrain. Roads are numbered and named, and the names of larger streams are also present. Features such as buildings and cemeteries are also shown on the map and are often helpful when locating the property of interest. Current indexes showing map coverage available in each of the 50 states may be obtained free by contacting the U.S. Geological Survey.

The national topographic map series includes quadrangles and other map series published by the Geological Survey. A map series is a family of maps conforming to the same specifications or having some common unifying characteristic such as scale. Adjacent maps of the same quadrangle series can generally be combined to form a single large map. The principal map series and their essential characteristics are tabulated below:

Map series	Scale	Standard quadrangle size (latitude-longitude)
7 ½ min	1:24,000	7 ½ × 7 ½ min
Puerto Rico 7 ½ min	1:20,000	7 ½ × 7 ½ min
15 min	1:62,500	15 × 15 min
Alaska 1:63,360	1:63,360	16 × 20 to 36 min
U.S. 1:250,000	1:250,000	1 × 2°
U.S. 1:1,000,000	1:1,000,000	4 × 6°

Maps of Alaska and Hawaii may vary from the foregoing standards. The first all-metric topographic maps published by the Geological Survey cover portions of Alaska. Map scale has been set at 1:25,000 and contour intervals are 5, 10, or 20 m. Distances, spot elevations, and similar data are shown in both metric and English units.

3-15   Geographic Information Systems   Much of the forest inventory information that was formerly maintained in the form of type maps, aerial photo overlays, or tabulated compartment data is now being stored, updated, and retrieved through what are commonly called geographic information systems (GISs). A GIS is a computerized data base for storing, manipulating, and displaying map (spatial) data and tabular (attribute) information. In a GIS, forest inventory information can be stored in a computer and directly linked to associated forest maps, which makes it both easier and faster to analyze and graphically display the results of forest inventories.

GISs can make forest inventory information more powerful by allowing resource managers to integrate it with other data commonly needed to make management decisions. For instance, managers may need to know the location of roads, streams, threatened or endangered species, or sensitive soils when developing management plans. Combining forest inventory data with other land-resource information allows managers to make more informed decisions. GISs have been, and continue to be, developed to provide essential information quickly and efficiently.

In a GIS, the basic types of data are separated into individual maps called themes (Fig. 3-7). The locations of each feature in each theme are stored in the computer data base, and the characteristics associated with that feature are linked to it. For example, the locations of stand boundaries are stored in the stands theme of the GIS, along with the inventory results (e.g., number of trees, volume) for each stand. Similarly, the locations and attributes (flow rate, pH, species present, etc.) of streams, soil types with their characteristics (soil series, drainage class, texture, erodability class, etc.), and other relevant resources are entered into the GIS. Each of these themes can be examined individually or in combination with others. Once maps of varying scales and formats have been digitized and stored, a GIS permits superimposition of map information for different themes. The combination of GISs with timely and efficient resource inventories will continue to play a central role in resource management decision making.

FIGURE 3-7
Data from diverse sources can be stored, manipulated, and displayed by use of geographic information systems.

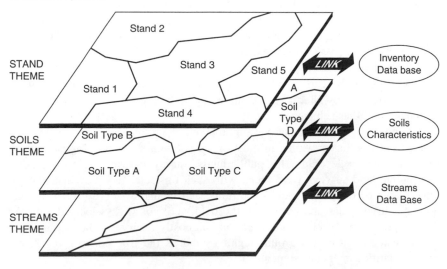

COLONIAL LAND SUBDIVISION

3-16   Metes and Bounds Surveys   A sizable segment of the United States, notably in the original 13 colonies, was subdivided and passed into private ownerships prior to the inauguration of a system for disposal of public lands in 1785. Many of these early land holdings were marked off and described by "metes and bounds," a procedure sometimes facetiously referred to as leaps and bounds.

The term *mete* implies an act of metering, measuring, and assigning by measure, and *bounds* refers to property boundaries or the limiting extent of an ownership. In some instances, older metes and bounds surveys may consist entirely of descriptions rather than actual measurements, e.g., "starting at a pine tree blazed on the east side, thence along a hedgerow to a granite boulder on the bank of the Wampum River, thence along the river to the intersection of Cherokee Creek . . . , etc." Fortunately, most metes and bounds descriptions are today referenced by bearings, distances, and permanent monuments. Even so, parcels of land are shaped in unusual and seemingly haphazard patterns, and a multitude of legal complexities can be encountered in attempting to establish the location of a disputed boundary along an old stone fence that disintegrated 50 years ago. Descriptions of metes and bounds surveys can ordinarily be obtained from plat books at various county court houses.

THE U.S. PUBLIC LAND SURVEY

3-17   History   Most of the United States west of the Mississippi River and north of the Ohio River, plus Alabama, Mississippi, and portions of Florida, has been subdivided in accordance with the U.S. Public Land Survey (Fig. 3-8). The first law governing public land surveys was enacted by Congress in 1785. That part of the Northwest Territory which later became the state of Ohio was the experimental area for the development of the rectangular system. The original intent was to establish *townships* exactly 6 miles square, followed by subdivision into 36 sections of exactly 1 mile square each. At first, no allowance was made for curvature of the earth, and numerous problems resulted. However, survey rules were revised by later acts of Congress, and the present system evolved as a culmination of these changes.

Adoption of a rectangular system marked the transition from metes and bounds surveys that prevailed in most of the colonial states to a logical and rational method for describing the public lands. Surveyors responsible for the earliest public land surveys were faced with such obstacles as crude instruments, unfavorable or dangerous field conditions, and changing survey rules. Consequently, survey lines and corners in the field were not always located with the desired precision. To eliminate litigation and costly resurveys, the original corners as established on the ground legally stand as the true corners, regardless of irregularities or inconsistencies.

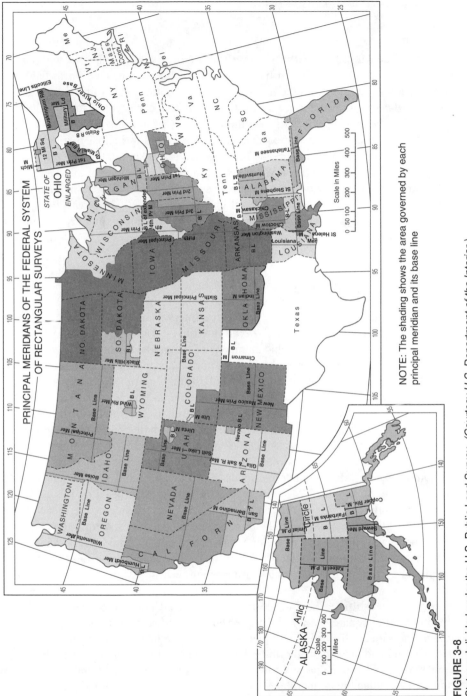

FIGURE 3-8
States subdivided under the U.S. Public Land Survey. *(Courtesy U.S. Department of the Interior.)*

NOTE: The shading shows the area governed by each principal meridian and its base line

47

3-18   The Method of Subdivision   The origin of a system begins with an *initial point,* usually established by astronomical observation. Passing through and extending outward from the initial point is a true north-south line known as a *principal meridian* and a true east-west *base line* that corresponds to a parallel of latitude. These two lines constitute the main axes of a system, and there are more than 30 such systems in existence (Fig. 3-8). Each principal meridian is referenced by a name or number, and the meridian is marked on the ground as a straight line. The base line is curved, being coincident with a geographic parallel. Starting at the initial point, the area to be surveyed is first divided into *tracts* approximately 24 miles square, followed by subdivision into 16 *townships* approximately 6 miles square and then into 36 sections approximately 1 mile square. An idealized system is shown in Fig. 3-9.

3-19   The 24-Mile Tracts   At intervals of 24 miles north and south of the base line, *standard parallels* are extended east and west of the principal meridian. These parallels are numbered north and south from the base line, as "first standard parallel north," and so on. At 24-mile intervals along the base line and along all standard parallels, *guide meridians* are run on *true north* bearings; these lines thus correspond to geographic meridians of longitude. Each guide meridian starts from a standard corner on the base line or on a standard parallel and ends at a closing corner on the next standard parallel to the north. Standard parallels are never crossed by guide meridians. Guide meridians are numbered east and west from the principal meridian, as "first guide meridian east," and so forth.

The tracts are 24 miles wide at their southern boundaries, but because guide meridians converge, they are less than 24 miles wide at their northern boundaries. As a result, there are two sets of corners along each standard parallel. *Standard corners* refer to guide meridians north of the parallel, while *closing corners* are those less than 24 miles apart which were established by the guide meridians from the south closing on that parallel. Convergence of meridians is proportional to the distance from the principal meridian; the offset of the second guide meridian is double that of the first, and that of the third guide meridian is 3 times as great. Of course, actual offsets on the ground may differ from theoretical distances because of inaccuracies in surveying.

3-20   Townships   The 24-mile tracts are divided into 16 townships, each roughly 6 miles square, by north-south *range lines* and east-west *township lines.* Range lines are established as true meridians at 6-mile intervals along each standard parallel and are run due north to the next standard parallel. Township lines are parallels of latitude that join township corners at intervals of 6 miles on the principal meridian, guide meridians, and range lines. Since range lines converge northward just as guide meridians do, the width of a township decreases from south to north, the shape is trapezoidal rather than square, and the area is always less than the theoretical 36 square miles.

The survey of townships within the 24-mile tract begins with the southwest township and continues northward until the entire west range is completed; then it moves to the next range eastward and again proceeds from south to north. Townships are numbered consecutively northward and southward from the base line and eastward and westward of the principal meridian. As illustrated in Fig. 3-9, T2N, R3W denotes a township that is 6 miles north of the base line and 12 miles west of the principal meridian.

FIGURE 3-9
Idealized subdivision of townships and sections

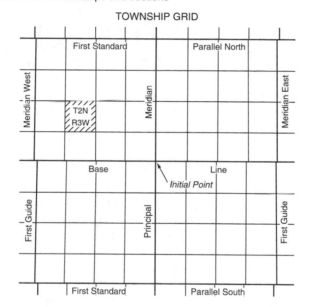

TOWNSHIP GRID

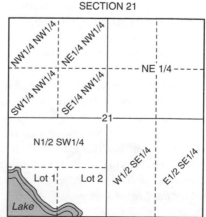

T2N   R3W

SECTION 21

3-21   Establishment of Sections and Lots Beginning in the southeast corner of a township, sections of approximately 640 acres are formed by running lines 1 mile apart parallel to eastern range lines and 1 mile apart parallel to southern township lines. By starting in the southeastern part of the township, irregularities are thrown into the northern and western tiers of sections in each township. Survey lines are first run around section 36, then 25, 24, 13, 12, and 1. The township subdivision thus starts at the eastern boundary and proceeds from south to north, establishing one tier of sections at a time. Sections are numbered as in Fig. 3-9.

Survey corners actually established on the ground include section corners and quarter corners, the latter being set at intervals of 40 chains for subdividing the sections into 160-acre tracts. These quarter sections may later be further divided into 40-acre parcels known as "forties." A complete land description begins with the smallest land parcel and covers each division in order on a size basis; the specific principal meridian involved is also part of the description. Thus the forty composing the most northwesterly portion of section 21 (Fig. 3-9) would be described as NW^1/$_4$ NW^1/$_4$ S.21, T2N, R3W, 5th P. M. To derive the approximate number of acres in a subdivision, the area of the section is multiplied by the product of the fractions in the legal description. From the previous example, ^{1}/$_4$ × ^{1}/$_4$ × 640 = 40 acres.

Accumulation of irregularities in northern and western tiers of sections often results in parcels of land that have an area considerably less than the 40 or 160 acres intended. Such subdivisions may be individually numbered as *lots.* Also, navigable streams and large bodies of water encountered on survey lines are meandered by running traverses along their edges. *Meander corners* set during such surveys may result in the recognition of additional irregularly shaped *lots* that commonly range from 20 to 60 acres in size.

3-22   Survey Field Notes Complete sets of field notes describing public land surveys can be obtained from the U.S. General Land Office in Washington, D.C., and from most state capitals. Field notes are public records, and only a nominal charge is made for copying them. They include bearings and distances of all survey lines, descriptions of corners, monuments and bearings objects, and notes on topography, soil quality, and forest cover types.

Field notes are essential for locating lost or obliterated survey corners from bearings objects or "witness trees." On original surveys, such objects were identified and located by recording a bearing and distance *from the corner to the object.* As a result, lost corners may be reestablished by reversing all bearings and chaining the specified distances from witnesses or bearings objects. Specific procedures for relocating original survey lines and corners are detailed in the U.S. Department of Interior's *Manual of Instructions for the Survey of the Public Lands of the United States.*

3-23 Marking Land Survey Lines In forested regions where land has been subdivided in accordance with the U.S. Public Land Survey, specific rules are sometimes formulated for marking trees so that different classes of land lines can be easily identified. Trees along township, range, and section lines may be marked with three bark blazes, preferably placed vertically on the tree stem near eye level. Quarter-section lines are referenced by two blazes, and finer subdivisions such as forty lines are indicated by a single blaze.

To avoid injury to trees, bark blazes 4 to 6 in. in diameter should be made with a drawknife and then painted with an appropriate color. When feasible, it is desirable to have blazing techniques and paint colors standardized in a given forest region, especially where numerous ownerships are represented. The following marking paint colors have been recommended by foresters in the Lake states:

Type of marking	Paint color
Property boundaries	Blue
Sale boundaries	Red
Cut trees	Yellow or orange
Leave trees	Light green
Research and inventory plots	White
Trails	Aluminum

PROBLEMS

3-1 Establish a pacing course and determine your (a) number of paces per chain, (b) number of paces required to measure the radii of $^1/_{10}$-, $^1/_5$-, and $^1/_4$-acre circular plots, and (c) number of feet per pace.

3-2 Why are divisions of a surveyor's chain called *links?* What is meant by the markings 1P, 2P, 3P, and 4P found on a surveyor's chain?

3-3 Change these magnetic bearings to true bearings, utilizing the declination specified for your locality: (a) N35°E, (b) S88°E, (c) S10°W, (d) N61°W. Express these bearings as azimuths, and then convert them to backsights.

3-4 Orient a staff compass with true north. Why are the positions of east and west reversed on the face of the compass?

3-5 What should be the sum of the interior angles for closed traverses having (a) three sides, (b) eight sides, (c) 12 sides?

3-6 Plot a closed traverse on cross-section paper at a scale of 20 chains per in. Determine the area by the graphical method and then by use of transects.

3-7 Assume you have a dot grid with 36 dots per sq in. What acreage will be represented by each dot at map scales of (a) 660 ft per in., (b) 25 chains per in., (c) 1 mile per in.?

3-8 Measure a given map area by means of the graphical method, dot grid, transects, and planimeter. Which technique is fastest? Which do you feel is most precise?

3-9 On standard USGS topographic quandrangles, how are contours and bench marks designated? What technique is used to denote woodland cover? What map symbols are used for (a) railroads, (b) power transmission lines, (c) churches, (d) schools, (e) airfields?

3-10 Visit the nearest property records repository in your locality. Prepare a facsimile of either (a) a plat and description of a metes and bounds survey or (b) a sample page of field notes from a GLO plat book. If feasible, supplement these data with a recent aerial photograph of the same locality.

REFERENCES

Aronoff, S. 1989. *Geographic information systems: A management perspective.* WDL Publications, Ottawa, Ontario. 294 pp.

Barrett, J. P., et al. 1979. Timber values of town forests. *Univ. of New Hampshire, N.H. Agr. Expt. Sta. Res. Report* 77. 44 pp.

Bolstad, P. V., and Smith, J. L. 1992. Errors in GIS. *J. Forestry* **90:**21–26, 29.

Brown, C. M. 1962. *Boundary control and legal principles.* John Wiley & Sons, New York. 275 pp.

Bryan, R. W. 1979. Sophisticated inventory system puts forest data on computer. *Forest Ind.* (January), 2 pp.

Congalton, R. G., and Green, K. 1992. The ABCs of GIS. *J. Forestry* **90:**13–20.

Davis, L. S. 1980. Strategy for building a location-specific, multi-purpose information system for wildland management. *J. Forestry* **78:**402–408.

Davis, R. E., Foote, F. S., and Kelly, J. W. 1966. *Surveying: Theory and practice,* 5th ed. McGraw-Hill Book Company, New York. 1096 pp.

U.S. Department of the Army. 1964. *Elements of surveying.* Tech. Manual TM 5-232, Government Printing Office, Washington, D.C. 247 pp.

U.S. Department of Commerce. 1962. Magnetic poles and the compass. Serial 726, Coast and Geodetic Survey, Government Printing Office, Washington, D.C. 9 pp.

U.S. Department of the Interior. 1973. *Manual of instructions for the survey of the public lands of the United States.* Government Printing Office, Washington, D.C. 333 pp.

Warren, B. J., and Cook, W. L. 1984. Surveying. Pp. 1089–1116 in *Forestry Handbook,* K. F. Wenger (ed.), John Wiley & Sons, New York.

Wilson, R. L. 1989. *Elementary forest surveying and mapping.* O.S.U. Bookstores, Inc., Corvallis, Oreg. 181 pp.

CUBIC VOLUME, CORD MEASURE, AND WEIGHT SCALING

4-1 Logs, Bolts, and Scaling Units When trees are cut into lengths of 8 ft or more, the sections are referred to as *logs*. By contrast, shorter pieces are called *sticks,* or *bolts*. The process of measuring volumes of individual logs is termed *scaling*. Logs may be scaled in terms of cubic feet, cubic meters, board feet, weight, and other units. The cubic foot is an amount of wood equivalent to a solid cube that measures 12 × 12 × 12 in. and contains 1728 cu in. The cubic meter, used in countries that have adopted the metric system, contains 35.3 cu ft. The board foot is a plank 1 in. thick and 12 in. square; i.e., it contains 144 cu in. of wood. The present discussion is concerned with log scales expressed in cubic volumes; board-foot volumes and scaling of sawlogs are described in Chap. 5.

Although tree cross sections rarely form true circles, they are normally presumed to be circular for purposes of computing cross-sectional areas. In measuring cubic-foot contents of logs, it is desirable to derive cross-sectional areas in square feet rather than in square inches; cubic volumes are then derived by multiplying average cross-sectional area times log length in feet. As diameters instead of radii of logs are measured, area in square inches may be derived by

$$\text{Area in square inches} = \frac{\pi D^2}{4}$$

where D is the log diameter in inches.

For this relationship, it is then necessary to divide the result by 144 to convert the area to square feet. This is most easily accomplished in one step by reducing the formula as follows:

$$\text{Area} = \frac{\pi D^2}{4(144)} = \frac{3.1416 D^2}{576} = 0.005454 D^2$$

By this simple conversion, cross-sectional areas are derived in square feet when diameters are measured in inches.

4-2 Log Volumes and Geometric Solids   If logs were perfectly cylindrical, cubic volumes would be derived by merely multiplying cross-sectional area in square units of length times log length. However, as logs taper from one end to another, only short sections of perhaps a few inches can logically be treated as cylinders. Still, there are several common geometric solids from which truncated sections can be extracted to approximate log forms (Fig. 4-1). Volumes of these solids of revolution are computed as follows:

Name of solid	Volume computation
Paraboloid	Area/2 × length
Conoid	Area/3 × length
Neiloid	Area/4 × length

As a rule, trees approximate the shape of truncated neiloids while the effects of butt swell are apparent. Logs from middle sections of tree stems are similar to truncated paraboloids, while upper logs approach the form of conoids.

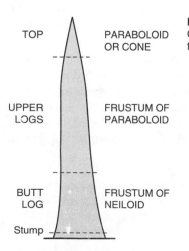

TOP

UPPER
LOGS

BUTT
LOG

Stump

PARABOLOID
OR CONE

FRUSTUM OF
PARABOLOID

FRUSTUM OF
NEILOID

FIGURE 4-1
Geometric shapes assumed by different portions of tree boles.

Cubic volumes for all solids of revolution are computed from the product of their *average* cross-sectional area and length. Thus in computing log volumes, the principal problem encountered is that of accurately determining the elusive average cross section. Three common formulas applied to this end are

$$\text{Huber's: Cubic volume} = B_{1/2}(L)$$

$$\text{Smalian's: Cubic volume} = \frac{(B + b)}{2} L$$

$$\text{Newton's: Cubic volume} = \frac{(B + 4B_{1/2} + b)}{6} L$$

where $B_{1/2}$ = cross-sectional area at log midpoint
$\quad B$ = cross-sectional area at large end of log
$\quad b$ = cross-sectional area at small end of log
$\quad L$ = log length

Areas and volumes are computed inside bark and may be expressed in either English or metric units. Huber's formula assumes that the average cross-sectional area is found at the midpoint of the log; unfortunately, this is not always true. The formula is regarded as intermediate in accuracy, but its use is limited because (1) bark measurements or empirical bark deductions are required to obtain mid-diameters inside bark and (2) the midpoints of logs in piles or ricks are often inaccessible and cannot be measured.

Smalian's formula, though requiring measurements at both ends of the log, is the easiest and least expensive to apply. It also happens to be the least accurate of the three methods, especially for butt logs having flared ends. Excessive butt swell must be allowed for by ocularly projecting a normal taper line throughout the log, or by cutting flared logs into short lengths to minimize the effect of unusual taper. Otherwise, log volumes may be overestimated by application of an average cross-sectional area that is too large. Because it has neither of the disadvantages cited for Huber's formula, Smalian's method of volume computation holds the greatest promise of the three for production log scaling.

Newton's formula necessitates the measurement of logs at the midpoint and at both ends. Although it is more accurate than the other two methods, the expense incurred in application limits its use to research, experimental techniques, and checks against other cubic volume determinations. It will be noted that for perfect cylinders, all three formulas provide identical results.

4-3 Commercial Scaling by the Cubic Foot   For species having a limited degree of natural taper, logs may be simply scaled as cylinders based on small-end diameters or cross sections. Such an approach has an obvious time and cost advantage over Smalian's method which requires measuring both ends of the log; furthermore, volume outside the scaling cylinder may be safely ignored for short log sections. Disregarding taper would be quite logical for items such as rotary-cut veneer logs, because little or no commercial veneer is produced until these logs have been reduced to cylindrical form.

Where log lengths are variable, taper cannot be completely disregarded. An allowance for volume outside the scaling cylinder may be made by applying a *fixed rate of taper* to all logs of a given species-group. For example, a taper rate of $1/2$ in. per 4 ft of length might be established for a certain species. The volume of a 16-ft log with a small-end diameter of 20 in. would be computed by 4-ft sections as follows:

Diameter, in.	Area, sq ft		Length, ft		Volume, cu ft
20.0	2.1817	×	4	=	8.7268
20.5	2.2921	×	4	=	9.1684
21.0	2.4053	×	4	=	9.6212
21.5	2.5212	×	4	=	10.0848
					37.6012 or 38

This computational procedure would not be followed by scalers themselves, of course. Instead, they would read the appropriate value from a special table showing volumes for many combinations of log diameters and lengths. Such tables are called *log rules.*

There have been numerous attempts to promote the cubic foot as the national log-scaling unit in the United States. However, the obvious advantages of a clearly defined measure which is independent of utilization standards, manufacturing efficiencies, and final product form have not completely prevailed. The U.S. Forest Service has adopted cubic-foot scaling of logs and timber for sales on national forests, but boardfoot scaling is still commonly used in many sectors of the forest products industry. Weight scaling of sawlogs also has possibilities for providing an objective standard of measurement.

4-4   Inscribed Square Timbers   It is sometimes necessary to determine quickly the dimensions of square timbers that can be cut from logs of various scaling diameters. Such information may be useful in measuring hewed products such as railroad ties or timbers when the outer portions of logs are wasted or ignored. The problem is basically one of fitting the largest possible square inside a circle of a specified size. A formula that will provide the length of this side S of an inscribed square from log diameter D is

$$\text{Side } S = \sqrt{\frac{D^2}{2}}$$

Cubic volumes of inscribed square timbers may be determined by merely squaring the length of the side and multiplying by log length. When the side is measured in inches and length is expressed in feet, the product must be divided by 144 for conversion to cubic feet.

MEASURING STACKED WOOD AND CHIPS

4-5   The Cord   A standard cord of wood is a rick that measures $4 \times 4 \times 8$ ft and contains 128 cu ft. Inasmuch as this space includes wood, bark, and sizable voids, the cord is more of an indication of space occupied than actual wood measured. Of course, cordwood is not necessarily cut into 4-ft lengths, and it is rarely stacked in rectangular ricks having 32 sq ft of surface area. Any stacked rick of roundwood may be converted to standard cords by

$$\frac{\text{Width (ft)} \times \text{height (ft)} \times \text{stick length (ft)}}{128}$$

When cordwood is cut into lengths shorter than 4 ft (e.g., firewood), a rick having 32 sq ft of surface area may be referred to as a *short cord*. If a similar rick is made up of bolts longer than 4 ft, it may be termed a *long cord*, or *unit*. In the United States, pulpwood is commonly cut into lengths of 5, 5.25, and 8.33 ft. When these stick lengths are multiplied by a cord surface area of 32 sq ft, the resulting units occupy 160, 168, and 266.6 cu ft of space, respectively. The *cunit* refers to 100 cu ft of *solid wood* rather than to stacked volume. Typical specifications for pulpwood purchased in the United States are as follows:

1 Bolts must be at least 4 in. in diameter inside bark at the small end and cut to the specified length.

2 Bolts must not exceed (18 to 24) inches in diameter outside bark at the large end.

3 Wood must be sound and straight.

4 Ends must be cut square and limbs trimmed flush.

5 No burned or rotten wood will be accepted.

6 All nails and other metals must be removed from bolts.

7 Mixed loads of pines and hardwoods are not acceptable.

Where the metric system is employed, stacked wood is measured in cubic meters; 1 m^3 is equivalent to 35.3 cu ft. A stacked rick of wood is converted to cubic meters by simply measuring all three dimensions of the rick in meters and obtaining the product of these dimensions:

$$m^3 = \text{width (m)} \times \text{height (m)} \times \text{bolt length (m)}$$

4-6 Solid Contents of Stacked Wood Purchasers of cordwood and pulp-wood are primarily interested in the amount of solid-wood volume contained in various ricks rather than in the total space occupied. The tabulation that follows provides a rough approximation of the volume of wood and bark for ricks with 32 sq ft of surface area and varying bolt lengths:

Size of rick, ft	Space occupied, cu ft	Solid wood and bark, cu ft
$4 \times 8 \times 4$	128	90
$4 \times 8 \times 5$	160	113
$4 \times 8 \times 5.25$	168	119
$4 \times 8 \times 6$	192	136
$4 \times 8 \times 8.33$	266.6	181

It should be borne in mind that the foregoing values are merely estimates that can vary greatly in individual situations. The species, method of piling, diameter and length of sticks, straightness, and freedom from knots can exert a significant influence on these average values, as illustrated in Figs. 4-2 to 4-4. Thus the cord method of measurement has been used in the past primarily because of simplicity and convenience rather than because of its accuracy.

When cubic-foot conversions are desired for solid wood alone, the volume of bark present must also be considered in establishing pulpwood values. For many coniferous species, bark may compose 10 to 30 percent of the total stick volume; in the Lake states, for example, the average is about 14 percent. Depending on bark volume and other factors outlined, the solid-wood content of a stacked cord may range from about 60 to 95 cu ft. A value of 79 has been widely used in the Lake states, and U.S. Forest Service studies in the South produced averages of

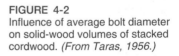

FIGURE 4-2
Influence of average bolt diameter on solid-wood volumes of stacked cordwood. *(From Taras, 1956.)*

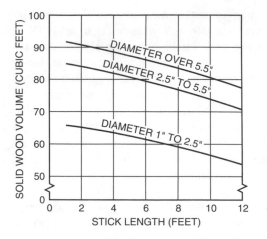

FIGURE 4-3
Influence of length of stick on solid-wood volume of a cord for conifers. *(From Taras, 1956.)*

72 cu ft per cord for southern pines and 79 for pulping hardwoods. Mills purchasing pulpwood have traditionally developed their own cubic-foot conversions based on the size and quality of wood being purchased; there is no single value that can be considered reliable in any given forest region.

Estimates of solid-wood content can be made with a steel tape graduated into 100 equal parts. The tape is stretched diagonally across the ends of the stacked bolts, and the number of graduations falling on wood (as opposed to bark and voids) provides the desired proportion. Several observations should be made for each rick to obtain reliable estimates.

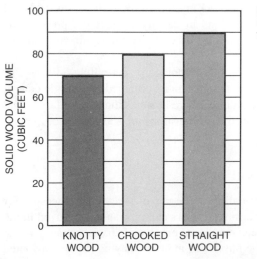

FIGURE 4-4
Effect of stick quality on solid-wood volume of 5-ft cordwood. *(From Taras, 1956.)*

4-7 An Ideal Measure The perfect unit of measurement would be one that is absolute, unambiguous, accurate, simple, and inexpensive to apply (Ker, 1962). If only peeled or debarked wood were purchased, and sticks could be individually measured, the cubic foot might be a logical unit for adoption. However, since low-value pulpwood bolts cannot be economically handled as separate items, this technique could be accurate but certainly not simple or inexpensive. The *xylometer,* or water-immersion method, has also been considered for determining cubic volumes, but few equipment models have progressed beyond the experimental stage. By this technique, volume of wood is derived through application of Archimedes' principle; i.e., the measured volume of water displaced is equivalent to the cubic volume of wood immersed.

4-8 Inventories of Chip Piles Large stockpiles of pulpwood chips, bark, or sawdust may be periodically measured for inventory and cost-accounting purposes. In the past, such inventories have been made by plane-table surveys or by ground surveys designed for computing stockpile cross sections. Today, cubic volumes of materials up to 100 ft high and covering 50 acres of land or more may be estimated efficiently by photogrammetric methods.

By this approach, stockpiles are first imaged on stereoscopic pairs of large-scale aerial photographs (e.g., 1:500). The stockpiles are then contoured at intervals of 1 ft by using a stereoscopic plotting instrument based on the floating-dot (parallax) principle as described in Chap. 13. After contours have been determined, the area of each contoured layer, or "slice," is determined by planimetry and the cubic volume of the pile computed, layer by layer.

If the stockpile consists of wood chips, the calculated cubic volume may be converted to weight units by periodic sampling of the chips to establish local weight/volume ratios. It has been determined, for example, that green southern pine chips weigh approximately 288 to 384 kg/m^3 (18 to 24 lb per cu ft). Local corrections should be made for variations in density in different piles of the same materials because settling and compaction will result in significant changes in weight/volume ratios.

A photogrammetric assessment has the advantage of permitting an easy cutoff time for inventories since all photographs can be obtained on a single date. It provides a permanent record of the stockpile size at a specific date and time, and volume can be rechecked at any later time if questions arise as to the accuracy of the estimate.

WEIGHT SCALING OF PULPWOOD

4-9 The Appeal of Weight Scaling The use of weight as a measure for purchasing wood is not a new idea; mine timbers and pine stumps utilized in the wood naval stores industry have been purchased on a weight basis for many years. The appeal of weight scaling in the pulpwood industry may be largely attributed to changes in the locale of measurement and purchases. Whereas wood was formerly scaled in the forest, measurements are now made at concentration yards or at the mill. Since 1955, a large segment of the pulp and paper industry has adopted weight scaling in lieu of linear measurements for stacked pulpwood (Fig. 4-5.)

Weight/price equivalents are usually based on studies of freshly cut wood. It is thus implicit that mills favor green wood with a high moisture content or that purchasers are prepared to assume the cost of carrying large wood inventories for seasoning purposes. As green wood is preferred in most instances, there is some incentive for the producer to deliver his wood immediately after cutting. While there are indications that many species lose very little moisture during the first 4 to 8 weeks in storage, the widespread belief that pulpwood seasons and loses weight rapidly works in favor of mills that desire freshly cut material. From the mill inventory viewpoint, the greener the wood delivered, the longer it can be stored at the yard without deterioration—an important consideration in warm and humid regions.

4-10   Variations in Weight   Most mills now utilizing weight scaling have developed their own local conversions by making paired weighings and cord-wood measurements of thousands of purchases. Weight equivalents may vary by species, mill localities, and points of wood origin. The influence of former

FIGURE 4-5
Weight scaling of truck-delivered pine pulpwood. *(U.S. Forest Service photograph.)*

measures in cords is reflected in the newer weight units; instead of wood being purchased by the ton or hundredweight, prices are usually based on average weights per standard cord.

The principal factors contributing to weight variations for a given species are wood volume, moisture content, and specific gravity. Variations in wood volume, or the actual amount of solid wood in a cord, are caused by differences in bolt diameter, length, quality, and bark thickness. Moisture content varies within species for heartwood versus sapwood. Specific gravity is affected by percent of summerwood and position in the tree; i.e., density tends to decrease from the lower to the top portion of the stem.

4-11   Wood Density and Weight Ratios   From a knowledge of moisture content and specific gravity (based on ovendry weight and green volume), the weight per cubic foot of any species may be computed by

$$\text{Density} = \text{specific gravity} \times 62.4 \left(1 + \frac{\% \text{ moisture content}}{100}\right)$$

The utility of this formula is illustrated by this example: assume that a weight/cord equivalent is required for loblolly pine, a species having a specific gravity of 0.46, moisture content of 110 percent, solid-wood volume per cord of 72 cu ft, and an estimated bark weight of 700 lb per cd. Substituting in the formula, we have

$$\text{Density} = 0.46 \times 62.4 \left(1 + \frac{110}{100}\right) = 60.3 \text{ lb per cu ft}$$

As there are 72 cu ft of wood per cord, the weight of solid wood per cord is 72×60.3, or 4342 lb. By adding the bark weight of 700 lb, the total weight is found to be 5042 lb per cd. Weight equivalents in Table 4-1 were derived by this procedure.

TABLE 4-1
COMPUTATION OF PULPWOOD WEIGHTS FOR SOUTHERN PINES[*]

Species	Specific gravity	Moisture content, %	Density, lb per cu ft	Solid volume per cord, cu ft	Weight of solid wood, lb	Estimated bark weight, lb	Total weight of wood and bark, lb
Loblolly pine	0.46	110	60.3	72	4342	700	5042
Longleaf pine	0.53	105	67.8	72	4882	650	5532
Shortleaf pine	0.44	120	60.4	72	4349	500	4849
Slash pine	0.52	120	71.4	72	5141	500	5641

*From Taras, 1956.

The foregoing technique of computing weight/cord ratios is valid only when both specific gravity and moisture content are accurately determined, because small variations in these factors can result in large weight changes. For each 0.02 change in specific gravity at a level of 100 percent moisture content, the weight of wood will change about 2.5 lb per cu ft. At the 100 percent level, a moisture content difference of 5 percent can cause a weight change of 1 to 2 lb per cu ft. In a cord having 72 cu ft of solid wood, the 0.02 change in specific gravity alone could mean a loss or gain of 180 lb of wood, while the stated moisture difference might involve 72 to more than 100 lb (Taras, 1956).

Table 4-2 was derived by solving the wood-density formula for a wide range of specific gravities and wood moisture contents. It may therefore be applied in developing approximate weight factors for a variety of tree species. The reader should remember, however, that most mills develop their price ratios by actual scaling of delivered wood rather than by this theoretical approach.

For those who wish to compute wood densities directly in metric units, the preceding relationship is modified to

$$\text{Density} = \text{sp gr} \times 1000 \left(1 + \frac{\%\text{ moisture content}}{100} \right)$$

TABLE 4-2
WEIGHT IN POUNDS PER CUBIC FOOT OF GREEN WOOD AT VARIOUS VALUES OF SPECIFIC GRAVITY AND MOISTURE CONTENT

Moisture content of wood, %	Weight in pounds per cubic foot for specific gravities* of:										
	0.30	0.34	0.38	0.42	0.46	0.50	0.54	0.58	0.62	0.66	0.70
30	24.3	27.6	30.8	34.1	37.3	40.6	43.8	47.0	50.3	53.5	56.8
40	26.2	29.7	33.2	36.7	40.2	43.7	47.2	50.7	54.2	57.7	61.2
50	28.1	31.8	35.6	39.3	43.1	46.8	50.5	54.3	58.0	61.8	65.5
60	30.0	33.9	37.9	41.9	45.9	49.9	53.9	57.9	61.9	65.9	69.9
70	31.8	36.1	40.3	44.6	48.8	53.0	57.3	61.5	65.8	70.0	74.3
80	33.7	38.2	42.7	47.2	51.7	56.2	60.7	65.1	69.6	74.1	78.6
90	35.6	40.3	45.1	49.8	54.5	59.3	64.0	68.8	73.5	78.2	83.0
100	37.4	42.4	47.4	52.4	57.4	62.4	67.4	72.4	77.4	82.4	87.4
110	39.3	44.6	49.8	55.0	60.3	65.5	70.8	76.0	81.2	86.5	91.7
120	41.2	46.7	52.2	57.7	63.1	68.6	74.1	79.6	85.1	90.6	96.1
130	43.1	48.8	54.5	60.3	66.0	71.8	77.5	83.2	89.0	94.7	100.5
140	44.9	50.9	56.9	62.9	68.9	74.9	80.9	86.9	92.9	98.8	104.8
150	46.8	53.0	59.3	65.5	71.8	78.0	84.2	90.5	96.7	103.0	109.2

*Based on weight when ovendry and volume when green. Values may be converted to kilograms per cubic meter by multiplying by 16.0185.
Source: "Wood Handbook," USDA, 1955.

The validity of this relationship can be verified from the earlier example. The previous result of 60.3 lb per cu ft multiplied by a conversion factor of 16.0185 is equivalent to 965.92 kg/m³. From the modified formula, we have

$$\text{Density} = 0.46 \times 1000 \left(1 + \frac{110}{100} \right) = 966 \text{ kg/m}^3$$

4-12   Advantages of Weight Scaling   The technique of weight scaling roundwood materials continues to gain in popularity because of these and other reasons:

1 It encourages delivery of freshly cut wood to the mill.

2 The method is fast, requires no special handling, and saves time for both buyer and seller. A greater volume of wood can be measured in a shorter time period and with fewer personnel.

3 Weight scaling is more objective than manual scaling, and positive records of all transactions are provided by automatically stamped weight tickets.

4 Incentive is provided for better piling of wood on trucks; this tends to increase the volume handled by the supplier.

5 Woodyard inventories are more easily maintained because of greater uniformity in record keeping.

PROBLEMS

4-1 From your instructor or a logging operation, obtain dimensions of three merchantable logs cut from the same tree. Determine cubic volumes of each by Huber's, Smalian's, and Newton's formulas. Tabulate results and explain reasons for differences noted.

4-2 Compute the volume (in cubic feet) for the following logs according to Huber's, Smalian's, and Newton's formulas:

a small-end diameter (ib) = 6.1 in.
midpoint diameter (ib) = 7.5 in.
large-end diameter (ib) = 9.0 in.
length = 16 ft

b small-end diameter (ib) = 24.0 in.
midpoint diameter (ib) = 26.4 in.
large-end diameter (ib) = 28.7 in.
length = 32 ft

c small-end diameter (ib) = 10.4 in.
midpoint diameter (ib) = 10.9 in.
large-end diameter (ib) = 11.5 in.
length = 8 ft

d small-end diameter (ib) = 32.0 in.
midpoint diameter (ib) = 32.8 in.
large-end diameter (ib) = 33.7 in.
length = 18 ft

4-3 Construct a working model of a xylometer. Use the model to determine the cubic volumes of 10 small pieces of roundwood. Compare with volumes computed by Smalian's and Huber's formulas for the same pieces. Explain reasons for the differences.

4-4 Determine the gross volume of pulpwood (in standard cords) on a railroad car or truck in your locality. Make notes on stick quality, length, range of diameters, and method of piling. Compute the total *value* of the wood at locale of measurement by application of current pulpwood prices in your vicinity.

4-5 Using gross pulpwood volume from problem 4-4, determine how much *solid wood* is contained in the load, assuming an average of 75 cu ft per cd. What is the *value* of solid wood per cubic foot?

4-6 Compute the volume, in standard cords, of the following stacks of wood:
  a 16 ft × 80 ft × 8 ft
  b 12 ft × 31 ft × 5 ft
  c 15 ft × 32 ft × 8.33 ft

4-7 Construct a display board of tree cross sections illustrating changes in wood specific gravity from stump to tree top. Use cross sections extracted at intervals of 2 or 3 ft for an important timber species in your locality.

4-8 Review articles in foreign journals and summarize studies of weight/volume relationships in a country that has adopted the metric system.

REFERENCES

Anonymous. 1953. Weights of various woods grown in the United States. *U.S. Dept. Agr., Forest Serv., Forest Prod. Lab. Tech. Note* 218. 8 pp.

Anonymous. 1955. Wood handbook. *U.S. Dept. Agr., Forest Serv., Forest Prod. Lab. Agr. Handbook* 72. 528 pp.

Enghardt, H., and Derr, H. J. 1963. Height accumulation for rapid estimates of cubic volume. *J. Forestry* **61:**134-137.

Hallock, H., Steele, P., and Selin, R. 1979. Comparing lumber yields from board-foot and cubically scaled logs. *U.S. Dept. Agr., Forest Service, Forest Prod. Lab. Res. Paper FPL* 324. 16 pp.

Ker, J. W. 1962. The theory and practice of estimating the cubic content of logs. *Forestry Chron.* **38:**168–172.

Miller, R. H. 1941. Measuring green southern yellow pine pulpwood by weight or by cord. *Paper Trade J.* **113** (July 17).

Taras, M. A. 1956. Buying pulpwood by weight as compared with volume measure. *U.S. Forest Serv., Southeast. Forest Expt. Sta., Sta. Paper* 74. 11 pp.

Zon, R. 1903. Factors influencing the volume of solid wood in the cord. *Forestry Quart.* **1:**125–133.

LOG RULES, SCALING PRACTICES, AND SPECIALTY WOOD PRODUCTS

5-1 The Board-Foot Anomaly The board foot is equivalent to a plank 1 in. thick and 12 in. (1 ft) square; it contains 144 cu in. of wood. Although the board foot has been a useful and fairly definitive standard for the measure of sawed lumber, it is an ambiguous and inconsistent unit for log scaling.

A *log rule* is a table or formula showing estimated volumes, usually in board feet, for various log diameters and lengths. During the past century, at least 100 board-foot log rules have been devised, and several have been widely adopted. However, none of these rules can accurately predict the mill output of boards, except when near-cylindrical logs are sawed according to rigid assumptions on which the rules are based. Although the scaler might employ any of several rules that indicate different log volumes, there is only one correct measure of the boards produced. Thus the terms *board-feet log scale* and *board feet of lumber* are rarely, if ever, synonymous.

The formula commonly used for determining the board-foot content of sawed lumber is

$$\text{bd ft} = \frac{\text{thickness (in.)} \times \text{width (in.)} \times \text{length (ft)}}{12}$$

Accordingly, a 1-in. × 12-in. × 12-ft plank contains 12 bd ft, and a 2-in. × 8-in. × 24-ft plank includes 32 bd ft. This method of computation is not entirely correct even for sawed lumber because of accepted dimensional differences between rough green boards versus finished (seasoned and planed) lumber. A green "two-by-four" may be originally cut to the nominal size of 2 × 4 in., but it can be acceptable in finished form and sold as a two-by-four if it measures only $1^1/_2 \times 3^1/_2$ in. The purchaser of 1000 bd ft of finished lumber is therefore likely to receive considerably less than the volume implied by rigid adherence to the formula cited.

5-2 General Features of Board-Foot Log Rules To be considered equitable to both buyer and seller, a log rule must be *consistent;* i.e., volumes should be directly correlated with log sizes over the entire range of dimensions encountered. Few log rules currently in use can meet this simple requirement. Most of the differences between board-foot log scale and the sawed lumber tally can be attributed to the inflexible assumptions that necessarily underlie such rules:

1 Logs are considered to be cylinders, and volumes are derived from the small ends of logs. Volume outside the scaling cylinder, resulting from log taper, is generally ignored. In a few instances, a fixed rate of taper is presumed to somewhat compensate for this volume loss.

2 It is assumed that all logs will be sawed into boards of a certain thickness (usually 1 in.) with a saw of a specified thickness, or "kerf."

3 A fixed procedure for sawing the log and allowing for slabs is postulated (Fig. 5-1).

4 There is a tacit implication that all sawmills operate at a uniform level of efficiency which provides equal lumber yields from similar logs. The fact that some mills may be able to cut and market shorter or narrower boards than others is disregarded.

As a corollary to the foregoing, the terms *minimum board width* and *maximum scaling length* are worthy of definition. Minimum board width refers to the narrowest board for which volume would be computed by a given log rule. For most rules, the minimum board width is not smaller than 4 in. or larger than 8 in. Maximum scaling length indicates the longest tree section that may be scaled as a single log. Such a limitation is essential where log rules include no taper allowance; otherwise an entire tree might be scaled from the top end as a 6-in. log. In the United States, local scaling practices usually limit the maximum scaling length to 16 ft in the east and 20 to 40 ft in the west.

Log rules have been constructed from empirical rules of thumb, sawmill lumber tallies, ratios of board feet to cubic feet, diagrams, mathematical formulas, and combinations of these techniques. The three most commonly used log rules in the United States are the Scribner, Doyle, and International $^1/_4$-in. All three are included in the Appendix.

FIGURE 5-1
One method of sawing a ponderosa pine log. Losses due to saw kerf, slabs, and shrinkage are apparent. Log diameter is about 15 in.

DERIVATION OF LOG RULES

5-3 Mill-Tally Log Rules Any sawmill may construct its own empirical log rule by keeping careful lumber tallies of boards cut from various-sized logs. Such rules may provide excellent indicators of log volume at the particular sawmills where they are compiled. However, as they represent only one example of manufacturing efficiency and utilization practice, they are rarely reliable for general use in other localities. The utility of mill-tally log rules is generally limited to mills performing "custom sawing," i.e., the production and sale of boards from logs supplied by customers.

5-4 Board Foot-Cubic Foot Ratios From a purely theoretical viewpoint, 1 cu ft of solid wood contains 12 bd ft of lumber. This mathematical conversion, however, presumes that the cubic foot is rectangular in shape and that twelve 1-in. boards can somehow be extracted without loss of saw kerf. When 1-in. boards must be sawed from round logs, with attendant losses in squaring the log and allowing for kerf, the conversion factor is more likely to range between 4 and 8 bd ft per cu ft. Adding to the difficulty of adopting a uniform conversion is the fact that the board foot-cubic foot ratio changes with log diameter, method of slabbing the log, saw thickness, and sizes of boards produced.

Other factors being constant, the board foot-cubic foot ratio increases with log diameter, because a smaller percentage of cubic volume is wasted in squaring up larger logs. Board foot-cubic foot relationships of 5:1 or 6:1 are sometimes recommended for rough conversions, but there is no single factor that is worthy of complete endorsement. Although the cubic foot alone might constitute an ideal scaling unit, its conversion to the less reliable board foot is unrealistic at best. (See Sec. 5-9.)

5-5 Scribner Log Rule Developed by J. M. Scribner around 1846, this rule was derived from diagrams of 1-in. boards drawn to scale within cylinders of various sizes. A saw kerf of $1/4$ in. is presumed. The exact minimum board width allowed is not definitely known, although it appears to have been 4 in. for at least some log diameters. No taper allowance was included, so the rule ignores all volume outside scaling cylinders projected from small ends of logs. Therefore, this rule will normally underestimate the mill output of lumber. The underestimate generally increases as the scaling length increases. When volumes of 16-ft logs are desired, the rule-of-thumb formula $0.8 (D - 1)^2 - D/2$ provides a close approximation of the Scribner log rule (Grosenbaugh, 1952).

In general, the Scribner rule is considered to be intermediate in accuracy, although it does not provide board-foot volumes that are entirely consistent with changing log diameters. A slight modification of the rule is the Scribner Decimal C log rule. Here, the original Scribner volumes are rounded off to the nearest 10 bd ft, and the last zero is dropped. This innovation is presumably an aid to the scaler who must record and total volumes for large numbers of logs. The Scribner Decimal C has been applied extensively in western United States. For eastern forests, the International $1/4$-in. rule is often used.

Those interested in constructing their own diagram log rules may do so by carefully drafting a series of circles representing the desired range of log diameters. It is then necessary to decide upon the minimum acceptable board width and the saw kerf to be employed. End views of boards are drawn to scale within each circle (Fig. 5-2). When all diagrams have been completed, the number of board feet in each log may be computed by

$$\text{bd ft} = \frac{\text{total number of sq in. of diagram boards}}{12} \times \text{log length (ft)}$$

5-6 Doyle Log Rule This rule, devised by Edward Doyle about 1825, is based on the mathematical formula

$$\text{bd ft} = \left(\frac{D - 4}{4}\right)^2 L$$

where D is the log diameter in inches and L is the log length in feet.

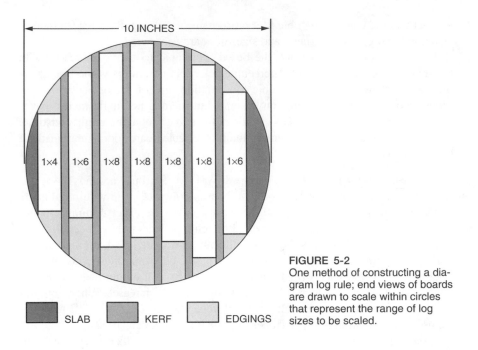

FIGURE 5-2
One method of constructing a diagram log rule; end views of boards are drawn to scale within circles that represent the range of log sizes to be scaled.

For 16-ft logs, the formula may be reduced to merely $(D - 4)^2$. Despite the fact that the formula is algebraically incorrect, use of the rule has persisted in southern and eastern United States. It was originally intended that the rule provide for a slabbing allowance of 4 in. and a saw kerf of $\frac{5}{16}$ in., or 25 percent. The 4-in. slab deduction is more than twice the amount ordinarily needed, and the kerf deduction is actually only about 4.5 percent. The net result is a highly inaccurate and inconsistent log rule that greatly underscales small logs because of the excessive slab deduction. Conversely, large logs are overscaled, for the insufficient kerf deduction is no longer absorbed by the heavy slab deduction.

The biggest fault of the Doyle rule lies in its inconsistency rather than its basic inaccuracy. The fact that volumes increase erratically with changing log diameters prohibits uniform adjustments in log prices to compensate for the abortive scale values. The rule can thus be considered a fair basis for transactions only when both buyers and sellers of logs are fully aware of its deficiencies. To provide a slight concession to the seller of small logs, some purchasers may either allow the inclusion of one bark thickness in measuring log diameters or record a scale equal to the log length when Doyle values are less than that amount. However, such local rules of thumb do little to alleviate the inherent inequalities of this anomalous rule.

In a few localities, the more erratic attributes of the Doyle and Scribner log rules are combined to form a diabolical yardstick called the *Doyle-Scribner* log rule. Doyle volumes are employed to underscale logs to about 24 to 28 in.; then the rule changes over to Scribner values to maintain the lowest possible board-foot values. If log prices are adjusted accordingly, the rule may be used without argument. All too often, however, it is the occasional seller of logs who may be unfamiliar with local scaling practices, and the buyer gains an undue advantage in the transaction.

5-7   Internati̲onal Log Rule   This rule, based on a reasonably accurate mathematical formula, is the only one in common use that makes an allowance for log taper. Devised in 1906 by Judson Clark, the International rule includes a fixed taper allowance of $1/2$ in. per 4 ft of log length. Thus scale values for a 16-ft log are derived by totaling board-foot volumes of four 4-ft cylinders, each $1/2$ in. larger in diameter than the previous one. In addition to the allowance for taper, the rule also provides rational deductions for slabbing and saw kerf. The original International $1/8$-in. rule assumed a $1/8$-in. saw kerf, plus $1/16$-in. allowance for board shrinkage, giving a total deduction of $3/16$ in. The International $1/4$-in. rule, devised from the original, provides for a $1/4$-in. saw kerf plus $1/16$ in. for shrinkage, or a total kerf deduction of $5/16$ in. Slabs are deducted in the form of an imaginary plank 2.12 in. thick and having a width equal to log diameter. It is assumed that all logs are cut into boards 1 in. thick.

Construction of the International $1/8$-in. rule begins with the simple cross-sectional area formula

$$\text{Area} = \frac{\pi D^2}{4} = 0.7854\, D^2$$

When log diameter D is substituted in inches, area is determined in square inches. Since 1 bd ft has a cross-sectional area of 12 sq in., the number of solid board feet in a cylinder is thus

$$\text{Solid bd ft} = \frac{0.7854\, D^2}{12}\, L, \text{ or } 0.06545\, D^2 L$$

where L is the log length in feet.

If the full cross-sectional area of a log could be cut into boards without loss of saw kerf, the foregoing formula would constitute a mathematically correct method of computing board-foot contents. As this is not the case, deductions must be made for kerf (sawing out the boards) and for slabs (squaring the log). The percentage p of log volume lost in saw kerf is derived by

$$p = \frac{k}{t + k}$$

where k is the saw kerf in inches and t is the board thickness in inches.

For the International $\frac{1}{8}$-in. rule ($\frac{3}{16}$-in. total deduction for kerf and board shrinkage), the kerf percentage for sawing 1-in. boards is

$$p = \frac{\frac{3}{16}}{1 + \frac{3}{16}} = \frac{3}{19} = 0.158 \text{ or about } 15.8 \text{ percent}$$

With a fixed maximum scaling length of 4 ft, the International $\frac{1}{8}$-in. rule now becomes

$$\text{bd ft} = (1 - p)0.06545\ D^2L, \text{ or } (1 - 0.158)0.06545\ D^2(4)$$

and

$$\text{bd ft} = (0.842)0.06545\ D^2(4) = 0.22\ D^2$$

The expression $0.22\ D^2$ includes the correct saw kerf allowance, but no deduction has been made for loss of log volume due to slabbing. Whereas kerf deductions are related to total log volume, slab allowances are closely associated with log diameter or circumference. Thus a correct slab deduction can be made by (1) decreasing log diameter or (2) extracting a plank with a width equal to log diameter. For the International rule, the latter method is used by removal of an imaginary plank 2.12 in. thick. The board-foot volume of the plank is computed for a 4-ft log section by

$$\text{Deduction for slabs and edgings} = \frac{2.12\ D}{12} \times 4 = 0.71\ D$$

When this final deduction is appended to the previously derived portion of the formula, the complete International $\frac{1}{8}$-in. rule for a 4-ft log section is

$$\text{bd ft} = 0.22\ D^2 - 0.71\ D$$

Some years after the International $\frac{1}{8}$-in. rule was published, it was modified to make it applicable for sawmills employing a $\frac{1}{4}$-in. kerf (total kerf and shrinkage allowance of $\frac{5}{16}$ in.). Instead of all scale values being recomputed by the process described here, the $\frac{1}{8}$-in. rule was reduced by the converting factor of 0.905. Thus, for 4-ft sections, the formula for the $\frac{1}{4}$-in. rule may be expressed as $0.905\ (0.22\ D^2 - 0.71\ D)$. For 16-ft log lengths, a simpler formula, $0.8\ (D - 1)^2$, will provide approximate volumes for the International $\frac{1}{4}$-in. rule (Grosenbaugh, 1952).

If the slab deduction for a formula log rule were made by reducing log diameter rather than by the International "plank method," a mathematically sound rule might be expressed as

$$\text{bd ft} = (1 - p)0.06545\ (D - s)^2\ L$$

where p = proportion of log volume lost as kerf
   D = log diameter, in.
   s = slab deduction, in.
   L = log length, ft

As no log taper allowance is automatically included in the rule, this factor must be controlled by placing limitations on the maximum scaling length.

Of the three principal log rules described here, the International is undoubtedly the most consistent, and it becomes quite accurate for mills producing mainly 1-in. boards with a $1/4$-in. saw thickness. The International $1/4$-in. rule has been officially adopted by several states and is widely used in various localities. In spite of its relative virtues, however, it has never gained the favor accorded such rules as the Scribner and even the Doyle for scaling work. Unrealistic as it may seem, many foresters are required to derive forest inventory data with the International rule and then handle log sales based on Scribner or Doyle volumes. As outlined in subsequent sections, conversions from one log rule to another are erratic and troublesome at best.

5-8 Overrun and Underrun   Comparisons of log rule values by various scaling diameters are presented in Table 5-1. The contrast in board-foot volumes for logs of identical sizes is even more strikingly illustrated by Fig. 5-3, where the International rule is used as a standard of comparison. For logs 10 in. in diameter, the Doyle value constitutes less than 60 percent of International scale, and the Scribner volume amounts to less than 85 percent. For 25- to 30-in. logs, all three log rules show a reasonable agreement. Beyond this point, the Scribner

TABLE 5-1
COMPARISON OF BOARD-FOOT LOG RULES FOR 16-FT LOGS

Log diameter, in.	Log rule				
	International $1/4$-in., bd ft	Scribner, bd ft	Scribner Decimal C, bd ft	Doyle, bd ft	Doyle-Scribner, bd ft
8	40	32	30	16	16
12	95	79	80	64	64
16	180	159	160	144	144
20	290	280	280	256	256
24	425	404	400	400	400
28	585	582	580	576	582
32	770	736	740	784	736
36	980	923	920	1024	923
40	1220	1204	1200	1296	1204

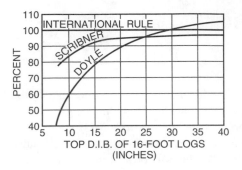

FIGURE 5-3
Relationships between three log rules for 16-ft logs. The International $^1/_4$-in. rule is used as the standard of comparison. *(Adapted from Schnur and Lane, 1948.)*

rule underscales logs, and the Doyle rule gives values much too large. The trends graphically shown here explain the reason for the devious combination log rule that utilizes Doyle volumes to about 24 in. and then abruptly switches to the Scribner scale.

The preceding comparison is not intended to convey the impression that the International log rule is faultless; nevertheless, it is a consistent rule that often closely approximates sawmill tallies. There is always some disparity between log scale and lumber yield. If the lumber output is greater, the excess difference is called *overrun*. When log scale values are larger than sawed output, an *underrun* occurs. Overrun and underrun are expressed as a percent of log scale by

$$\text{percent of overrun or underrun} = \frac{\text{mill tally} - \text{log scale}}{\text{log scale}} \times 100$$

When log sizes and sawmilling practices are equal, the amount of overrun or underrun is primarily dependent on the log rule used for scaling. This fact can be verified at any mill by checking a random selection of logs, as shown in Table 5-2. Here, scale values and lumber tallies were recorded for 15 logs at a hardwood mill in Arkansas. All logs were cut into 1-in. boards with a saw kerf loss of $^1/_4$ in. Overrun averaged 3.3 percent for the International $^1/_4$-in. rule and nearly 37 percent for the Doyle rule. Discrepancies of this magnitude are acceptable only when prices paid for logs are adjusted according to the scaling rule locally applied.

5-9 Board-Foot Volume Conversions In spite of the difficulties of equating various log rules and volume units such as board feet versus cords, conversions are occasionally desirable. Graphs similar to Fig. 5-3 may be used to change board-foot scales from one log rule to another, or individuals may develop their own factors that reflect local sizes of logs handled. In western Oregon, Washington, and Alaska, for example, the U.S. Forest Service considers

TABLE 5-2
SCALE AND OVERRUN COMPARISON OF DOYLE AND INTERNATIONAL $1/4$-IN.
LOG RULES*

Log no.	Scaling diameter, in.	Log length, ft	Doyle scale, bd ft	International scale, bd ft	Lumber tally, bd ft
1	13	10	50	70	83
2	11	14	43	70	70
3	16	12	108	130	127
4	11	16	49	80	107
5	11	16	49	80	70
6	15	12	91	115	112
7	18	12	147	170	174
8	11	12	37	55	55
9	10	12	27	45	45
10	13	12	61	85	82
11	10	14	32	55	55
12	16	12	108	130	124
13	11	12	37	55	65
14	19	12	169	190	190
15	12	12	48	70	87
Totals, bd ft			1056	1400	1446
Overrun, percent			+36.9	+3.3	

*Based on data collected by the senior author at a circular sawmill in southeast Arkansas.

that 1000 bd ft scaled by the Scribner Decimal C rule is roughly equal to 1400 bd ft by the International $1/4$ -in. rule.

In most regions of the United States, factors have been developed for converting stacked cordwood to board feet and vice versa. A sample tabulation of such factors for the Lake states is shown in Table 5-3. Although it is theoretically possible to convert cordwood of any size to board-foot units, results are questionable unless bolts are large enough to have been scaled as *bona fide* logs.

BOARD-FOOT LOG SCALING

5-10 Scaling Straight, Sound Logs The scaling of a straight and sound log is simply a matter of determining its length and average diameter inside bark (dib) at the small end. Lengths may be estimated or measured with a tape. Diameters are commonly determined with a *scale stick,* i.e., a rule graduated in inches and imprinted with log rule volumes for varying lengths. The "average" log diameter to be scaled is ocularly selected in most cases. However, on unusually elliptical logs the two extreme diameters may be measured for computing an average value.

TABLE 5-3
NUMBER OF ROUGH CORDS PER THOUSAND BOARD FEET,
LAKE STATES*

Bolt top dib, in.	Cords per MBF			No. of 8-ft bolts per cord
	Doyle	Scribner Decimal C	International $1/4$-in.	
6	11.1	4.4	2.2	45
7	6.7	3.0	3.0	33
8	5.0	4.0	2.7	25
9	4.2	2.5	2.5	20
10	3.5	2.1	2.1	16
11	3.2	2.5	2.2	13
12	2.8	2.3	2.0	11
13	2.5	2.0	1.8	10
14	2.5	2.1	1.9	8
15	2.4	2.0	1.9	7

*Adapted from Ralston, 1956. Conversions assume an average of 79 cu ft of solid wood per standard cord.

Depending on local scaling practices, the minimum scaling diameter is ordinarily set at 6 to 8 in. Smaller logs are given zero scale or *culled,* i.e., disregarded and eliminated from the scale record. When log diameters fall exactly halfway between scale-stick graduations (such as 12.5 in.), it is customary to drop back to the lower value—12 in. in this instance. Scaling diameters definitely above the halfway mark are raised to the next largest graduation; thus a 12.6-in. log would be scaled as 13 in.

Log lengths are usually taken at 2-ft intervals, although 1-ft intervals are used for certain species. All logs should have a trim allowance of 2 to 6 in. When logs are cut "scant" (without sufficient trim allowance) or in odd lengths, the scale is ordinarily based on the next shortest acceptable length. When long logs or tree-length sections are being scaled, the locally adopted maximum scaling length should be observed to avoid loss of volume due to excessive taper.

5-11   Log Defects   If a log is straight and free from defects, the gross scale (as read from the scale stick) is also the *net,* or *sound,* scale. From the standpoint of log scaling, defects include only those imperfections that will result in losses of wood *volume* in sawing the log. By contrast, those imperfections affecting log *quality,* or *grade,* only are not regarded as scaling defects. Thus scale deductions are made for such items as rot, wormholes, ring shake, checks, splits, and crook but not for sound knots, coarse grain, light sap stain, or small pitch pockets.

Making scale deductions for log defects is basically a matter of (1) determining the type and extent of the defect and (2) computing the board-foot volume that will be lost as a result. When the defect volume is subtracted from gross log scale, the usable volume remaining is the net, or sound, scale. Although certain guides or rules can be developed to somewhat standardize deduction techniques, the extent of many interior log defects can be learned only by working with experienced scalers and seeing defective logs sawed into boards on the mill carriage.

A point worthy of mention is that no deductions are made for defects outside the scaling cylinder or for those that penetrate 1 in. or less into the scaling cylinder. Defects outside the scaling cylinder are disregarded because this volume is ordinarily excluded from the original log scale (except for the International rule). Defects that penetrate the scaling cylinder 1 in. or less may be ignored because this portion of the log is normally lost in slabbing anyhow. If, for example, an exterior defect penetrates 3 in. into the cylinder of a log scaled by the Scribner rule, only the last 2 in. of penetration would be considered in making a scale deduction.

The principal forms of quantitative log defects encountered are:

1 Interior defects, such as heartrot or decay, hollow logs, and ring shake (mechanical separation of annual rings)
2 Exterior, or peripheral, defects, such as sap rot, seasoning checks, wormholes, catface, and fire or lightning scars
3 Crook defects, such as excessive sweep, crook, and forked or "crotched" logs
4 Operating defects, such as breakage, splits, and end brooming

5-12 Board-Foot Deduction Methods Defect deductions can be accomplished by at least three approaches, viz., by reducing log diameters, by reducing log lengths, or by diagramming defects for mathematical computations. Exterior, or peripheral, defects (checks, sap rot) are best handled by diameter reductions. Butt rot and many crook defects are accommodated by reducing log lengths. For internal and partially hidden defects, the diagram-formula method is suitable. By this method, interior defects are enclosed by an imaginary solid and the board-foot contents computed for subtraction from gross log scale. Deductions are made as 1-in. boards, and that part of the defective section that would normally be lost as saw kerf is not deductible. For the Scribner and other cylinder log rules assuming 1-in. boards and a $1/4$-in. kerf, the standard deduction formula is

$$\text{bd ft loss} = \frac{w \times t \times l}{15}$$

where w = width of defect enclosure, in.
$\quad\quad t$ = thickness of defect enclosure, in.
$\quad\quad l$ = length of defect enclosure, ft

One inch is usually added to both the width and the thickness of the defect in calculating the deduction. For defects that run from one end of a log to the other, measurements are taken at the larger defect exposure. It will be recognized that this is the basic board-foot formula for lumber (Sec. 5-1), except that the denominator has been changed from 12 to 15. This reduction to 80 percent of the solid board-foot content effectively removes the 20 percent deduction due to a $1/4$-in. saw kerf, because this portion would be lost anyway. For the International $1/4$-in. rule, where the kerf-shrinkage allowance is actually $5/16$ in., a denominator of 16 rather than 15 has been suggested for the formula. Several common log defects are illustrated in Fig. 5-4. In three instances, the standard deduction formula has been used for determining net log scales.

All nine logs illustrated have scaling diameters of 24 in. and lengths of 16 ft; gross scale by the Scribner Decimal C log rule is 40 (400 bd ft). For purposes of illustration here, it is assumed that log and board lengths are acceptable only in 2-ft multiples, with a minimum length of 8 ft. Minimum board width is 4 in. Where deductions have been computed, they are rounded off to the nearest 10 bd ft and converted to decimal scale. In most cases, deductions are made in accordance with the *National Forest Log Scaling Handbook* (USDA, 1985).

Log 1   As all defective material is outside the scaling cylinder, no deduction is necessary.

Log 2   The 4-ft crotched portion is deemed unusable; thus the length is reduced, and the scale for a sound 12-ft log is recorded.

Log 3   Because of rotten sapwood, only the heartwood portion of this log is scaled. The deduction is automatically made in scaling by diameter reduction.

Log 4   For surface or sun checks that penetrate along the radii of the log, it is common practice to drop back to one-half the depth of the checks to obtain the scaling diameter. If the checks here were 4 in. deep on all sides, the scaling diameter would be $24 - (2 + 2) = 20$ in. The reason for not scaling entirely inside the checks is that the loss due to checks is not usually as great in the interior of a log as it is at the ends.

Log 5   Sweep results in a deduction only when it causes a deviation that exceeds the top taper. For the log illustrated, a sweep of 6 in. was established by projecting the scaling cylinder straight through the log. As losses due to sweep are related to log size, the deduction may be approximated by expressing the sweep measurement as a percent of scaling diameter. In this example, the deduction is $6/24 \times 400 = 100$, or 10 decimal scale. Both sweep and crook can be minimized by careful log bucking. When logs are excessively crooked, deductions are made by merely reducing the length as in log 2.

Log 6   Sector or V-shaped defects bear the same relationship to log volume as the sector bears to a circle. For the spiral lightning scar affecting one-fourth of the log circumference, the deduction is one-fourth of the gross scale. Wormholes and frost cracks may also be handled by this method of deduction.

SCRIBNER DECIMAL C LOG RULE

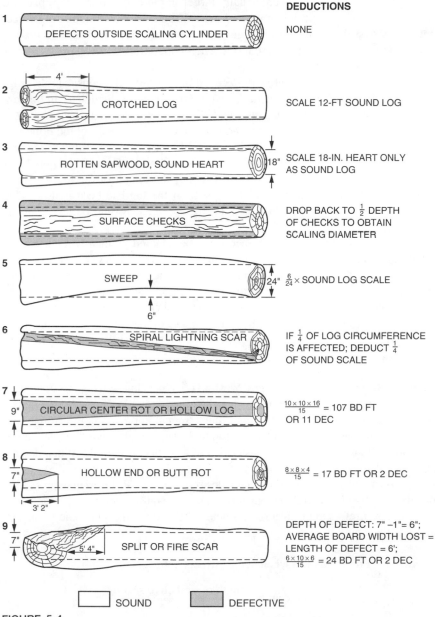

DEDUCTIONS

1. DEFECTS OUTSIDE SCALING CYLINDER — NONE

2. CROTCHED LOG — SCALE 12-FT SOUND LOG

3. ROTTEN SAPWOOD, SOUND HEART — SCALE 18-IN. HEART ONLY AS SOUND LOG

4. SURFACE CHECKS — DROP BACK TO $\frac{1}{2}$ DEPTH OF CHECKS TO OBTAIN SCALING DIAMETER

5. SWEEP — $\frac{6}{24} \times$ SOUND LOG SCALE

6. SPIRAL LIGHTNING SCAR — IF $\frac{1}{4}$ OF LOG CIRCUMFERENCE IS AFFECTED; DEDUCT $\frac{1}{4}$ OF SOUND SCALE

7. CIRCULAR CENTER ROT OR HOLLOW LOG — $\frac{10 \times 10 \times 16}{15} = 107$ BD FT OR 11 DEC

8. HOLLOW END OR BUTT ROT — $\frac{8 \times 8 \times 4}{15} = 17$ BD FT OR 2 DEC

9. SPLIT OR FIRE SCAR — DEPTH OF DEFECT: 7" −1"= 6"; AVERAGE BOARD WIDTH LOST = LENGTH OF DEFECT = 6'; $\frac{6 \times 10 \times 6}{15} = 24$ BD FT OR 2 DEC

☐ SOUND   ▨ DEFECTIVE

FIGURE 5-4
Typical log defects and methods of computing deductions by the Scribner Decimal C log rule.

Log 7   The standard deduction formula is applied for hollow logs and those with center rot; such defects are usually larger at the butt end of a log, because they tend to follow the configuration of annual rings. In this example, the defect encompasses a 9-in. circle at its larger end. One inch must be added in boxing the defect to allow for sawing around it. The deduction is thus (10 in. × 10 in. × 16 ft)/15, or 11 decimal scale.

Log 8   When logs are hollow or decayed at one end only, considerable judgment and experience is required to determine the depth of the defect. In this example, the defect is 7 in. in diameter and slightly more than 3 ft deep. Again, 1 in. is added to allow for sawing around the defect and the depth of the hollow is increased to 4 ft, because boards are acceptable only in 2-ft multiples. The deduction is thus (8 in. × 8 in. × 4 ft)/15, or 2 decimal scale. It should be noted that if the defect had penetrated more than 8 ft into the length of the log, it would have been deducted for the full 16 ft—just as if the log had been hollow. This would be necessary because boards opposite the defect would be less than the minimum acceptable length of 8 ft.

Log 9   Splits, fire scars, or catfaces may be handled by merely reducing log length or by applying the deduction formula. In the latter case, it is necessary to determine (1) the length of the defect with respect to the projected scaling cylinder, (2) the depth of the defect into the scaling cylinder, and (3) the average board width lost because of the defect. As shown here, the length is 5 ft 4 in., or 6 ft. The depth of 7 in. is reduced by 1 in. because of slabbing loss, and the average board width lost is estimated as 10 in. The deduction is therefore (6 in. × 10 in. × 6 ft)/15, or 2 decimal scale.

5-13   Cull Percent Deduction Methods   When log defects are computed in terms of board feet as in Sec. 5-12, a different deduction formula is required for each log rule used. A simpler and more logical approach is to estimate the defect volume as a *percent* of total log volume, thereby avoiding deduction methods that are tied to the inconsistencies of a particular log rule. When defects are computed as cull percents, the volume to be deducted can be easily translated into board feet, cubic feet, or other desired units. L. R. Grosenbaugh (1952) has devised five basic cull percent formulas for handling common log defects.[1] In all cases, *d* refers to the average diameter of the log at the small end in inches, and *L* is the length in feet.

Rule 1   Proportion lost when defect affects entire section:

$$\text{Cull percent} = \frac{\text{length of defective section}}{L}$$

---

[1]In reality, Grosenbaugh's formulas provide cull proportions rather than cull percents; they may be regarded as percents if multiplied by 100.

Rule 2 Proportion lost when defect affects wedge-shaped sector:

$$\text{Cull percent} = \frac{\text{length of defective section}}{L} \times \frac{\text{central angle of defect}}{360°}$$

Rule 3 Proportion lost when log sweeps (or when its curved central axis departs more than 2 in. from an imaginary chord connecting the centers of its end-areas; ignore sweep less than 2 in.):

$$\text{Cull percent} = \frac{\text{maximum departure minus 2 in.}}{d}$$

Rule 4 Proportion lost when log crooks (or when a relatively short section deflects abruptly from straight axis of longer portion of log):

$$\text{Cull percent} = \frac{\text{length of deflecting section}}{L} \times \frac{\text{maximum deflection}}{d}$$

Rule 5 Proportion lost when average cross section of interior defect is enclosed in ellipse (or circle) with major and minor diameters measurable in inches:

$$\text{Cull percent} = \frac{\text{(major) (minor)}}{(d-1)^2} \times \frac{\text{length of defect}}{L}$$

When rule 5 is applied, a defect in the peripheral inch of log (slab collar) can be ignored, but the ellipse should enclose a band of sound wood at least $1/2$ in. thick. When it is necessary to use a rectangle instead of an ellipse to enclose the defect, the cull percent will be five-fourths as much as for an ellipse with the same diameters as the rectangle. An obvious modification when a ring of rot surrounds a sound heart with average diameter H (in inches) is to estimate the sound proportion as $(H-1)^2/(d-1)^2$ and the defective proportion as $1 - [(H-1)^2/(d-1)^2]$.

In the rare case when cubic scale for products other than sawlogs is being used, sweep ordinarily is not considered to cause loss and $(d+1)^2$ is used instead of $(d-1)^2$ as a divisor for interior defect deduction.

Applications of the preceding formulas are illustrated in Fig. 5-5. Cull percents are multiplied by gross log volume to derive the defect volume in terms of desired units. These values are then subtracted from gross log volumes to arrive at net, or sound, scales.

5-14 Merchantable versus Cull Logs Logs are considered merchantable (valuable enough for utilization) if they can be profitably converted into a

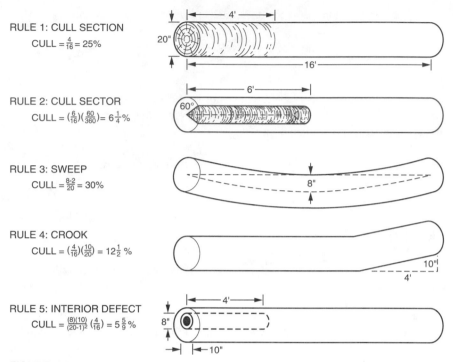

RULE 1: CULL SECTION

$$CULL = \tfrac{4}{16} = 25\%$$

RULE 2: CULL SECTOR

$$CULL = (\tfrac{6}{16})(\tfrac{60}{360}) = 6\tfrac{1}{4}\%$$

RULE 3: SWEEP

$$CULL = \tfrac{8-2}{20} = 30\%$$

RULE 4: CROOK

$$CULL = (\tfrac{4}{16})(\tfrac{10}{20}) = 12\tfrac{1}{2}\%$$

RULE 5: INTERIOR DEFECT

$$CULL = \tfrac{(8)(10)}{(20-1)^2}(\tfrac{4}{16}) = 5\tfrac{5}{9}\%$$

FIGURE 5-5
Application of cull percent deductions for log defects. *(Adapted from Grosenbaugh, 1952.)*

salable product such as lumber. Nonmerchantable logs are referred to as *culls.* If minimum dimensional requirements are met, the distinction between merchantable and cull logs is usually determined by the amount of defect encountered. In many localities, logs are considered merchantable only if they are at least 50 percent sound. The exact percentage applied, of course, is dependent on log size and species. A high-value, black walnut veneer log might be acceptable if only 30 percent sound, but a yellow pine log having a comparable defect would probably be culled. Thus merchantability limits vary with locality, kind of log, and changing economic conditions.

5-15 Scaling Records Log-scaling data are recorded on specially printed forms or in scale books. A complete scaling record includes the individual log number, species, diameter, length, gross scale, type and amount of defect, and net scale. When few log defects are encountered, the essential tally may occasionally be limited to species, log length, and gross scale. Log diameters are normally needed only for calculating defect deductions.

To conserve writing space and time, the type of defect can be indicated by locally accepted letter codes. Suggested designations are rot, R; sweep, S; wormholes, W; crack, C; catface or fire scar, F; and so on. The completed scaling record should additionally show the location or name of purchaser, scaler's initials, date, and log rule used. When gross scales are in terms of a decimal scale, defect deductions should be computed accordingly.

Standardized scaling records are essential when such tallies are the basis for log sales and purchases. When complete records of log dimensions and defects are required, scalers are more likely to make all measurements carefully. Furthermore, "check scaling" by supervisory personnel is most effective when specific data for each log are clearly noted in scale books.

5-16   Automated Log Scaling   While log scaling may continue to be a responsibility of foresters in the field, there have been promising developments in the design and application of automated scaling devices at large industrial sites. For example, the "autoscaler" illustrated in Fig. 5-6 measures the average diameter and length of each log (at each end) on a conveyor. At regular intervals, the processing unit prints a log inventory, by diameter/length classes, that summarizes such items as piece count, linear measure, and volumes by selected log rules. A similar installation near Longview, Washington, has indicated that diameters are reliably measured within ±0.25 in. (0.6 cm) and lengths within ±1 in. (2.5 cm).

FIGURE 5-6
The LC-310 Autoscaler log-scaling system. *(Courtesy Atmospheric Sciences, Inc.)*

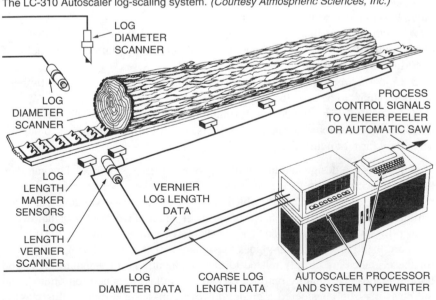

LOG GRADING

5-17   Need for Log Grading   In addition to being scaled, which involves estimating the board-foot content, logs may also be graded. Grading entails placing logs in quality classes. Since logs of different quality (i.e., different grades) vary in value per unit volume, both scale and grade are needed for determining the value of a given assortment of logs.

Defects in logs fall into two main categories: (1) those that result in losses of wood volume and (2) those that reduce strength or appearance or otherwise limit utility of the sawn products. The first are scaling defects, and the second are grading defects.

Log grades, and by extension grading rules, are related to the quality of the products to be produced from logs. Even with the product specified, the problem of developing meaningful grades remains elusive. For instance, a given log may produce lumber of several grades. Furthermore, since the objective is to maximize the value of each log, the proportion of grades will vary with manufacturing technology and market conditions. Hence, it is not surprising that grading practices have not been widely standardized. However, there are general principles and procedures of grading that have been found useful and that should be kept in mind.

5-18   Hardwood Log Grading   Four log-use classes have been designated to cover current hardwood utilization practices. These classes are (Rast et al., 1973):

1 *Veneer class.* Includes high-value logs as well as some relatively low value logs that can be utilized as veneer logs.

2 *Factory class.* Includes logs adapted to the production of boards that later can be remanufactured so as to remove most defects and obtain the best yields of clear face and sound cuttings.

3 *Construction class.* Includes logs suitable for sawing into ties, timbers, and other items to be used for structural purposes.

4 *Local-use class.* Includes logs suitable for products not usually covered by standard specifications: crating, pallet parts, mine timbers, industrial blocking, secondary farm buildings, etc.

After logs are designated in one of the four log-use categories, they can then be graded according to their size, abnormalities, and surface characteristics. In most lumber mills, veneer logs (which are typically the most valuable class) are withdrawn and shipped to a hardwood veneer plant. Two grades of hardwood veneer logs—*prime veneer* and *veneer*—are recognized. Logs in the factory class are then placed into log grades (designated F1, F2, F3). The specifications for these grades are closely related to the specifications for standard hardwood lumber grades. The quality range of construction-class logs is often limited, and the class may be considered a grade; however, in some instances this class is further divided into grades. Requirements for classification into the local-use category are limited. In general, if the log meets minimum size requirements (8 in. scaling diameter, 8 ft scaling length) and is one-third or more sound, it can be classified in the local-use category. Grades are usually not recognized within the local-use class.

To summarize, the typically applied log-use classes and the grades recognized within each class of hardwood logs are

	Log-use class			
	Veneer	Factory	Construction	Local use
Grades:	Prime veneer, Veneer	F1, F2, F3	No grades	No grades

5-19 Softwood Log Grading   There are numerous grade specifications for softwood logs. However, two log-use classes cover most softwood utilization practices:

1 *Veneer class.* Includes high-value logs as well as some relatively low value logs that can be utilized as veneer logs.

2 *Sawlog class.* Includes logs adapted to the production of yard and structural lumber.

Grading factors used for softwood logs are log diameter, log length, sweep, and cull. At a typical sawmill, veneer logs are removed and sold separately, if markets are available. Grading specifications for veneer logs vary by species and region and are generally complex.

A variety of grading systems have been developed for softwood sawlogs. While some of the grading rules are rather detailed and involved, a relatively simple system developed by the U.S. Forest Service for southern pine sawlogs that will be sawed into standard yard and structural lumber has been found to be quite effective. The clear-face log grading rules for southern pine logs (Schroeder et al., 1968) serve as an example of a softwood log grading system.

Logs are graded in three steps. First, one must determine the number of clear faces of the log. (A face is one-fourth the circumference of the log surface and extends the full length of the log.) One attempts to include as many defects as possible in one face so that the maximum number of clear faces are obtained. On the basis of the number of clear faces, the log is tentatively classified into one of three grades:

Grade 1 Logs with three or four clear faces
Grade 2 Logs with one or two clear faces
Grade 3 Logs with no clear faces

After a tentative log grade is established from the face examination, the log is reduced one grade for each of the following:

1 *Sweep.* Degrade any tentative grade 1 or 2 log one grade if sweep is 3 in. or more *and* equals or exceeds one-third the scaling diameter of the log.

2 *Heartrot.* Degrade any tentative grade 1 or 2 log one grade if evidence of advanced heartrot is found.

Note: No log is degraded below grade 3 if its net scale is at least one-third the gross log scale after deductions have been made for sweep and/or rot. Logs with total scaling deductions for sweep and rot exceeding two-thirds the gross scale of the log are classified cull.

The clear-face grading system for southern pine logs is easy and quick to use, and it has been found to be as accurate for predicting lumber-grade yields as more complex log grading methods.

WEIGHT SCALING OF SAWLOGS

5-20 Advantages and Limitations In general, the advantages of weight scaling pulpwood as cited in Chap. 4 apply equally well to transactions involving sawlogs. The chief difference is that price adjustments must be made in weight scaling sawlogs to take care of variations in log quality and size. Without such adjustments, crooked or defective logs might command the same price as straight, clear logs, and small-diameter logs (yielding less lumber per ton) could bring as much as larger logs (Page and Bois, 1961).

Circumstances most favorable to weight scaling of sawlogs exist when truck-loads are made up of a single species and when there is a relatively narrow range of log diameters present on any given load. It is therefore not surprising that numerous experiments in sawlog weight scaling have been conducted with southern pine logs. Such logs are fairly uniform in size and quality, with few defect deductions being required. On the other hand, mixed hardwood logs of varying quality, degree of soundness, and log size present severe obstacles to effective weight scaling.

Changing from stick scaling to weight scaling raises many questions, the primary one being, What log weight provides an equivalent for 1000 bd ft of lumber? There is no single or ready answer to this query, of course. Approximations for several species are presented in Table 5-4, but it must be recognized that these values are affected not only by log size and quality but also by such items as moisture content, wood density, and proportion of heartwood versus sapwood. As in measuring pulpwood, volume-weight conversions are preferably based on local measurements by purchasers of logs.

5-21 Volume-Weight Relationships for Sawlogs Because of the in-grained custom of using board-foot log rules, volume-weight conversions are likely to be initially based on predicted log scales rather than on expected lum-

TABLE 5-4
AVERAGE GREEN WEIGHTS OF LOGS PER THOUSAND BOARD FEET
OF LUMBER

Species	Weight, lb	Species	Weight, lb
Ash, white	11,100	Maple, red	11,900
Aspen (Popple)	10,800	Maple, sugar	12,900
Basswood	9,500	Oak, red	14,800
Beech	12,700	Oak, white	14,400
Birch, yellow	13,200	Pine, jack	11,500
Cedar, western red	6,200	Pine, loblolly	12,400
Cherry, black	10,500	Pine, longleaf	11,100
Chestnut	12,600	Pine, Norway (red)	9,700
Cottonwood	10,700	Pine, pitch	12,400
Cypress, southern	11,800	Pine, shortleaf	10,400
Elm, slippery	12,600	Pine, slash	12,200
Elm, white	11,300	Pine, sugar	11,500
Fir, balsam	10,400	Pine, white	9,000
Fir, Douglas	8,700	Pine, yellow (western)	11,300
Gum, black	10,400	Poplar, yellow (tulip)	8,800
Gum, red (sweet)	10,600	Redwood	8,900
Hackberry	11,300	Spruce, black	7,700
Hemlock, eastern	11,200	Sycamore	12,000
Hickory	14,700	Walnut, black	11,900
Locust, black	13,400	Willow, black	11,800

Source: U.S. Forest Service. The presumed range of log diameter is 10 to 16 in.

ber yields. A series of 50 to 100 paired weighings and stick-scaled truckloads will provide a basis for determining the number of pounds or tons per thousand board feet (MBF), according to a particular log rule. Recoverable volume varies by log size; consequently, both the total weight and the number of logs must be known when log sizes encompass a wide diameter range. Equations based on truckload weights and the number of logs on the load have been successfully employed to predict sawlog volumes.

As an example, Bower (1962) derived the following equation for loblolly pine:

$$Y = -3.954N + 0.0925W$$

where Y = total board-foot volume, International $1/4$-in. rule for a truckload of logs
N = number of 16-ft logs on load
W = total load weight, lb

Because diameter is a fair indication of log quality, the log count per ton is also useful as a rough grading index or as a basis for premium payments.

Ideally, weight-volume relationships for sawlogs would either be derived on the basis of an expected mill tally of lumber or be computed independently of *any* presumed product. Except for custom and apathy, there is no reason why roundwood materials cannot be purchased and sold strictly on the basis of weight—without an implied conversion back to cords or board feet. Such a changeover might well be initiated when the United States adopts the metric system for measuring primary wood products.

SPECIALTY WOOD PRODUCTS

5-22 Specialty Products Defined As arbitrarily applied in this book, specialty wood products encompass an agglomeration of logs, bolts, roundwood, timbers, and stumps that are distinctive because of their shapes, sizes, quality, measurement standards, or intended use. Aside from veneer logs, the items included here may be additionally classified as products purchased in individual units (i.e., piece products such as poles or railroad ties) or products purchased in bulk form (e.g., mine timbers and fuel wood).

5-23 Veneer Logs Illogical as it may appear, veneer logs are ordinarily measured and purchased in terms of board-feet log scale. To compensate for the fact that size and quality standards are more stringent than for most sawlogs, a premium price is paid for logs of veneer quality (Fig. 5-7). This price may sometimes amount to two or three times the price paid for logs that are sawed into yard lumber. Although grading specifications for veneer logs vary widely, quality requirements are based largely on species, log diameter, and freedom from defects such as crook, knots, bird peck, worm holes, ring shake, stains, and center rot.

Instead of scaling veneer logs in terms of board feet, it would be more realistic to compute their contents in cubic feet or calculate expected yield in terms of veneer sheets of a given thickness. For rotary-cut veneers obtained from sound logs, output can be closely estimated from the difference between two cylinders—one based on the dib of the veneer bolt at the small end and the other based on a presumed core diameter. Thus the maximum surface area of rotary-cut veneer to be expected from a sound wood cylinder may be computed by

$$\text{Veneer yield in square feet} = \frac{B - b}{t} w$$

where B = cross-sectional area of log at small end, sq ft
  b = cross-sectional area of residual core, sq ft
  t = veneer thickness, thousandths of a foot
  w = sheet width (log length), ft

FIGURE 5-7
Eastern hardwood logs purchased for the manufacture of rotary-cut veneers. *(U.S. Forest Service photograph.)*

For excessively tapered logs, actual yields may be greater than that indicated, because some veneer is obtained from material outside the presumed cylinder. On the other hand, yields may be less for logs having interior defects. Nevertheless, for sound logs the formula will provide predictions that are much more reliable and realistic than scale methods based on board feet. The relationship can also be easily adapted to metric units.

5-24 Poles and Piling   These are roundwood products selected primarily for strength, durability (or capability for preservative treatment), and resistance to exposure and mechanical stresses. Along with several other piece products, they are grouped according to distinct classes and price grades on the basis of species, dimensions, straightness, and freedom from defects (Fig. 5-8).

The principal species utilized for poles are southern pines, western red-cedar, western hemlock, northern white cedar, lodgepole pine, and Douglas-fir. As a rule, poles are marketed under specifications compiled by the American Standards Association (Williston, 1957). Poles are grouped into one of 10 size classes, depending on length, minimum top circumference, and minimum circumference 6 ft from the butt end.

Wood piling may be of any species that will withstand driving impact and support the loads imposed. Specifications are based on intended use, straight-

FIGURE 5-8
Peeled pine poles awaiting preservative treatment at a mill in
Mississippi. *(U.S. Forest Service photograph.)*

ness, uniformity of taper, soundness, and dimensions (length, minimum top diameter, and both minimum and maximum diameters at 3 ft from the butt end).
Most piling, especially if used in salt water, is pressure-treated with creosote as
protection against shipworms.

Piling is purchased at a stated price per linear foot for specified dimensions,
and value accrues rapidly with increasing length and desired taper characteristics. Standard specifications for round timber piling are available from the
American Society for Testing and Materials (Williston, 1957). In general, poles
and piling must be peeled at the time of cutting. Therefore, the woodland owner
anticipating the sale of such roundwood should carefully study the purchaser's
requirements before trees are severed from the stump.

5-25 Fence Posts Posts are round, split, or sawed piece products ranging
from about 3 to 8 in. in diameter. Lengths are usually 7 to 8 ft, though some
posts are as long as 20 ft. If posts are split and untreated, they are preferably
made from durable species such as various cedars, redwood, white oak, or catal-
pa. Those that are peeled, seasoned, and treated with preservatives are commonly made from red oak, southern pines, western pines, and Douglas-fir.

Posts may be cut from trees too small for efficient utilization as pulpwood or from the top sections of pulpwood and sawlog trees. If preservative treatment is required, posts are peeled either when cut or at concentration yards. The worth of posts as stumpage (standing trees) may be minimal, but this value (and retail price) is increased considerably by seasoning and preservative treatment. Nondurable species are commonly pressure-treated with a wood preservative.

5-26 Railroad Ties   The principal species used for railroad crossties are red and white oaks, Douglas-fir, gums, and southern pines. Today, most ties utilized by leading railroads are sawed rather than hand-hewn and pressure-treated with preservatives to prolong service life. Red oak is a preferred species, because it is dense, strong, and easily treated, and possesses superior resistance to mechanical wear. There are seven standard classes of crossties, based on width. Ties for standard gauge railroads are 8, 8½, or 9 ft long.

Before felling trees intended for conversion into railroad ties, the forester should check with local buyers or railroad agents about acceptable timber species and quantities that can be marketed. Trees of hardwood species are preferably cut in fall or winter when seasoning progresses slowly and there is less chance for end-checking of logs, sap stains, and incipient decay. Ties must be straight-grained, with ends cut square and all bark removed. Defects such as bark seams, decay, splits, shakes, holes, and unsound knots are not permitted in high-grade ties.

Expected yields of crossties from standing trees or logs can be computed by diagramming various tie dimensions within circles representing cross-sectional areas of logs. The technique is analogous to that of determining the size of inscribed square timbers (Sec. 4-4) or constructing a diagram log rule (Sec. 5-5). Tabulations of predicted yields by tie class and log size are referred to as *tie log rules*. Such rules are of considerable aid in evaluating alternative product uses for standing timber.

5-27 Mine Timbers   More than three-fourths of the mine timbers used in the United States and Canada are cut from assorted hardwoods such as beech, maple, hickory, ash, poplar, gum, and oak. Depending on local custom and requirements, mine timbers may be round, split, hand-hewn, or sawed. Because dimensions vary from place to place, the following specifications are merely indicative of size ranges encountered.

Mine props are round timbers used as supports for roofs and sides of tunnels; they range from 4 to 14 in. in diameter and 3 to 12 ft long. *Lagging* is round timber about 3 in. in diameter and 7 ft long; it is used behind props and caps to form the sides and roofs of tunnels. *Caps* are hewn or sawed timbers of various sizes that are placed across the tops of paired props as a support for roof lagging. *Sills* are hewn or sawed foundations for props, ranging from 8 to 12 in. across the widest face and of varied lengths. *Mine ties* are ordinary track ties 4 to 5 in. wide on the face and 3 to 5 ft long.

Mine timbers are commonly purchased on a green-weight basis, though certain sawed items (e.g., mine ties) may be handled as piece products or measured in terms of board feet. Round and untreated mine props may be sold without removal of bark. Mine timbers are used primarily in areas where coal, iron ore, copper, lead, zinc, and silver are extracted. Thus the principal markets are found in regions producing large quantities of these minerals.

5-28   Stumps for the Wood Naval-Stores Industry   In southeastern United States, residual stumps of old-growth longleaf and slash pines are utilized for the extraction of turpentine, rosin, and various pine oils. Only heartwood (sometimes referred to as *lightwood*) is valuable; hence wood producers prefer older stumps from which all sapwood has been removed by weathering and decay. Stumps and taproots are "pushed" out of the ground with large bulldozers equipped with coarsely toothed blades. Then they are loaded onto trucks or railroad cars for transport to one of more than a dozen steam-and-solvent extraction plants in Alabama, Florida, Georgia, Louisiana, or Mississippi.

Pine stumpwood is purchased by the ton, and where stumps are readily accessible, profitable removal operations may be conducted for yields as low as 1 ton (five to eight average stumps) per acre. In general, stump lands are classed as "operable" if the area will support harvesting equipment during average seasons, if the area can be worked without undue damage to live trees, and if the stumpwood tract size is 25 acres or larger.

Inventories or cruises of pine stumps are accomplished by use of sample strips or plots similar to those designed for tallies of live timber (Chap. 10). Tracts are designated as operable, timber-locked, inaccessible, or nonproductive, and stumps are tallied by three to five groundline diameter categories. Foresters concerned with inventories or leases of stumpwood may obtain prices and additional specifications from wood naval-stores extraction plants.

5-29   Bolts and Billets   Bolts are short sections of logs, usually less than 8 ft long (e.g., veneer bolts). When bolts are split or sawed lengthwise, they are called billets. Collectively, bolts and billets are used for such products as cooperage, excelsior, handles, vehicle parts, shingles, baseball bats, pencils, and matches. A variety of hardwood and coniferous species are utilized. For example, white oak is used for tight cooperage, hickory for handles, western redcedar for shingles, ash for baseball bats, and white pine for matches.

In general, bolts and billets are shorter than 8 ft in length, are less than 12 in. in diameter, and must be made up of high-quality materials. However, exact size and grade requirements are so diversified that accurate specifications must be obtained locally. Bolts under 12 in. in diameter are usually measured and sold in terms of stacked cords; those 12 in. and larger may be scaled in board feet. Billets may be sold by piece counts, short cords, or standard cords.

5-30 Fuel Wood With increased emphasis on energy conservation, firewood has once again become an important forest product in the United States and Canada. Firewood is commonly measured and sold in terms of stacked "short cords" or "face cords," e.g., a 4-ft × 8-ft stack of wood with stick lengths that range from about 16 in. to 2 ft or more. Thus a face cord of 16-in.-long firewood is actually one-third of a standard cord. As with pulpwood, the amount of *solid* wood contained depends on stick diameters, straightness, and care in stacking.

In some parts of Anglo-America, firewood is sold by the ton. Persons who purchase wood by weight instead of volume should seek out the driest wood available; the weight of unseasoned firewood is greatly influenced by its moisture content. And although bark is rarely removed from firewood, sticks over 8 in. in diameter should be split to facilitate seasoning.

Firewood is usually cut and marketed locally. As a result, a wide variety of species are utilized, with preference given to dense hardwoods such as oak, hickory, beech, birch, maple, ash, and elm. The heavier species weigh about 2 tons per cord and in a dry condition will provide about as much heat as a short ton of coal, 175 gal of domestic fuel oil, or 24,000 cu ft of natural gas (USDA, 1978).

PROBLEMS

5-1 Compute the total volume and total value for these items of lumber:
   a 139 pieces 3 in. × 6 in. × 16 ft @ $228.50 per MBF
   b 254 pieces 2 in. × 4 in. × 18 ft @ $232.45 per MBF
   c 346 pieces 2 in. × 8 in. × 20 ft @ $336.00 per MBF
5-2 Devise a diagram log rule for scaling diameters of 10 through 40 in., by 2-in. classes. Assume a saw kerf of $1/4$ in., board thickness of 1 in., minimum board width of 6 in., and a uniform log length of 16 ft. Compare diagram volumes with those listed for the Scribner log rule.
5-3 Compute cubic-foot contents for each cylinder diagrammed in problem 5-2. Derive board foot-cubic foot ratios for each diameter class. Explain possible reasons for the consistency or variability of these ratios.
5-4 Compile a log rule based on the formula in Sec. 5-7. Assume a $5/16$-in. kerf, and make the slab deduction in the form of a 1-in. collar around the circumference of the log. Using a taper allowance of 1 in. per 10 ft of log length, compute log rule values for diameters of 12 to 40 in. and a log length of 20 ft.
5-5 Visit a sawmill in your own locality, and conduct a simple study of mill overrun based on two different log rules. Tabulate results as shown in Table 5-2.

5-6 The following tabulation of mill overrun or underrun for southern pine logs was compiled by the U.S. Forest Service (Campbell, 1962):

	Percent of overrun or underrun			
Log dib, in.	Doyle	Scribner Decimal C	International $^1/_4$-in.	No. of logs
6	+400	+28	−2	89
7	200	26	−2	102
8	130	23	−3	134
9	90	21	−3	162
10	70	19	−4	155
11	50	17	−4	132
12	42	14	−5	167
13	32	12	−5	119
14	26	10	−6	128
15	20	8	−6	85
16	16	5	−7	74
17	12	3	−8	43
18	8	1	−8	42
19	4	−2	−9	22
20	0	−4	−9	16
21	−2	−6	−10	8
22	−4	−8	−11	8
23	−6	−10	−11	3
24	−8	−13	−12	2
Total				1491

Plot percent of overrun (+) or underrun (−) over log diameter for each of the three log rules. Draw smooth curves (in different colors) for each set of plotted points. Label curves as in Fig. 5-3, and title the graph. Then prepare a brief written report on advantages and disadvantages of the three log rules.

5-7 Select 10 to 20 defective sawlogs for a sample scaling project. Tally by log number, species, diameter, length, gross scale, type and amount of defect, and net scale. Record gross volumes in terms of the Scribner Decimal C log rule. Compute defect deductions by using (a) board-foot deduction methods described in Sec. 5-12 and (b) Grosenbaugh's cull percent formulas outlined in Sec. 5-13.

5-8 Investigate sawlog weight-scaling practices in your own locality. Determine which species are involved, average green log weights per MBF, and average number of logs per ton. Prepare a written report on your findings.

5-9 Use the data in Table 5-2 on log-scaling diameter, log length, and lumber tally to develop a mill-tally log rule.

    a Plot lumber tally *(Y)* versus log diameter *(D)*, log length *(L)*, and log diameter squared times length *(D²L)*. Which variable *(D, L,* or *D²L)* is more closely related to lumber tally?

b Develop a mill-tally log rule by fitting the simple linear regression

$$Y = b_0 + b_1 X$$

where Y = lumber tally (bd ft)

$$X = D^2 L$$

c On a single graph, plot curves of Doyle, International $1/4$-in., and your mill-tally log rule from part (b) for 16-ft logs with scaling diameters 10 through 20 inches. Which log rule, Doyle or International $1/4$ in., corresponds more closely to your fitted mill-tally log rule?

5-10 Compile a veneer log rule by use of the formula presented in Sec. 5-23. Assume a bolt length of 4, 6, or 8 ft, and compute veneer yields for logs 14 to 36 in. in diameter, by 2-in. classes. Base your table on a standard veneer thickness of $1/16$ in.

5-11 Determine minimum log diameters that are required to produce crossties of the following dimensions: (a) 6×6 in., (b) 6×8 in., (c) 7×9 in.

5-12 Investigate current retail prices for fuel wood in your locality. What is the average price per cubic foot or per standard cord? Estimate the stumpage value of fuel wood per standard stacked cord, and express this as a percent of market price. Then compare the price paid for a standard stacked cord of pulpwood (fob mill) in your locality with the retail price of the equivalent amount of fuel wood. Can you supply rational reasons for the differences noted?

REFERENCES

Amateis, R. L., Burkhart, H. E., Greber, B. J., and Watson, E. E. 1984. A comparison of approaches for predicting multiple-product yields from weight-scaling data. *Forest Sci.* **30:**991–998.

Anonymous. 1972. System scales logs automatically; uses infrared scan, digital control. *Forest Industries* (Sept.). 4 pp.

Avery, T. E., and Herrick, A. M. 1963. *Field projects and classroom exercises in basic forest measurements.* Univ. of Georgia Press, Athens, Ga. 151 pp.

Bower, D. R. 1962. Volume-weight relationships for loblolly pine sawlogs. *J. Forestry* **60:**411–412.

Campbell, R. A. 1962. Overrun—southern pine logs. *U.S. Forest Serv., Southeast. Forest Expt. Sta. Res. Note* 183. 2 pp.

———. 1964. Forest Service log grades for southern pine. *U.S. Forest Serv., Southeast. Forest Expt. Sta. Res. Paper SE*-11. 17 pp.

Curtis, A. B. 1978. Wood for energy: An overview. *U.S. Forest Serv.,* State and Private Forestry, Atlanta, Ga. 4 pp.

Fasick, C. A., Tyre, G. L., and Riley, F. M., Jr. 1974. Weight-scaling tree-length timber for veneer logs, saw logs, and pulpwood. *Forest Prod. J.* **24:**17–20.

Freese, F. 1973. A collection of log rules. *U.S. Dept. Agr., Forest Service, Forest Prod. Lab. Gen. Tech. Rep.* FPL 1. 65 pp.

Grosenbaugh, L. R. 1952. Shortcuts for cruisers and scalers. *U.S. Forest Serv., Southern Forest Expt. Sta. Occas. Paper* 126. 24 pp.

Ker, J. W. 1966. The measurement of forest products in Canada: Past, present and future historical and legislative background. *Forestry Chron.* **42:**29–38.

Mann, C. N., and Lysons, H. H. 1972. A method of estimating log weights. *U.S. Forest Serv., Pacific Northwest Forest and Range Expt. Sta. Res. Paper* PNW-138. 75 pp.

McKinley, T. W. 1953. Veneer tables: Surface feet tabulations for rotary cut and special veneers. *J. Forestry* **51:**826–827.

Page, R. H., and Bois, P. J. 1961. Buying and selling southern yellow pine saw logs by weight. *Georgia Forest Res. Council Rept.* 7. 9 pp.

Ralston, R. A. 1956. The break-even point for rough 8-foot bolts merchantable as sawlogs or cordwood. *U.S. Forest Serv., Lake States Forest Expt. Sta. Tech. Note* 469. 2 pp.

Rast, E. D., Sonderman, D. L., and Gammon, G. L. 1973. A guide to hardwood log grading. *U.S. Forest Serv., Northeast. Forest Expt. Sta. Gen. Tech. Rep.* NE-1. 31 pp.

Sander, G. H. 1970. Measuring trees. Pacific Northwest Cooperative, Oregon State University, Corvallis, Oreg. PNW Bull. 31. 26 pp.

Schnur, G. L., and Lane, R. D. 1948. Log rule comparison: International $1/4$-inch, Doyle, and Scribner. *U.S. Forest Serv., Central States Forest Expt. Sta.* 6 pp.

Schroeder, J. G., Campbell, R. A., and Rodenback, R. C. 1968. Southern pine log grades for yard and structural timber. *U.S. Forest Serv., Southeast. Forest Expt. Sta. Res. Paper* SE-39. 9 pp.

Smith, N. 1981. Wood: An ancient fuel with a new future. Worldwatch Institute, Washington, D.C. Paper no. 42. 48 pp.

U.S. Department of Agriculture. 1978. Firewood for your fireplace. Government Printing Office, Washington, D.C. Leaflet 559. 2 pp.

U.S. Department of Agriculture. 1985. *National Forest Log Scaling Handbook.* U.S. Forest Service, Government Printing Office, Washington, D.C. 247 pp.

Williston, H. L. 1957. Pole grower's guide. *U.S. Forest Serv., Southern Forest Expt. Sta. Occasional Paper* 153. 34 pp.

MEASURING STANDING TREES

6-1 Tree Diameters The most frequent tree measurement made by foresters is diameter at breast height. In the United States, diameter at breast height (dbh) is defined as the average stem diameter, outside bark, at a point 4.5 ft above ground as measured from the uphill side of the stem. In countries that use the metric system, dbh is usually taken 1.30 m above ground. Direct measurements are usually made with a diameter tape, tree caliper, or Biltmore stick. Collectively, instruments employed in determining tree diameters are referred to as *dendrometers*.

With a diameter tape, tree circumference is the variable actually measured (Fig. 6-1). The tape graduations, based on the relationship between the diameter and circumference of a circle, provide for direct readings of tree diameter, usually to the nearest 0.1 in. If a steel diameter tape is level and pulled taut, it is the most *consistent* method of measuring dbh. However, as tree cross sections are rarely circular, taped readings of irregular trees are likely to be positively biased.

Wooden or steel tree calipers provide a quick and simple method of directly measuring dbh. For ordinary cruising work, a single caliper measurement will usually suffice (Fig. 6-2). Directional bias can be minimized by measuring all diameters from the tree face closest to a cruise plot center. If stem cross sections are decidedly elliptical, two caliper readings at right angles should be made and the average diameter recorded. When caliper arms are truly parallel and in correct adjustment, the instrument gives reliable measures of dbh to the nearest 0.1 in. Calipers are ideal for trees to about 18 in. in diameter. The diameter tape is often preferred for bigger stems, because large calipers are bulky and awkward to handle in thick underbrush.

FIGURE 6-1
Measuring tree dbh with a steel diameter tape. *(U.S. Forest Service photograph.)*

A modification of the conventional caliper is the diameter "fork," a two-pronged instrument that can be held in one hand while measuring small trees. One prong of the fork is movable and spring-loaded, resulting in an automatic adjustment to the sides of the stem. Diameters are read from a built-in-arc type scale on the fork.

The Biltmore stick is a straight wooden stick specially graduated for direct readings of dbh. Based on a principle of similar triangles, the stick must be held horizontally against the tree dbh at a predetermined distance from the observer's eye (Fig. 6-3). The cruiser's perspective view is compensated for by the dbh graduations; i.e., the inch units get progressively shorter as tree diameters increase. Thus it is possible to measure a 40-in.-diameter tree with a stick about 25 in. long.

Scale graduations for the Biltmore stick may be computed by

$$\text{dbh graduation} = \sqrt{\frac{AD^2}{A + D}}$$

where A is the fixed distance from the eye to the stick in inches and D is any selected tree diameter in inches.

FIGURE 6-2
Measuring tree dbh with wooden calipers. *(Photograph by Rick Griffiths.)*

On commercially manufactured Biltmore sticks, diameter graduations are usually based on a fixed distance of 25 in. from the observer's eye to dbh. However, foresters may construct sticks based on a different arm reach by use of the preceding formula (Avery, 1959). Because of the difficulty of maintaining the proper distance from eye to tree, the Biltmore stick must be regarded as a rather crude measuring device. With care, diameters of small trees can be read to the nearest full inch, but accuracy tends to decrease for larger trees, because of the shortened intervals between inch graduations. The Biltmore stick is handy for occasional cruising work, but tree calipers or the diameter tape should be used for more reliable measurements of individual trees.

6-2   Diameter at Breast Height for Irregular Trees   Whatever the type of dendrometer used, constant care must be exercised to measure trees exactly at dbh—or at a rational deviation from this point when irregular stems are encountered. For trees growing on slopes, for example, it is recommended that dbh be measured 4.5 ft above ground from the *uphill* side of the tree. Figure 6-4 illustrates suggested methods of maintaining consistency in obtaining diameter measurements.

FIGURE 6-3
Measuring tree dbh with a Biltmore stick. *(Photograph by Rick Griffiths.)*

When swellings, bumps, depressions, or branches occur 4.5 ft above ground, tree diameters are usually taken just above the irregularity at a point where it ceases to affect normal stem form. If a tree forks immediately above dbh, it is measured below the swell resulting from the double stem. Stems that fork below dbh are considered as two separate trees, and diameters are measured (or estimated) approximately 3.5 ft above the fork. Cypress, tupelo gum, and other swell-butted species are measured 1.5 ft above the pronounced swell or "bottleneck," if the swell is more than 3 ft high. Such measurements are usually referred to as *normal diameters* and are abbreviated dn.

When there is heavy snow cover on the ground or when diameters are measured under floodwater conditions, a pole should be used as a probe to locate true ground level; otherwise, the point of diameter measurement may be made too high up on the tree stem.

If successive diameter measurements are taken on the same trees (as on permanent sample plots), relative accuracy can be improved by marking the exact dbh point on each tree. And when calipers are used, measurements should be made *in the same direction* each time.

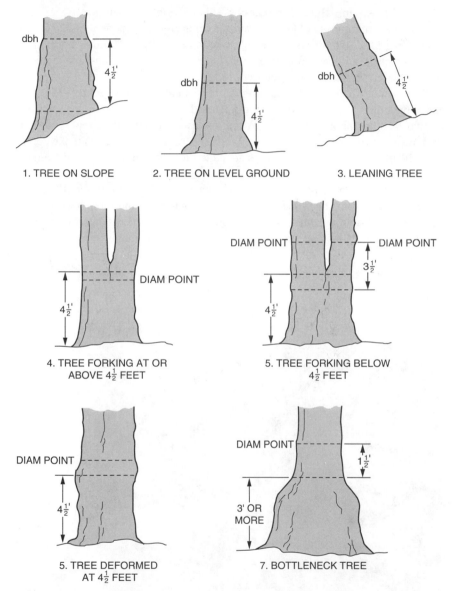

1. TREE ON SLOPE

2. TREE ON LEVEL GROUND

3. LEANING TREE

4. TREE FORKING AT OR ABOVE $4\frac{1}{2}$ FEET

5. TREE FORKING BELOW $4\frac{1}{2}$ FEET

5. TREE DEFORMED AT $4\frac{1}{2}$ FEET

7. BOTTLENECK TREE

FIGURE 6-4
Points of dbh measurement for sloping ground or irregular tree stems. *(Source: U.S. Forest Service.)*

6-3 Measuring Bark Thickness Although dbh measurements are made outside bark, a common objective for computing tree volume is the diameter inside bark (dib). Reliable measures of bark thickness are essential, because the breast-height ratio of dib/dob is often applied to estimate inside-bark diameters for inaccessible points on the tree stem.

The standard measurement tool employed is called a *bark gauge* (Fig. 6-5). Since bark thickness tends to vary from one side of a tree to another, a minimum of two readings should be taken. When dob is obtained with calipers, the two bark measurements should be made exactly where the caliper arms make contact with the tree stem. The two readings are added together and subtracted from dob to obtain dib.

Where dbh is determined with a diameter tape, bark thickness should be measured radially from the wood surface to the contour of the tape. Two or more thicknesses should be measured, depending on the eccentricity of the cross section. By this technique, bark thickness is regarded as the difference in diame-

FIGURE 6-5
Measuring bark thickness at breast height.

ters of two concentric circles, one defined by the bark surface, the other by the interior wood surface (Mesavage, 1969b).

When commercially manufactured bark gauges are not available, a fair substitute can be improvised by filing graduations on the steel bit of a sharpened screwdriver, or by using a carpenter's brace and auger bit.

6-4   Tree Diameter Classes   Although tree diameters are commonly measured to the nearest 0.1 in., it is often expedient to group such measurements into diameter classes. If 1-in. classes are used, it is customary to drop back to the lower value when diameters fall exactly halfway between inch graduations—just as in log scaling (Sec. 5-10). Thus the class boundaries or true class limits for 8-in. trees are 7.6 to 8.5 in.; the 9-in. class ranges from 8.6 to 9.5 in., and so on. Two-inch class boundaries are commonly defined for 8-in. trees as 7.1 to 9.0 in., for 10-in. trees as 9.1 to 11.0 in., and so on.

6-5   Basal Area and Mean Diameter   Tree-stem diameter measurements are often converted to cross-sectional areas. The cross-sectional area at breast height is called *basal area*. To compute tree basal area, one commonly assumes that the tree stem is circular in cross section at breast height. Thus the formula for calculating basal area is (Sec. 4-1):

$$BA(ft^2) = \frac{\pi dbh^2}{4(144)} = 0.005454 \ dbh^2$$

when basal area is desired in square feet and dbh is measured in inches. If metric units are used, and basal area is desired in m^2 and dbh is measured in cm, the formula for basal area is

$$BA(m^2) = \frac{\pi dbh^2}{4(10,000)} = 0.00007854 \ dbh^2$$

Average dbh of a stand of trees is a highly informative statistic for forest managers. Two average dbh values are in common use. The arithmetic mean dbh ($\overline{dbh}$) is simply the sum of the tree dbh values divided by the number of trees; that is,

$$\overline{dbh} = \frac{\sum\limits_{i=1}^{n} dbh_i}{n}$$

Alternatively, the quadratic mean diameter (also termed the diameter of the tree of mean basal area) may be computed. If basal area (BA) is in square feet and dbh is in inches, one notes from the foregoing expression for basal area that

$$\text{dbh}^2 = \frac{\text{BA}}{0.005454}$$

The quadratic mean dbh ($\overline{\text{dbh}}_q$) in inches is computed by determining the mean basal area in square feet per tree ($\overline{\text{BA}}$) and substituting into the following formula:

$$\overline{\text{dbh}}_q = \sqrt{\frac{\overline{\text{BA}}}{0.005454}}$$

Computation of quadratic mean dbh in metric units follows by analogy. It should be noted that, for typical tree-diameter data, the two mean dbh values will not be equal, but rather the quadratic mean dbh will be slightly larger than the arithmetic mean dbh.

6-6 Upper-Stem Diameters Out-of-reach diameters are frequently required in studies of tree form, taper, and volume. Although such diameters are best obtained by direct measurement, the use of ladders and climbing irons is time-consuming, awkward, and often hazardous. As a result, a number of diverse upper-stem dendrometers have been proposed. These include such items as calipers attached to a pole, binoculars with a mil scale in one eyepiece, telescopic stadia devices, and split-image rangefinders. A comprehensive investigation of optical dendrometers has been conducted by Grosenbaugh (1963).

Most upper-stem dendrometers are limited in usefulness because either they do not provide sufficient accuracy or they are prohibitively expensive. Some are also quite complex in operation. An ideal upper-stem dendrometer would be simple to use, portable, relatively inexpensive, accurate at all tree heights, and operable independently of distance from point of measurement. Although it may be unrealistic to expect all these attributes in a single instrument, several are incorporated in the pentaprism tree caliper (Wheeler, 1962).

In effect, the pentaprism caliper may be compared to an imaginary giant caliper that can be clamped on a tree stem at any point and from any distance without special calibration. Two pentaprisms, one fixed and the other movable, are mounted so that extended parallel lines of sight may be viewed simultaneously (Fig. 6-6). Prisms are oriented so that the right side of the tree stem is brought into coincidence with the left side, which is viewed directly. A scale is provided so that dob may be read through the fixed (left-hand) prism at the point of coincidence.

FIGURE 6-6
Measuring an upper-stem diameter with the Wheeler pentaprism tree caliper. *(U.S. Forest Service photograph.)*

Tests of the Wheeler pentaprism caliper indicate that upper-stem diameters as high as 50 ft above ground may be read to an accuracy of 0.2 to 0.5 in. Greater accuracy may be feasible if an optical-lens system is used to replace the sighting tube on original models of the instrument.

A popular instrument for measuring upper-stem diameters is the Barr and Stroud optical dendrometer. Mounted on a tripod, this instrument is a split-image, coincident-type magnifying rangefinder for estimating inaccessible diameters, heights, and distances. The last model manufactured, model FP 15, is designed to the following accuracy specifications:

Diameter   0.1 in. for tree diameters from 1.5 to 10 in., and 1 percent for diameters from 10 to 200 in.

Height $1^1/_2$ percent at all heights above 10° elevation.

Distance   0.2 percent at 45 ft, 0.6 percent at 90 ft, 1.2 percent at 300 ft, 2.2 percent at 600 ft, and 6.8 percent at 2000 ft.

The instrument readings are nonlinear for both distance and diameter; scale readings must therefore be transformed by using tables supplied by the manufacturer or by using special computer programs. Such conversions can be tedious when a large number of readings are involved.

Field tests of this instrument have shown that the manufacturer's claims for accuracy can be substantiated under good sighting conditions. Extensive use of such optical dendrometers *may* provide the forester with a means of eliminating the conventional tree volume table. With reliable dendrometer readings, volume growth for the upper stem may be determined on standing trees by repeated measurements at specified time intervals.

Where reliable instruments are available for measuring upper-stem diameters of standing trees, log volumes may be determined by "height accumulation," a concept of tree measurement developed by Grosenbaugh (1954). In essence, this technique consists of selecting tree diameters above dbh in a specified progression; then tree height to each diameter is estimated, recorded, and accumulated. In contrast to usual tree measurement procedures, outside bark diameter at selected intervals is treated as the independent variable, and heights are recorded at irregular intervals. This technique permits individual trees to be broken down into various product uses and recombined with similar sections from other trees.

TREE HEIGHTS

6-7 Height Measurement Principles Instruments used for measuring tree heights are collectively referred to as *hypsometers*. Many types of height-measuring devices and instruments have been evolved, but only a few have gained wide acceptance by practicing foresters. Thus only two of the more common designs are discussed here. The basic trigonometric principle most frequently embodied in hypsometers is illustrated in Fig. 6-7. The observer stands at a fixed horizontal distance from the base of the tree, usually 50, 66, or 100 ft. Tangents of angles to the top and base of the tree are multiplied by horizontal distance to derive the

FIGURE 6-7
Principle of height measurement with the Abney level and similar hypsometers.

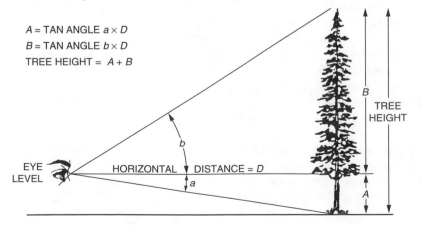

A = TAN ANGLE $a \times D$
B = TAN ANGLE $b \times D$
TREE HEIGHT = $A + B$

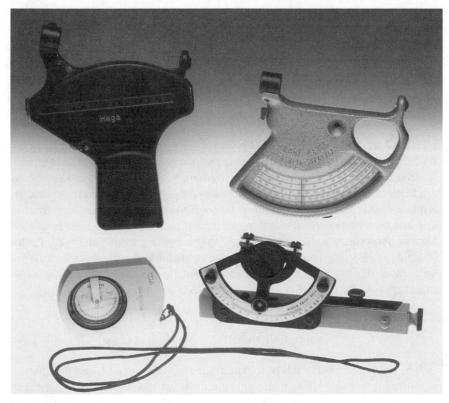

FIGURE 6-8
Examples of hypsometers (upper left, Haga altimeter; upper right, Blume-Leiss; lower left,
Suunto clinometer; lower right, Abney level) based on the tangents of angles. *(Photograph
by Rick Griffiths.)*

height of each measured section of the stem. The Abney level and several *cli-
nometers* or *altimeters* operate on this principle (Fig 6-8), yielding height readings
directly in feet or meters at fixed horizontal distances from the tree.

For accurate results, trees must not lean more than 5° from the vertical,
and the fixed horizontal distance must be determined by taped measurement or
careful pacing. Instruments equal in caliber to the Abney level will provide readings within 2 to 5 percent of true heights, provided both points of tree measurement are clearly discernible. Leaning trees should be measured at right angles to
the direction of lean to minimize height errors.

When using an instrument such as the Abney level on gentle terrain, a level
line of sight from the observer's eye will usually intercept the tree stem some-
where between stump height and the tree top. As a result, angular readings to the
base and the top of the tree will appear on *opposite* sides of the zero point on the

graduated instrument scale. In such instances, the two readings must be *added* together to obtain the desired height value.

In mountainous terrain, the observer's hypsometer position may be below the base of the tree or occasionally above the desired upper point of measurement. If a level line of sight from the observer fails to intercept the tree stem, both angular readings will then appear on the *same side* of the instrument zero point. Tree height is derived by taking the *difference* between the two readings.

6-8   Merritt Hypsometer   This linear scale, often imprinted on one face of a standard Biltmore stick, is based on a principle of similar triangles. It is normally used for determining merchantable log heights rather than total heights, and graduations are placed at 16-ft log intervals or 8-ft half-log intervals. As with the Biltmore stick, the hypsometer must be positioned at a fixed distance from the eye and the observer must stand a specified distance from the tree. In use, the Merritt hypsometer is held vertically with the lower end of the stick on a line of sight to tree-stump height. With the stick held plumb, the observer then glances up to note the log height at the desired point on the upper stem. Improvised rules may be calibrated for any desired arm reach and specified distance by

$$\frac{\text{Arm reach (in.)}}{\text{Distance from tree (ft)}} = \frac{\text{scale interval (in.)}}{\text{log height (ft)}}$$

The foregoing ratio is solved to determine the scale interval, and this distance is uniformly marked off on a straight rule to define the desired log-height spacings. The Merritt hypsometer is a useful aid for estimating tree heights by log intervals, but it is not reliable for precise work. Where heights must be recorded to 1- or 2-ft intervals, an Abney level or clinometer of comparable accuracy should be used. For trees less than 50 ft tall, *direct* linear measurement of tree heights may be feasible by using jointed light-weight tubing having brightly colored stripes at 1- or 2-ft intervals. This technique may be particularly applicable in coniferous plantations where vehicular access is possible.

6-9   Total versus Merchantable Heights   Total tree height is the linear distance from ground level to the upper tip of the tree crown. The tip of the crown is easily defined when trees have conical shapes, but it may not be readily discernible for deciduous trees having irregular or round-topped crowns. Thus the measurement of total height is more applicable to coniferous trees having *excurrent* branching characteristics than to broad-leaved deciduous trees with *deliquescent* branching patterns. Recording of total heights is preferred to merchantable lengths on permanent sample plots when tree-growth measurements are based on periodic remeasurements of the same trees. Here, measurement of the entire stem is likely to be more objective and less subject to errors of judgment than heights measured to an ocularly selected merchantable top.

Merchantable tree height refers to the usable portion of the tree stem, i.e., the part for which volume is computed or the section expected to be utilized in a commercial logging operation. For smooth, straight stems, merchantable height may be simply defined as the length from an assumed stump height to an arbitrarily fixed upper-stem diameter. Exact location of the upper-diameter limit may require considerable proficiency in ocular estimation, perhaps including occasional checks with an upper-stem dendrometer.

When upper limits of stem merchantability are not dictated by branches, crook, or defect, minimum top diameters may be chosen as a percentage of dbh. With sawtimber-sized trees, for example, minimum top diameters may be set at approximately 60 percent of dbh for small trees, 50 percent of dbh for medium-sized trees, and 40 percent of dbh for large trees. This procedure, more often applied to conifers than to hardwoods, rationally presumes that the larger the dbh, the rougher the upper stem of a tree. Thus top-log-scaling diameters will be larger for mature or old-growth trees than for smaller, second-growth stems. When merchantable heights are tallied for inventory purposes, minimum top diameters must be selected in accordance with the particular volume table to be used. Failure to observe this precaution will result in highly inaccurate estimates of individual tree volume.

6-10 Sawlog Merchantability for Irregular Stems   For many hardwood species, minimum top diameters and sawlog merchantability are regulated by tree form, branches, stem roughness, or defect (rotten cull material). Some typical stem forms that may be encountered in sawlog height determination are illustrated in Fig. 6-9. Bole sections designated as *upper stem* refer to sound portions unsuitable for sawlogs but usable for lower-grade products such as fence posts or pulpwood. Limbs and sound cull material are considered unmerchantable because of roughness, form, or size. Following is a brief description prepared by the U.S. Forest Service for each tree pictured:

A: Sawtimber tree   Sawlog length terminates at 9-inch top dob. Meets minimum qualifications of a 12-foot sawlog. Upper stem portion contains no cull and terminates at 4 inches dob. Sawlog length is recorded as 12 feet; bole length as 21 feet.

B: Sawtimber tree   Sawlog portion terminated by limbs at 13 inches dob. Contains no cull and meets minimum grade specifications. Both bole length and sawlog length are 14 feet. Portion between whorls of limbs is large enough in diameter but not in length to qualify as upper stem (i.e., it is less than 4 feet long).

C: Rotten cull tree   Although sawlog portion is 20 feet long, a 13-foot section of rotten cull prevents utilization of a log meeting minimum grade specifications; thus the entire sawlog portion is culled. Because more than half the volume in that portion is rotten, the tree is classed as a rotten cull.

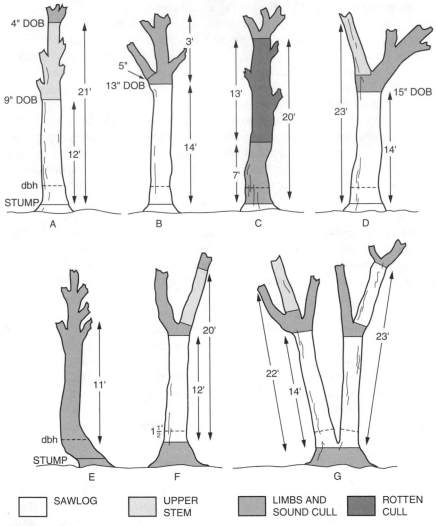

SAWLOG   UPPER STEM   LIMBS AND SOUND CULL   ROTTEN CULL

FIGURE 6-9
Merchantable height limits for irregular bole forms. *(Source: U.S. Forest Service.)*

D: Sawtimber tree   Sawlog portion terminates because of branching at 15-inch top dob. Right-hand fork is too limby to qualify as upper stem, but 7 feet of left-hand fork qualifies as upper stem.

E: Rough tree   Sawlog top terminates by branches 11 feet above crooked butt. No sawlog meeting minimum requirements present.

F: Sawtimber tree   Despite sound cull in the sawlog portion due to butt swell, a 12-foot sawlog is present. Seven feet of right-hand fork qualifies as upper stem. Left-hand fork does not qualify because of crook.

G: Two sawtimber trees Because lowest fork is below dbh, each fork is appraised and recorded as a separate tree. The lower 14-foot section in the left-hand fork meets requirements for a sawtimber tree. A 6-foot portion of the largest stem in upper fork qualifies as upper stem material. In the main right-hand fork, a 13½-foot sawlog plus a 9-foot sawlog (with an intervening 1-foot section of sound cull) is recorded as 23 feet of sawlog length.

TREE FORM EXPRESSIONS

6-11 Taper Tables and Functions Because trees taper, often irregularly, from stump to top, it is common to make some evaluation of stem form in the construction and application of tree volume tables. The rate of tree taper varies not only by species but also by age, dbh, and tree height. If a series of diameter measurements are taken at intervals along the bole, average taper rates may be derived for groups of trees characterized by a particular shape or form category. Such tabulations are referred to as *taper tables,* and they are commonly compiled according to generalized form classes, irrespective of tree species.

TABLE 6-1
UPPER-LOG TAPER (INCHES) BY 16-FT LOGS

dbh, in.	2-log tree 2d log	3-log tree 2d log	3d log	4-log tree 2d log	3d log	4th log
10	1.4	1.2	1.4			
12	1.6	1.3	1.5	1.1	1.4	1.9
14	1.7	1.4	1.6	1.2	1.5	2.0
16	1.9	1.5	1.7	1.2	1.6	2.1
18	2.0	1.6	1.8	1.3	1.7	2.2
20	2.1	1.7	1.9	1.4	1.8	2.4
22	2.2	1.8	2.0	1.4	2.0	2.5
24	2.3	1.8	2.2	1.5	2.2	2.6
26	2.4	1.9	2.3	1.5	2.3	2.7
28	2.5	1.9	2.5	1.6	2.4	2.8
30	2.6	2.0	2.6	1.7	2.5	3.0
32	2.7	2.0	2.7	1.7	2.5	3.1
34	2.8	2.1	2.7	1.8	2.5	3.3
36	2.8	2.1	2.8	1.8	2.6	3.4
38	2.9	2.1	2.8	1.9	2.6	3.4
40	2.9	2.2	2.8	1.9	2.7	3.4

Source: Mesavage and Girard, 1946. (Table abridged.)

An example of upper-log taper rates is shown in Table 6-1. For specified dbh and merchantable height classes, average stem taper is shown for each successive 16-ft log above the butt section. It will be noted that both tree age and species are ignored. This particular table is the basis for construction of widely adopted Mesavage-Girard form-class volume tables discussed in Chap. 7. Improbable as it may seem, these taper rates were originally compiled from *ocular* estimates of about 2000 trees of assorted species in the anthracite region of Pennsylvania. They were later modified and extended on the basis of ocular estimates of about 20,000 additional trees throughout the South.

In contrast to the preceding technique, taper tables are preferably constructed by complete stem analyses of felled trees or from optical dendrometer readings of standing trees. With such information, taper curves may be defined for trees of any size, thus permitting the calculation of tree volumes for any degree of stem utilization.

A number of attempts have been made to derive rates of stem taper for various species by means of mathematical functions. Although no universal expression for taper curves has been adopted, Kozak et al. (1969) have shown that for certain coniferous species, upper-stem diameters (dib) can be reliably predicted from this parabolic function:

$$d^2/\text{dbh}^2 = b_0 + b_1(h/H) + b_2(h^2/H^2)$$

and
$$d = \text{dbh}\sqrt{b_0 + b_1(h/H) + b_2(h^2/H^2)}$$

where d = stem diameter at any given height h above ground
H = total tree height
b_0, b_1, b_2 = regression coefficients

6-12   Form Factors and Quotients   The comparison of tree bole forms with various solids of revolution (cylinders, paraboloids, etc.) may be expressed in numerical terms as *form factors*. Such ratios are derived by dividing stem volume by the volume of a chosen solid. However, as form factors cannot be computed until essential stem diameters are obtained, it has become customary to express stem configuration in terms of *form quotients* derived directly from the diameter ratios themselves.

A *form quotient* is the ratio of some upper-stem diameter to dbh. The value is always less than unity and is usually expressed as a percentage. Higher form quotients indicate lower rates of stem taper and correspondingly greater tree volumes. For a given species, form quotients are lowest for open-grown trees with long live crowns and highest for forest-grown trees with relatively short crowns. Thus for given soil and site conditions, stand density has an indirect effect on tree-taper rates. The primary expression of form that has been used in the United States is known as *Girard form class*.

FIGURE 6-10
The points of diameter measurement for determining Girard form class are shown by tree bands at breast height and at the top of the first 16-ft log. This ponderosa pine has a form class of 82 and a total height of 69 ft.

6-13   Girard Form Class   This form quotient is computed as the ratio between stem diameter, *inside bark,* at the top of the first 16-ft log and dbh, *outside bark.* With a log trimming allowance of 0.3 ft and a 1-ft stump, the upper-stem measurement is taken 17.3 ft above ground. As an example, a tree with a first-log-scaling diameter of 16 in. and a dbh of 20 in. has a Girard form class of $16 \div 20 = 0.80$, or 80 percent (Fig. 6-10).

Sawtimber volume tables based on Girard form class assume that trees having the same diameter and merchantable height will have similar, though not necessarily identical, rates of taper in the sawlog portion *above the first log.* It is thereby implied that all volume differences in trees of the same diameter and merchantable height may be attributed largely to taper variations occurring *in the first log.* Girard form-class tables are *composite* volume tables; i.e., they are compiled independently of tree species and are applicable to both coniferous and broad-leaved trees.

For swell-butted species such as cypress and tupelo gum, diameter measured 1.5 ft above the pronounced swell should be substituted for dbh in computing Girard form class. Measurement of the scaling diameter remains at 17.3 ft above ground; thus the two diameters may be only 6 to 10 ft apart in some cases. For trees deformed by chipped turpentine faces at breast height, the normal diameter should be measured just above the highest face. Although the diameter here will be smaller, form class will be higher, resulting in a compensating volume increase.

6-14   Importance of Form Measurements   Most foresters prefer form expressions based on relatively accessible measurements, a factor that has probably contributed to the popularity of Girard form class. Even here, the diameter measurement at 17.3 ft can rarely be ocularly estimated with precision and consistency; thus it is usually better to have a carefully *measured sample* of a few stems rather than rough estimates of form class for each tree tallied. Obtaining reliable inside-bark measurements at the top of the first 16-ft log implies the use of ladders or climbing irons—an expensive and time-consuming task. As an alternative, however, dob at the top of the first log can be determined with an optical dendrometer, and the corresponding dib computed from a breast-height ratio of dib/dob. Or the form class can be ocularly estimated by using a simple sighting device such as that suggested by Wiant (1972). When applying Girard form-class values, the difference between one class and another (e.g., 79 versus 80) amounts to approximately 3 percent in terms of merchantable tree volume.

TREE AGE

6-15   Definitions   The age of a tree is defined as the elapsed time since germination of the seed or the time since the budding of the sprout or cutting from which the tree developed. The age of a plantation is commonly taken from the year it was formed, i.e., exclusive of the age of the nursery stock that may have been planted.

The terms *even-aged* and *uneven-aged* are often applied to forest stands; therefore, it is appropriate to define these expressions. *Even-aged stands* are those in which tree ages do not differ by more than 10 to 20 years. In stands where the harvesting or rotation age is 100 years or more, however, age differences up to 30 percent of the rotation age may be allowed (Society of American Foresters, 1971).

Uneven-aged stands are those where age differences exceed the stated limits or where three or more age classes are represented. *All-aged stands* are rarities that are virtually nonexistent. In theory, they include trees of all ages from minute seedlings to the harvest or rotation age.

The selection of sample trees for age and site index determinations in even-aged stands requires an evaluation of relative dominance or crown levels for various trees. Four crown classes are recognized:

1 *Dominant.* Trees with crowns extending above the general level of the crown cover and receiving full light from above and partly from the side; larger than the average trees in the stand, with crowns well developed but possibly somewhat crowded on the sides.

2 *Codominant.* Trees with crowns forming the general level of the crown cover and receiving full light from above, but comparatively little from the sides; usually with medium-sized crowns more or less crowded on the sides.

3 *Intermediate.* Trees shorter than those in the two preceding classes, but with crowns either below or extending into the crown cover formed by codominant and dominant trees, receiving little direct light from above and none from the sides; usually with small crowns considerably crowded on the sides.

4 *Suppressed.* Trees with crowns entirely below the general level of the crown cover, receiving no direct light either from above or from the sides.

6-16   Age from Annual Rings   Many tree species found in northern temperate zones grow in diameter by adding a single and distinctive layer of wood each year. The formation of this layer starts at the beginning of the growing season and continues well through it. Earlywood (or springwood) is more porous and lighter in color than latewood (or summerwood). The combination of one springwood and one summerwood band comprises a year's growth. On a stem cross section, these bands appear as a series of concentric rings. A count of the number of rings at any height gives the number of years' growth that the tree has undergone since attaining that height; thus, a ring count made on the cross section at ground level provides the total age of the tree.

If the annual ring count is made on a stump cross section or higher up on the stem, the count provides the age of the tree from that point upward. It is therefore necessary to add the number of years required for the tree to attain the height of measurement to derive total tree age.

Although the most reliable ring counts are made on complete cross sections, it is obvious that except on logging operations, this method involves destructive sampling. Ages of standing trees are therefore determined by extracting a radial core of wood with an instrument called an *increment borer*. The hollow auger of the increment borer is pressed against the standing tree (usually at breast height) and turned until the screw bit reaches the center of the tree. A core of wood is forced into the hollow auger; the borer is then given a reverse turn to snap the core loose and permit its removal with an extractor. The number of annual rings on the core gives the age of the tree from the point of the boring upward (Fig. 6-11).

The reliability of annual ring counts depends on the species and the environmental conditions under which a particular tree may be growing. Fast-growing coniferous species in northern temperate zones usually provide the easiest counts. Difficulties are encountered when there is little contrast between springwood and summerwood, as in the case of some diffuse-porous, deciduous broadleaved species.

Trees growing under adverse environmental conditions may produce extremely narrow or almost nonexistent rings that are difficult to count except on sanded cores or cross sections. A catastrophe such as an extended drought or tree defoliation by insects, followed by favorable growth conditions, can lead to the formation of *false rings*. Such rings, which often display an incomplete circumference, may be distinguishable only when cross sections are available for analysis.

FIGURE 6-11
Extraction of a core of wood with an increment borer (left) and measurement of the core (right). Annual rings are discernible in the right-hand view. *(U.S. Forest Service photograph.)*

6-17 Age without Annual Rings In many forest regions of the world (e.g., Australia, New Zealand, and numerous other countries in tropical or southern temperate zones), tree growth is generally *not* characterized by annual rings. Annual rings are usually absent in tropical conifers and in diffuse-porous, evergreen broad-leaved species, unless they are growing in subalpine or alpine conditions. Therefore, except where plantations are established, it may only be possible to approximate individual tree ages.

For a limited number of species around the world, e.g., *Pinus strobus,* the age of young trees may be estimated from branch whorls. The seasonal height growth of the tree begins with the bursting of the terminal bud, which lengthens to form the leader. At the base of the leader, a circle of branchlets grows out at the same time, thus marking the height of the tree as it was before the season's growth started. The following year, the process is repeated, and so on. A count of these branch whorls will give the approximate age of the tree. For some species, 2 to 4 years must be added to account for the early growth years when no whorls were produced.

When neither annual rings nor branch whorls can be relied upon, age determination for trees in natural stands is extremely difficult. *Relative* tree age can be roughly gauged by the relative size of the tree, shape and vigor of the crown, and texture or color of the bark. Occasionally, the age of trees in naturally regenerated even-aged stands can be closely approximated from the year of the logging operation that resulted in the establishment of the stand. In similar fashion, trees whose germination followed a catastrophic event such as a fire or hurricane can also be dated. And in a few instances, counting the annual rings of "indicator species" may help to ascertain the ages of nearby trees that do not exhibit annual rings. For example, ring counts on certain scrub species have been used to assist in determining the age of associated eucalyptus trees in Tasmania (Carron, 1968).

PROBLEMS

6-1 Number 20 standing trees of varying diameters. Ocularly estimate each dbh to the nearest inch. Then, in order, remeasure diameters with (a) a Biltmore stick, (b) a diameter tape, and (c) calipers. Tabulate all measurements according to tree number on a single tally sheet. Using the average of two caliper readings as a standard, obtain plus or minus deviations for the other three diameter estimates. Discuss your findings and preferences in a brief written report.

6-2 Design and construct a one-handed tree caliper of the "fork" type.

6-3 Design and construct a simple bark gauge.

6-4 Construct an upper-stem dendrometer based on twin prisms, a rangefinder, or the stadia principle. Test on 10 or more trees of known dimensions.

6-5 Number 20 standing trees of varying total or merchantable heights. In order, obtain heights by (a) ocular estimation, (b) Merritt hypsometer, (c) Abney level, and (d) any other available clinometer. Tabulate and analyze findings as in problem 6-1, using Abney readings as the measurement standard.

6-6 Construct a Biltmore stick and Merritt hypsometer for your own arm reach. Graduate the rule by 1-in.-diameter classes and for height intervals most commonly used in your locality.

6-7 Number 5 to 10 standing trees (preferably mature conifers), and determine for each (a) Girard form class and (b) another expression of tree form suggested by your instructor. Use an upper-stem dendrometer for obtaining out-of-reach diameters. Which form expression is most easily derived in the field and which is most commonly used in your region?

6-8 Select 50 to 100 felled trees on a logging operation and obtain dib measurements at 4-ft intervals from dbh to top. Construct a taper curve for the species, or summarize average values in a taper table.

6-9 Select several eccentric tree stumps at a recent harvesting site and obtain complete tree cross sections from each. Also obtain two increment cores from each stump—one across the longer radius and one across the shorter radius. Have several persons count the rings on the cores; then compare results with stump counts. For the species evaluated, how many years must be added to stump age to obtain total tree age? Can you detect any false rings on the cross sections? Does the eccentricity of cross sections tend to be predictable; i.e., is the long axis of cross sections aligned in a fairly constant compass direction? If so, can you give possible explanations for this growth pattern?

6-10 Attempt to determine the ages of several mature trees in natural stands by methods other than annual ring counts. Prepare a brief report on the favored estimation technique for a species in your locality.

6-11 Determine the arithmetic mean and the quadratic mean diameter for the following set of dbh values (in inches):

   6, 11, 9, 15, 19, 7, 8, 12

Convert the two means from inches to centimeters.

REFERENCES

Avery, T. E. 1959. An all-purpose cruiser stick. *J. Forestry* **57:**924–925.

Biging, G. S., and Wensel, L. C. 1988. The effect of eccentricity on the estimation of basal area and basal area growth. *Forest Sci.* **34:**338–342.

Brickell, J. E. 1970. More on diameter tapes and calipers. *J. Forestry* **68:**169–170.

Bruce, D. 1975. Evaluating accuracy of tree measurements made with optical instruments. *Forest Sci.* **21:**421–426.

Carron, L. T. 1968. *An outline of forest mensuration.* Australian National University Press, Canberra, A. C. T. 224 pp.

Gregoire, T. G., Zedaker, S. M., and Nicholas, N. S. 1990. Modeling relative error in stem basal area estimates. *Can. J. For. Res.* **20:**496–502.

Grosenbaugh, L. R. 1954. New tree-measurement concepts: Height accumulation, giant tree, taper, and shape. *U.S. Forest Serv., Southern Forest Expt. Sta. Occasional Paper* 134. 32 pp.

———. 1963. Optical dendrometers for out-of-reach diameters: A conspectus and some new theory. *Forest Sci. Monograph* 4. 47 pp.

———. 1981. Measuring trees that lean, fork, crook, or sweep. *J. Forestry* **79:**89–92.

Howe, G. T., and Adams, W. T. 1988. Clinometer versus pole measurement of tree heights in young Douglas-fir progeny tests. *West. J. Appl. For.* **3:**86–88.

Hunt, E. V., Jr. 1959. A time and accuracy test of some hypsometers. *J. Forestry* **57:**641–643.

Jackson, M. T., and Petty, R. O. 1973. A simple optical device for measuring vertical projection of tree crowns. *Forest Sci.* **19:**60–62.

Jonsson, B. 1981. An electronic caliper with automatic data storage. *Forest Sci.* **27:**765–770.

Kozak, A., Munro, D. D., and Smith, J. H. G. 1969. Taper functions and their application in forest inventory. *Forestry Chron.* **45:**278–283.

Kozlowski, T. T. 1971. *Growth and development of trees.* Vol. II: *Cambial growth, root growth, and reproductive growth.* Academic Press, Inc., New York. 514 pp.

Matern, B. 1990. On the shape of the cross-section of a tree stem. An empirical study of the geometry of mensurational methods. *Swedish Univ. of Agr. Sci. Report* 28. 46 pp.

Mesavage, C. 1965. Definition of merchantable sawtimber height. *J. Forestry* **63:**30–32.

———. 1969a. New Barr and Stroud dendrometer, model FP 15. *J. Forestry* **67:**40–41.

———. 1969b. Measuring bark thickness. *J. Forestry* **67:**753–754.

———, and Girard, J. W. 1946. Tables for estimating board-foot content of timber. U.S. Forest Service, Government Printing Office, Washington, D.C. 94 pp.

Rennie, J. C. 1979. Comparison of height measurement techniques in a dense loblolly pine plantation. *So. J. Appl. For.* **3:**146–148.

Society of American Foresters. 1971. *Terminology of forest science, technology, practice and products.* W. Heffer and Sons, Ltd., Cambridge, England. 349 pp.

Wheeler, P. R. 1962. Penta prism caliper for upper-stem diameter measurements. *J. Forestry* **60:**877–878.

Wiant, H. V. 1972. Form class estimates—A simple guide. *J. Forestry* **70:**421–422.

VOLUMES AND WEIGHTS
OF STANDING TREES

7-1 Purpose of Volume and Weight Tables A volume table is a tabulation that provides the average contents for standing trees of various sizes and species. Volume units most commonly employed are board feet, cubic feet, cords, or cubic meters. Volumes may be listed for some specific merchantable portion of the stem only, or for the total stem. In some instances volumes are tabulated for both the sawlog material and the pulpwood top sections.

Board-foot volume tables are usually based on existing log rules; thus they can never be more reliable than the log rule selected as a basis for their construction. The principal objective in compiling such tables is to obtain a board-foot estimate for standing trees that would correspond with the volume obtained if the same trees were felled, bucked, and scaled as logs. Thus such tables are used in timber estimating as a means of ascertaining the volume and value of standing trees in a forested tract.

In modern practice, equations are generally used to predict tree volumes rather than obtaining the values from tables. However, the term *volume table* has persisted in forestry usage as a generic term meaning tabulations *or* equations that show the contents of standing trees.

Tree weight tables are analogous to volume tables except that weights (green or dry) of standing trees are predicted rather than volumes. Weight tables (or equations) are generally expressed in terms of pounds or kilograms. For brevity, this text will refer to tree volume and weight tables or equations simply as volume tables when it is clear that volume or weight estimates for standing trees are intended.

7-2   Types of Tree Volume and Weight Tables   The principal variables ordinarily associated with standing tree volume or weight are diameter at breast height (dbh), tree height, and, perhaps, tree form. Height may be expressed as total tree height or length of the merchantable stem. Although merchantable stem length is more highly correlated with merchantable volume than is total tree height, it is a difficult variable to define precisely and to measure consistently in the field. Consequently, for species that are typically single-stemmed, total height is generally used when predicting both total and merchantable volumes.

Tree volume tables that are based on the single variable of dbh are commonly referred to as *local* volume tables; those that require the user to also obtain tree height and possibly form or taper are referred to as *standard* volume tables. These labels are often misleading, for they tend to imply that local volume tables are somehow inferior to standard volume tables. Such an assumption is not necessarily true, particularly when the local table in question is derived from a standard volume table. More appropriate labels for these two types of volume or weight tables would be *single-entry* and *multiple-entry* tables.

Volume tables, whether of the single-entry or multiple-entry variety, may also be classified as *species* tables or *composite* tables. In the first instance, separate tables are constructed for each important timber species or groups of species that are similar in terms of tree form. On the other hand, composite tables are intended for application to diverse species, often including both conifers and hardwoods. To compensate for inherent differences in stem taper and volume between various species groups, provision is usually made for additionally measuring tree form, or correction factors are developed for various species. Otherwise, composite tables will overestimate volumes of some trees while underestimating volumes of others.

The main disadvantage of species tables is the large number of species encountered in most regions. When it is not feasible to construct separate tables for each species, those of similar taper and shape may be grouped together. To avoid such difficulties, composite tables utilizing some measurement of tree form in lieu of species differentiation have been adopted in several regions.

MULTIPLE-ENTRY VOLUME TABLES

7-3   Form-Class versus Non-Form-Class Tables   Multiple-entry volume tables provide an estimate of individual tree volume based on dbh, height, and, sometimes, a measure of tree form. Tree form is a difficult variable to describe, and there is often a high degree of variability in form, both within and between species. Nevertheless, many form-class volume tables have been prepared, and those based on Girard form class are typical of the tables still in use.

The theory and measurement of Girard form class were discussed in the previous chapter (Sec. 6-13). Form-class volume tables based on this concept of butt-log taper are among the most widely accepted standard tables in eastern United

States. The biggest disadvantages in using these tables are (1) the general tendency toward rough estimates of form class rather than actual measurements and (2) the wide variations in upper-stem form that cannot be adequately accommodated by measuring butt-log taper only. The fact that each change in form class (as from 77 to 78) accounts for about 3 percent of merchantable tree volume should serve as a precaution against purely ocular estimates of this independent variable.

Multiple-entry board-foot volume tables have been compiled from International, Scribner, and Doyle log rules for form classes of 65 to 90 (Mesavage and Girard, 1946). Table 7-1, based on form class 80 and the International rule, provides an abridged example of the format employed. The user must first determine the form class of the tree; then only dbh and sawlog length in 16-ft logs and half logs are needed to derive merchantable board-foot volumes. In most published versions, these tables cover a dbh range of 10 to 40 in. (by 1-in. classes) and a height range of one to six logs.

7-4   Compilation of Mesavage-Girard Form-Class Tables   In the original tabulations of the Mesavage-Girard form-class tables, the scaling diameter of the butt log is derived from the estimated or measured form class. Thus for a 20-in.-dbh three-log tree of form class 80, the scaling diameter of the first log is 0.80×20, or 16 in. Scaling diameters for all upper logs are derived from a single

TABLE 7-1
STANDARD VOLUME TABLE, INTERNATIONAL
$1/4$-IN. RULE, FORM CLASS 80*

dbh, in.	Volume by 16-ft logs, bd ft			
	1	2	3	4
10	39	63	80	
12	59	98	127	146
14	83	141	186	216
16	112	190	256	305
18	144	248	336	402
20	181	314	427	512
22	221	387	528	638
24	266	469	644	773
26	315	558	767	931
28	367	654	904	1096
30	424	758	1050	1272
32	485	870	1213	1480
34	550	989	1383	1691
36	620	1121	1571	1922
38	693	1256	1772	2167
40	770	1403	1977	2432

*From Mesavage and Girard, 1946.

taper table, an abridged version of which appears in the preceding chapter (Table 6-1). For the tree in question, a taper rate of 1.7 in. is assigned to the second log and 1.9 in. to the third log.

Reference to the International $1/4$-in. rule for 16-ft logs indicates a scale volume of 181 bd ft for the 16-in. butt log. For the second log, the scaling diameter is 16 minus 1.7, or 14.3 in., and the log scale is 142 bd ft. For the third log, the scaling diameter is 14.3 minus 1.9, or 12.4 in., and the log scale is 104 bd ft. By totaling the scale values for the three logs (181 + 142 + 104), the standing tree volume of 427 bd ft is derived (Fig. 7-1). This calculated tree volume may be verified by reference to Table 7-1.

The Mesavage-Girard form-class tables have enjoyed a long and useful life, but sizable volume errors can occur when the upper-stem taper for a particular species differs appreciably from rigidly assumed taper rates or when form class is not accurately determined in the field. There is a general trend away from the use of tabulations of tree volumes and toward the direct computation of volumes from mathematical formulas. Before proceeding ahead to the next section, read ers who desire can obtain a review of regression analysis, the statistical technique commonly used to estimate coefficients in tree volume and weight equations, in Chap. 2.

7-5 Constructing Multiple-Entry Volume Tables The preferred method for constructing tree volume tables is by regression analysis. By this approach, a number of independent variables can be analyzed to determine their relative value in predicting the dependent variable of tree volume. And regression equations involving several independent variables and hundreds of sample observations can be efficiently solved by use of electronic computers.

Although many independent variables have been incorporated into regression equations for predicting tree volume, measurements of stem diameter and height tend to account for the greatest proportion of the variability in volume. Thus tree volumes for a given species may be predicted from the "combined variable" method described by Spurr (1952):

$$V = b_0 + b_1 \, \mathrm{dbh}^2 H$$

This formula is, of course, identical to the equation $Y = b_0 + b_1 X$. One merely substitutes the combined variable of "diameter squared times height" for the quantity X in the basic equation for a straight-line relationship. Solution of the equation is by simple linear regression techniques.

The diameters, heights, and volumes required to develop the volume function are ideally obtained by direct stem measurements of felled trees. It is important that a representative sample of trees spanning the full range of sizes (dbh and heights) of interest be obtained. If felled trees are not available, volumes may be computed from optical dendrometer measurements of standing trees. By this approach, diameter readings are made at intervals along the stem, and sectional

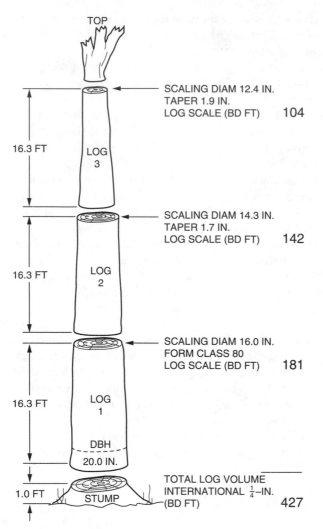

TOP

SCALING DIAM 12.4 IN.
TAPER 1.9 IN.
LOG SCALE (BD FT)   104

16.3 FT   LOG 3

SCALING DIAM 14.3 IN.
TAPER 1.7 IN.
LOG SCALE (BD FT)   142

16.3 FT   LOG 2

SCALING DIAM 16.0 IN.
FORM CLASS 80
LOG SCALE (BD FT)   181

16.3 FT   LOG 1

DBH
20.0 IN.

TOTAL LOG VOLUME
1.0 FT   STUMP   INTERNATIONAL $\frac{1}{4}$–IN.
(BD FT)   427

FIGURE 7-1
Derivation of merchantable tree volume for Mesavage-Girard
form-class tables.

volumes are computed by Smalian's or another suitable formula. These sectional volumes are then summed to produce an estimate of the portion of stem volume of interest.

Although form is not explicitly included as an independent variable in this volume function, differences in tree taper tend to be accounted for by the employment of separate prediction equations for each species or species group. Also, form is implicitly considered through the fitting of the intercept term b_0.

Assuming that total height is employed to estimate merchantable volume, the intercept for a straight line would typically be negative. This is because there is zero merchantable volume Y for positive values of diameter squared times height X. Thus when the straight-line relationship is extended to $X = 0$, it crosses the Y axis below the origin, i.e., at a negative value. This negative constant b_0, which is subtracted regardless of tree size, is a large portion of the b_1 dbh^2H value for a small tree but a small portion for a large tree, thus implying poorer form for small trees than for large trees.

The volume table shown in Table 7-2 was derived by fitting the combined-variable equation to data from ponderosa pine in the Front Range of Colorado.

TABLE 7-2
MERCHANTABLE CUBIC-FOOT VOLUME TABLE FOR PONDEROSA PINE*

dbh, in.	Total height above ground, ft								
	20	30	40	50	60	70	80	90	100
5	0.9	1.5	2.2	2.8					
6	1.4	2.3	3.2	4.1					
7	2.0	3.2	4.4	5.6	6.8				
8	2.7	4.2	5.8	7.4	8.9	10.5			
9	3.5	5.4	7.4	9.3	11.2	13.2			
10	4.3	6.7	9.1	11.5	13.8	16.2			
11	5.3	8.1	11.0	13.8	16.7	19.5	22.4		
12	6.3	9.7	13.1	16.4	19.8	23.2	26.6		
13	7.4	11.4	15.3	19.2	23.2	27.1	31.0		
14		13.2	17.7	22.3	26.8	31.3	35.9		
15		15.1	20.3	25.5	30.7	35.9	41.1		
16		17.2	23.1	29.0	34.8	40.7	46.6	52.5	
17		19.4	26.0	32.6	39.2	45.9	52.5	59.1	
18		21.7	29.1	36.5	43.9	51.3	58.7	66.1	
19		24.2	32.4	40.6	48.8	57.0	65.3	73.5	
20		26.8	35.9	44.9	54.0	63.1	72.2	81.2	
21			39.5	49.5	59.5	69.4	79.4	89.4	
22			43.3	54.2	65.2	76.1	87.0	98.0	108.9
23			47.3	59.2	71.1	83.1	95.0	106.9	118.8
24				64.4	77.3	90.3	103.3	116.2	129.2
25				69.8	83.8	97.9	111.9	126.0	140.0
26				75.4	90.6	105.7	120.9	136.1	151.2
27				81.2	97.6	113.9	130.2	146.6	162.9
28					104.8	122.4	139.9	157.5	175.0
29					112.3	131.1	149.9	168.7	187.5
30					120.1	140.2	160.3	180.4	200.5

*Cubic-foot volume, inside bark, from a stump height of 1 ft to a 4-in. top diameter (inside bark). Computed from $V = -0.44670 + 0.00216$ dbh^2H; coefficient of determination = 0.9744; standard error of estimate ±3.0 cu ft (14.29% of mean). Dbh classes represent full inches (e.g., 5-in. class includes 5.0 to 5.9), and midpoints were used when generating the table (i.e., 5.5 in. was used for the 5-in. class, etc.). From Edminster et al., 1980.

The resultant equation (from Edminster et al., 1980) is

$$V = -0.44670 + 0.00216 \text{ dbh}^2 H$$

where V = cubic feet, inside bark, from stump height of 1 ft to 4-in. top diameter
(inside bark)
dbh = diameter at breast height, in.
H = total tree height, ft

In the construction of tables for total tree volume, the combined-variable equation, as shown previously, may be used or the equation may be conditioned to pass through the origin

$$V = b_1 \text{ dbh}^2 H$$

Conditioning through the origin may be desirable for *total* tree volume prediction when predicted values for very small trees are desired. Such conditioning is not recommended for *merchantable* volume prediction, however, because merchantable volume equations should logically have negative intercepts. The equation $V = b_1 \text{dbh}^2 H$ is called the "constant form factor" equation because form is not explicitly or implicitly included. All trees, regardless of size, are assumed to be of similar form.

In addition to the merchantable cubic-foot volume equation shown previously, Edminster et al. (1980) also developed an equation for total cubic-foot volume from ponderosa pine data. To predict total volume, the constant form factor model was employed:

$$V = 0.00226 \text{ dbh}^2 H$$

where V is cubic feet, inside bark, of the entire stem, including stump and top, and dbh and H are as previously defined.

Another equation form that has been widely applied in past studies for predicting tree volumes is the model originally suggested by Schumacher and Hall (1933):

$$V = a \text{ dbh}^b H^c$$

The Schumacher and Hall equation, shown in its nonlinear form, can be transformed to a linear equation:

$$Y = b_0 + b_1 X_1 + b_2 X_2$$

and the coefficients estimated by standard linear regression methods by applying a logarithmic transformation, i.e.,

$$\log V = b_0 + b_1 \log \text{dbh} + b_2 \log H$$

where $Y = \log V$
$\quad X_1 = \log \text{dbh}$
$\quad X_2 = \log H$

The Schumacher and Hall equation, as shown here, passes through the origin and is appropriate for predicting *total* volume. When predicting *merchantable* volume, however, the equation should be conditioned to pass through an appropriate point by translating the axes.

Logarithmic volume equations have the advantage of more nearly satisfying the homogeneity of variance assumption of ordinary regression, but suffer from the disadvantage that a transformation bias is introduced. The transformation bias results from the fact that the transformed regression equation passes through the arithmetic means of the logarithms of the X and Y variables—which are the geometric means of the original variables. Geometric means are always less than arithmetic means, unless all values in a set of numbers are identical, and hence an underprediction bias is introduced. This bias is generally not large, however, and various bias correction factors have been proposed (Flewelling and Pienaar, 1981). Past experience has shown that the logarithmic tree volume expression proposed by Schumacher and Hall generally provides results similar to those from arithmetic tree volume expressions such as the combined-variable equation.

Equations which involve form class in addition to dbh and height have been fitted to tree data by regression techniques. One model that has been used, the "combined-variable form-class formula," is

$$V = b_0 + b_1 F + b_2 \text{ dbh}^2 H + b_3 F \text{ dbh}^2 H$$

where F is a measure of form, generally Girard form class. This formula is sometimes simplified to the "short-cut form-class formula" (Spurr, 1952):

$$V = b_0 + b_1 F \text{ dbh}^2 H$$

7-6 Selecting a Multiple-Entry Volume Table For most timber inventories and forest management plans, the forester has a wide selection of multiple-entry volume tables available for possible use. The choice of a reliable table requires careful scrutiny and an objective evaluation. To determine whether a particular table is suitable for a given inventory project, these and other questions might be appropriately asked:

1 For what species and locality was the table developed?
2 How many sample trees formed the basis for table construction?
3 Who is the author or publisher of the table?
4 What type of height and form measurements are required?
5 Are merchantability limits and units of volume suitable for the project at hand?
6 How were tree volumes originally obtained in deriving the table?
7 What method of table construction was used?
8 What evidence of table accuracy and reliability is available?

If feasible, it is desirable to visit a harvesting operation on timberlands similar to that where the volume table will be applied. Then one can measure the dbh, height (according to the definition used in the volume table construction), and form (if needed) for the range of sizes of trees that are being cut. Felled-tree volumes should be obtained according to the utilization standards used to construct the volume table being evaluated. Volumes for each of the sample trees can then be predicted from the independent variables (dbh, height) and compared with the measured volumes. Although individual trees will vary considerably from predicted averages in volume tables, the mean difference between actual and predicted volumes for samples of 50 to 100 trees should be near zero. If the mean difference is not near zero, bias is indicated. One might also examine the trend of differences across dbh and height classes to determine whether the table provides acceptable accuracy for the desired range of tree sizes.

For a given species or species group, foresters often have an array of multiple-entry volume tables from which to choose. These tables are commonly constructed from sample data obtained from a restricted portion of the region where the species occurs. Consequently, there has been much interest in "locality effects" on tree volume tables. How much variation might one expect due to the location from which the sample data were obtained? Can a volume table, based on tree data from one locality, be safely applied in another area? After a thorough study of factors affecting the total volume, Spurr (1952) concluded that "the locality, type of growth, and site where a tree grows apparently do not affect total cubic-foot volume sufficiently to justify the development of more than one volume table for a given species." Since that time, there has been additional evidence to support Spurr's conclusion. For example, in a study involving four data sets for plantation-grown loblolly pine that were similar in age but widely separated geographically, Van Deusen et al. (1981) concluded that a single volume-prediction equation can be used through much of the range of a given species.

7-7   Allowing for Various Utilization Standards   Multiple-entry volume tables commonly provide estimates of the contents of tree boles from stump height to a fixed top diameter. With multiple products being cut from tree boles and rapidly changing utilization standards, it is important to have the option of obtaining volume estimates to various top diameters and between given top diameters. These volume estimates should be consistently and logically related so that they sum to the total stem volume. Furthermore, for given combinations of dbh and total height, the volume to a 4-in. top, for example, should always be less than the volume to a 3-in. top.

One approach to providing flexibility in volume tables, while maintaining logical relationships, is to apply volume-ratio equations. In this approach, the ratio of merchantable volume to total volume is predicted; multiplying the ratio times the total volume gives the desired merchantable volume. Volume tables have been developed for several species by using this general approach (e.g., Honer, 1967).

Equations developed by Burkhart (1977) will be used to illustrate the volume-ratios procedure. First, total stem volume is predicted by using the combined-variable model. The equation for total cubic-foot volume, outside bark, of plantation-grown loblolly pine is

$$V = 0.34864 + 0.00232 \text{ dbh}^2 H$$

where dbh is in inches and H is total tree height in feet. Next, the ratio of merchantable stem volume divided by total stem volume is predicted from dbh and top diameter. The equation for outside-bark volume ratios with outside bark top diameters is

$$R = 1 - 0.32354 \, (d_t^{3.1579}/\text{dbh}^{2.7115})$$

where R = merchantable cubic-foot volume to top diameter d_t/total stem volume,
      cu ft
      d_t = top dob, in.

The reader should note that this equation is conditioned so that when $d_t = 0$, i.e., when one is at the tip of the tree, the ratio R equals 1, and multiplying R times total volume thus gives total volume.

To demonstrate the application of these prediction equations, it may be assumed that merchantable volumes to 4- and 6-in. tops are desired for a tree that measures 10 in. dbh and 75 ft total height. One first substitutes the tree dimensions into the total-volume equation

$$V = 0.34864 + 0.00232(10^2)(75)$$
$$= 17.75 \text{ cu ft}$$

Cubic-foot volume to a 4-in. top diameter is computed by substituting in the ratio equation and then by multiplying the estimated ratio times the total volume. Substituting the selected top limit and measured dbh gives

$$R = 1 - 0.32354(4^{3.1579}/10^{2.7115})$$
$$= 0.950$$

which is multiplied times the total volume to compute cubic-foot volume to a 4-in. top. Therefore,

$$V_4 = (17.75)(0.950)$$
$$= 16.86 \text{ cu ft}$$

Similarly, cubic-foot volume to a 6-in. top is computed as

$$R = 1 - 0.32354(6^{3.1579}/10^{2.7115})$$
$$= 0.820$$
$$V_6 = (17.75)(0.820)$$
$$= 14.56 \text{ cu ft}$$

Cubic-foot volume between top limits of 4 and 6 in. can be computed by subtraction $(16.86 - 14.56 = 2.30 \text{ cu ft})$.

7-8 Tree Volumes from Taper Equations Taper curves allow for changing utilization standards; if tree profiles can be accurately described, then volume for any merchantability limit or segment can be computed. Much work has been done on developing taper equations, and some of the results are quite complex. The relatively simple parabolic function presented by Kozak et al. (1969) will be used to illustrate the use of taper equations. This function has been found to fit well over about 85 percent of a tree bole, with lack of fit occurring primarily near the butt and top sections of the tree:

$$d^2/\text{dbh}^2 = b_0 + b_1(h/H) + b_2(h^2/H^2)$$

where d = diameter at any given height h above ground
$\quad\quad H$ = total tree height
b_0, b_1, b_2 = regression coefficients

Estimated upper-stem diameters are obtained by rearranging the function as

$$d = \text{dbh}\sqrt{b_0 + b_1(h/H) + b_2(h^2/H^2)}$$

To obtain volumes for any desired portion of the tree bole, diameters for short segments can be predicted. The predicted diameters can be used to compute volumes of the segments, and the segment volumes summed to obtain volume for the desired portion of the tree bole. More precise estimates of volume can be obtained, however, through mathematical integration of the taper equation (Sec. 7-9).

Taper curves can also be used to estimate the height at a specified diameter. By applying the quadratic formula to the parabolic taper equation used for illustrative purposes here, one can obtain an expression for height h to any specified diameter d for trees of given dbh and total height values:

$$h = \frac{-b_1 H - \sqrt{(b_1 H)^2 - 4b_2 \left(b_0 H^2 - \frac{d^2 H^2}{\mathrm{dbh}^2} \right)}}{2b_2}$$

Such equations allow estimation of merchantable heights to specified top diameters and computation of lineal footage for specific products.

Taper curves provide maximum flexibility for computing volumes of any specified portions of tree boles. Tree boles are highly variable, however, and it is difficult to describe their shapes over entire lengths without resorting to rather complex functions. Solutions for volume and height to specified diameters naturally become more difficult computationally when complex taper equations are used. However, modern electronic computers have largely eliminated the computational difficulties associated with taper equations.

Estimated volumes from taper equations are biased, because the sum of squared deviations about diameter (or some function of diameter) is minimized in the regression fitting process rather than the sum of squared deviations about volume. This bias is generally not large, however, and techniques have been developed for fitting "compatible" taper equations (Demaerschalk, 1972; Reed and Green, 1984). A compatible taper equation, when integrated, produces an identical estimate of total volume to that given by an existing total-volume equation.

7-9 Integrating Taper Functions   For those readers with a background in calculus, this section describes how to obtain tree volumes through mathematical integration of taper equations. The expression for stem cross-sectional area is integrated over the length desired. If one assumes that tree cross sections are circular in shape, then the area in square feet when diameter d is measured in inches would be

$$\text{Area} = \frac{\pi d^2}{4(144)} = 0.005454 d^2$$

Integrating the expression for area in square feet over the length desired in feet will give the volume in cubic feet for that segment. That is,

$$V \text{ (cu ft)} = 0.005454 \int_{h_1}^{h_2} d^2 dh$$

where h_1 and h_2 denote the limits of integration.

Assuming that the Kozak et al. (1969) taper model is employed,

$$d^2/\text{dbh}^2 = b_0 + b_1(h/H) + b_2(h^2/H^2)$$

and
$$d^2 = \text{dbh}^2[b_0 + b_1(h/H) + b_2(h^2/H^2)]$$

Substituting the expression for d^2 into the integral for volume results in

$$V = 0.005454\text{dbh}^2 \int_{h_1}^{h_2} [b_0 + b_1(h/H) + b_2(h^2/H^2)]dh$$

Solving the integral results in

$$V = 0.005454\text{dbh}^2 [b_0(h) + (b_1/2)(h^2/H) + (b_2/3)(h^3/H^2)]\Big|_{h_1}^{h_2}$$

Coefficients for coastal Douglas-fir from Kozak et al. (1969) are

$$b_0 = 0.85458$$
$$b_1 = -1.29771$$
$$b_2 = 0.44313$$

Suppose that the volume of the 32-ft segment from 32 to 64 ft above ground is desired for a Douglas-fir tree measuring 30 in. dbh and 150 ft in total height. Substituting into the expression for volume results in

$$V = \{0.005454(30)^2[0.85458(64) - (1.29771/2)(64^2/150)$$
$$+ (0.44313/3)(64^3/150^2)]\} - \{0.005454(30)^2[0.85458(32)$$
$$- (1.29771/2)(32^2/150) + (0.44313/3)(32^3/150^2)]\}$$
$$= 76.4 \text{ cu ft (inside bark)}$$

If the taper equation is integrated over the total bole length (i.e., from 0 to H, where H denotes total tree height), an expression for total tree volume will result. Following through with the same parabolic taper function, we have

$$0.005454\text{dbh}^2 [b_0(h) + (b_1/2)(h^2/H) + (b_2/3)(h^3/H^2)]\Big|_0^H$$
$$= \{0.005454\text{dbh}^2 [b_0(H) + (b_1/2)(H^2/H) + (b_2/3)(H^3/H^2)]\} - 0$$

Noting that the above expression can be written as

$$0.005454[b_0 + (b_1/2) + (b_2/3)]\text{dbh}^2H$$

and setting $0.005454[b_0 + (b_1/2) + (b_2/3)]$ equal to b, we note that the implied volume equation is

$$V = b(\text{dbh}^2H)$$

which is a logical model for total tree volume, i.e., the constant form factor equation.

SINGLE-ENTRY VOLUME TABLES

7-10 Advantages and Limitations Volume tables based on the single variable of dbh may be constructed from existing multiple-entry volume tables or from the scaled measure of felled trees. Such tables are particularly useful for quick timber inventories, because height and form estimates are not required and trees can be tallied by species and dbh only. Elimination of height and form determinations also tends to assure greater uniformity in volume estimates, particularly when two or more field parties are cruising within the same project area.

Construction of volume tables based on dbh alone presumes that a definitive height-diameter relationship exists for the species under consideration, i.e., that trees of a given diameter class tend to be of similar height and form. If this is true, all trees in a given dbh class can be logically assigned the same average volume. Height-diameter or volume-diameter relationships can often be established for hardwood or coniferous species growing under relatively uniform site and stand density conditions. When soils and topography are notably varied, it is usually necessary to construct single-entry tables for each broad site class encountered.

Volume tables based on dbh alone are sometimes compiled for inventories of relatively small areas, but this is not an essential condition; in some instances, "local" tables may be as widely applicable as "standard" volume tables. Thousands of sample trees may be represented by some so-called local tables. The exact number of sample measurements required depends upon characteristics of the tree species involved, variability of soil-site conditions, and the desired geographic area of application. From 30 to 100 samples are usually considered a minimum number for small tracts, depending on the range of diameter classes to be included in the final table.

7-11 Constructing a Single-Entry Table from Measurements of Felled Trees To obtain tree volumes essential for this procedure, measurements may be obtained from felled trees on logging operations or by "scaling" standing trees with a reliable upper-stem dendrometer (Sec. 6-6). Sample trees should be selected in an unbiased manner and a sufficient number of measurements made to span the desired range of dbh classes for each species involved.

For each sample tree, measurements should be obtained of (1) dbh to the nearest 0.1 in., (2) tree volume in desired units, and (3) total tree height. The last item, though not actually needed for constructing the table, serves as a useful indication of the sites or geographic areas to which the table may be applied. To illustrate the procedure of table construction, the data for yellow-poplar in Table 7-3 were supplied by the Southeastern Forest Experiment Station, U.S. Forest Service. Cubic-foot volumes are outside bark for the main stem to a 4-in. top dob.

The felled-tree volumes are related to dbh through regression analysis. Tree volumes have a curvilinear relationship with dbh but are approximately linearly related to dbh squared (Fig. 7-2). In general form, the local volume equation relationship can be expressed as

$$V = b_0 + b_1 \, \text{dbh}^c$$

The cubic-foot volume data for this example follow an approximately straight-

TABLE 7-3
YELLOW-POPLAR DATA USED IN CONSTRUCTING A
SINGLE-ENTRY VOLUME TABLE

dbh, in.	Volume, cu ft	Total ht., ft	dbh, in.	Volume, cu ft	Total ht., ft
6.0	4.7	65	16.1	59.5	98
6.3	5.3	63	16.2	48.3	86
7.2	7.0	69	16.8	76.2	105
7.4	7.4	63	17.2	58.7	98
8.0	12.5	78	17.6	75.7	106
8.5	10.5	66	18.4	78.9	101
9.3	14.4	74	19.3	89.1	111
8.6	13.4	80	18.7	85.4	102
10.2	21.7	83	20.4	104.3	109
9.8	17.5	77	19.8	92.5	103
11.5	23.0	74	20.7	102.9	108
11.4	24.6	84	21.3	113.1	101
12.2	38.1	98	22.4	115.1	106
12.0	31.8	98	22.2	135.3	120
13.4	41.6	96	23.0	125.6	108
12.8	35.0	90	23.4	152.4	128
14.0	43.1	95	24.3	167.9	115
14.1	41.5	91	23.8	153.6	124
14.9	45.8	87	25.3	138.5	107
15.4	55.0	98	25.8	177.3	118

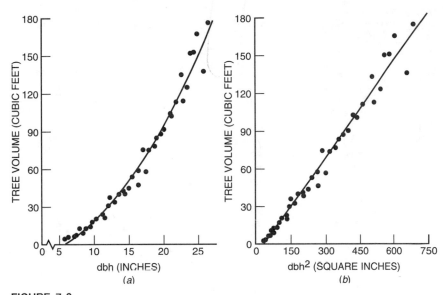

FIGURE 7-2
(a) Curvilinear relationship of tree volume to dbh and (b) the same relationship transformed to a straight line. Based on measurements of 40 felled yellow-poplar trees in the southern Appalachians.

line pattern when plotted over dbh squared, indicating that c can be set equal to 2 and simplifying the model to

$$V = b_0 + b_1 \, \text{dbh}^2$$

Applying simple linear regression techniques to the yellow-poplar data results in

$$V = -8.4166 + 0.2679 \, \text{dbh}^2$$

Table 7-4, a single-entry, or local, volume table, was compiled by substituting diameter class midpoint values into the regression equation to obtain predicted average tree volumes by dbh class. Average total heights, by dbh classes, in the sample data are shown to aid users in determining whether the table is appropriate for a given inventory situation. The average total heights were determined by substituting diameter class midpoint values in a height-dbh regression fitted to the yellow-poplar data.

This method of constructing a single-entry volume table works reasonably well when felled trees of representative sizes are available for measurement. However, felled trees from harvesting operations rarely make up a typical sample of standing trees, because they may represent a different population or have

TABLE 7-4
SINGLE-ENTRY VOLUME TABLE FOR
YELLOW-POPLAR*

dbh, in.	Total volume, cu ft	Average total height, ft
6	1.2	58
7	4.7	66
8	8.7	72
9	13.3	77
10	18.4	82
11	24.0	86
12	30.2	89
13	36.9	92
14	44.1	95
15	51.9	97
16	60.2	99
17	69.0	101
18	78.4	103
19	88.3	105
20	98.8	106
21	109.7	107
22	121.3	109
23	133.3	110
24	145.9	111

*Based on measurements of 40 felled trees. Tree volumes are in cubic feet (outside bark) to a 4-in. top dob. Equation: $V = -8.4166 + 0.2679 \, dbh^2$; coefficient of determination = 0.978.

distinctive characteristics that influenced their volume and, thus, caused them to be cut. When this is the case and felled trees are nonrepresentative samples, single-entry volume tables derived from such data would be biased and unreliable. As an alternative, tree volumes might be obtained from random samples of standing trees, or tables might be constructed from height-diameter relationships as described in the next section.

7-12 Construction from a Multiple-Entry Table This method of deriving a single-entry volume table is dependent on a well-established height-diameter relationship and the existence of a reliable "standard" table from which volumes may be interpolated. Field measurements of 50 to 100 merchantable or total heights, spanning the desired range of tree dbh classes, should be obtained from the selected project area. If the multiple-entry volume table to be used is based on merchantable heights, field measurements must be carefully taken to identical top diameters or merchantability limits. An example of tree data for 54 Sierra redwood trees is shown in Table 7-5.

TABLE 7-5
HEIGHT-DIAMETER DATA FOR 54 YOUNG-GROWTH SIERRA
REDWOOD TREES

dbh, in.	Total ht., ft	dbh, in.	Total ht., ft	dbh, in.	Total ht., ft
12.5	62	12.0	60	45.0	148
13.0	65	30.6	115	52.4	153
46.8	135	22.4	95	30.7	110
30.7	120	38.0	128	38.4	132
56.5	145	56.5	145	25.3	100
14.0	61	44.1	133	18.0	90
44.0	133	60.0	130	35.1	122
59.9	160	60.0	160	24.0	111
16.5	75	30.3	107	43.5	140
45.0	121	15.2	76	24.9	89
56.0	140	36.0	137	42.0	137
26.2	103	51.4	144	51.9	131
36.4	116	21.1	95	59.6	145
58.7	160	34.2	129	17.5	72
36.8	120	20.7	81	54.0	135
45.7	133	48.6	143	40.0	130
18.4	82	57.3	153	23.5	91
27.5	108	13.5	71	45.0	127

Numerous height-diameter regression models have been proposed and used in the past. Because the basic height-diameter relationship is not linear, but is sigmoid in shape over the full range of diameters (Fig. 7-3), a transformation must be made to apply linear regression methods for estimating coefficients in the relationship. One model that has been found satisfactory for a wide range of species and for both total and merchantable tree heights is

$$\log H = b_0 + b_1 \, \mathrm{dbh}^{-1}$$

When height values are transformed to logarithms and dbh measurements to reciprocals of the original values, a straight-line relationship exists between the transformed values (Fig. 7-3); thus simple linear regression techniques can be used to solve for b_0 and b_1 in the transformed model. The application of linear regression analysis to the data in this example results in

$$\ln H = 5.21909 - 14.32872 \, \mathrm{dbh}^{-1}$$

where ln H is the natural logarithm of total tree height in feet and dbh is in inches. Figure 7-3 shows the height-diameter data and the fitted regression line on a transformed and an arithmetic scale.

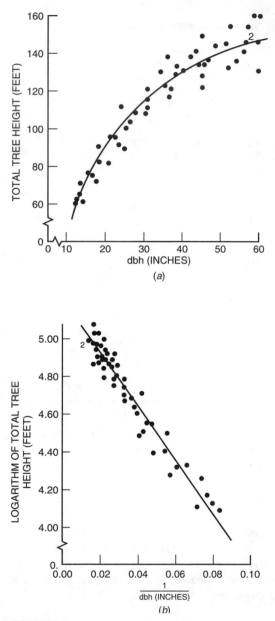

FIGURE 7-3
(a) Curvilinear relationship of tree height to dbh for
second-growth Sierra redwood trees and (b) the same
relationship transformed to a straight line.

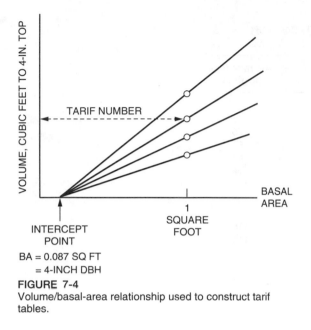

FIGURE 7-4
Volume/basal-area relationship used to construct tarif
tables.

Average heights were predicted for 2-in. dbh classes by substituting into the
height-diameter equation. The midpoint of each dbh class and the predicted
height were then substituted into the multiple-entry volume equation derived by
Wensel and Schoenheide (1971):

$$\ln V = -6.66790 + 1.54423 \ln \text{dbh} + 1.29808 \ln H$$

to obtain average predicted gross cubic-foot volumes by dbh class. These average values were tabulated into a single-entry volume table (Table 7-6). The
completed table includes average predicted heights from the height-diameter regression as an indication of the conditions to which the table may be safely applied.

7-13   Tarif Tables   The term *tarif* is of Arabic origin and simply means
tabulated information. In the context of tree volume estimation, tarif tables are
collections of local volume tables (Table 7-7). Tarif tables are based on the assumption that volume has a linear relationship to diameter squared or basal area.
When English units are used and volume to a 4-in. top limit is the dependent
variable, the straight-line relationship intercepts at a basal area of 0.087 sq ft, the
basal area of a 4-in. tree (Fig. 7-4). The volume of the tree with a basal area of
1.0 sq ft is called the tarif number. With two points fixed (basal area = 0.087 sq
ft, volume = 0 cu ft; and basal area = 1.0 sq ft, volume = tarif number in cu ft) on

TABLE 7-6
SINGLE-ENTRY VOLUME TABLE FOR
YOUNG-GROWTH SIERRA REDWOOD*

dbh class, in.	Tree volume, cu ft	Average total height, ft
12	11.0	56
14	17.4	66
16	25.2	75
18	34.4	83
20	44.8	90
22	56.5	96
24	69.4	102
26	83.3	106
28	98.3	111
30	114.3	115
32	131.3	118
34	149.2	121
36	168.0	124
38	187.7	127
40	208.2	129
42	229.5	131
44	251.6	133
46	274.5	135
48	298.1	137
50	322.5	139
52	347.5	140
54	373.3	142
56	399.7	143
58	426.9	144
60	454.6	146

*Height-diameter equation: $\ln H = 5.21909 - 14.32872$ dbh^{-1}. Volume is gross cubic feet; multiple-entry volume equation: $\ln V = -6.66790 + 1.54423 \ln dbh + 1.29808 \ln H$. Volume equation from Wensel and Schoenheide, 1971.

a straight line, the entire local volume table for a given tarif number is specified. The tarif numbers, which represent different tree forms, are used to specify which local volume or tarif table to use.

Users must select an appropriate tarif, or local volume, table for the area of interest. One means of determining a tarif number is to use felled-tree measurements. The method proceeds as follows:

1 Select representative trees in the area of interest (typically 20 to 30 trees are selected).

2 Fell the trees, measure the volume of each tree, and, for each tree, find the table giving the closest match to the measured volume for a tree of the measured dbh.

TABLE 7-7
EXAMPLE OF A TARIF TABLE*

dbh, in.	Volume in cubic feet (ib, to 4-in. top dib) for tarif numbers of				
	20	25	30	35	40
6	2.4	3.0	3.6	4.2	4.8
7	3.9	4.9	5.9	6.9	7.9
8	5.7	7.2	8.6	10.0	11.5
9	7.8	9.7	11.7	13.6	15.5
10	10.0	12.5	15.1	17.6	20.1
11	12.5	15.7	18.8	22.0	25.1
12	15.3	19.1	22.9	26.8	30.6
13	18.3	22.9	27.4	32.0	36.6
14	21.5	26.9	32.3	37.6	43.0
15	25.0	31.2	37.5	43.7	50.0

*Abridged from Turnbull et al., 1980.

3 Average the tarif numbers from the felled trees and use the mean tarif number when selecting the volume table to use for inventory purposes in that stand.

Felled-tree information is time-consuming and expensive to obtain. As an alternative, volume can be computed from diameter and height data obtained with optical dendrometer (Sec. 6-6) measurements on standing sample trees. Although some loss in information is inevitable when using standing-tree rather than felled-tree data, it can be an attractive alternative when time and cost constraints prohibit tree felling.

Another alternative to felled-tree data is to use tarif access tables. For a species, these tables list the tarif numbers estimated for a tree of measured dbh and height. When the access table approach is taken, the procedure is

1 Select representative trees in the area of interest (in practice, 20 to 30 trees are usually chosen).

2 Measure the dbh and height of the sample trees. For each dbh-height sample tree, use the species-specific access tables to obtain a tarif number.

3 Average the tarif numbers from the sample trees and use the mean tarif number to determine the local volume (tarif) table to use for estimating volume in that stand.

Tabulations termed *comprehensive* tarif tables have been developed which provide tree volumes in several units of measure and utilization limits, as well as volume/basal-area ratios and growth multipliers, all within a related system. In addition to the tabular approach, equations have been developed to estimate tarif numbers and volumes for various top limits and units of measure. Equations are more readily implemented in computing systems than are tables.

TREE WEIGHT TABLES

7-14   Field Tallies by Weight   The continued emphasis on weight scaling as a basis of payment for pulpwood and sawlogs has spurred interest in field tallies based on tree weights rather than on various units of volume. It is logical that standing trees should be measured in the same units as those on which log purchases and sales are transacted.

Any ordinary volume table can be converted to a weight basis if weight-volume equivalents can be reliably established. For example, if it has been determined that a given species has a mean specific gravity of 0.47 and a mean moisture content of 110 percent, the weight per cubic foot would be (see Sec. 4-11)

$$\text{lb per cu ft (green)} = 0.47 \times 62.4 \left(1 + \frac{110}{100}\right)$$
$$= 61.6$$
$$\text{lb per cu ft (dry)} = 0.47 \times 62.4 \left(1 + \frac{0}{100}\right)$$
$$= 29.3$$

A standing tree having a volume of 20 cu ft inside bark would then be assigned a green weight (without bark) of

$$(20)(61.6) = 1232 \text{ lb}$$

or a dry weight of

$$(20)(29.3) = 586 \text{ lb}$$

The same technique can be used to convert volume tables expressed in board feet or cords, provided that acceptable weight equivalents are derived. As previously outlined, Tables 4-1 and 5-4 provide pulpwood and sawlog weight approximations for a number of commercial timber species.

7-15   Weight Tables for Tree Boles   In lieu of converting volume tables to weight tables, it is preferable to weigh a series of sample trees and to relate tree weight to tree dimensions (dbh and height) via regression models similar to those used for volume prediction. Green weight of sycamore trees from an 11-year-old plantation in Georgia was predicted by employing the combined-variable model (Belanger, 1973). The resulting equation is

$$\text{Green bole weight (3-in. top dob)} = -32.35109 + 0.15544 \text{ dbh}^2 H$$

where dbh is in inches and H is total tree height in feet. Table 7-8 was compiled from this equation.

TABLE 7-8
GREEN-WEIGHT TABLE FOR PLANTATION-GROWN SYCAMORE, IN POUNDS*

dbh, in.	Total height, ft							
	45	50	55	60	65	70	75	80
4	80	92	104	117	129			
5	143	162	181	201	220	239		
6		247	275	303	331	359	387	
7			387	425	463	501	539	
8			515	565	614	664	714	764
9				723	786	849	912	975
10					978	1056	1133	1211

*Equation: $W = -32.35109 + 0.15544 \, dbh^2 H$, where W is green weight (including bark) to a 3-in. top dob; coefficient of determination = 0.99. From Belanger, 1973.

Although most transactions are in terms of green weight, dry weight is more appropriate when the wood will be utilized for pulped products. Dry weight is more highly correlated with the final yield of pulped products than is green weight, which is influenced by variations in moisture content. Dry weight, as green weight, can be assessed in standing trees through use of appropriate tables or equations. Belanger (1973) also developed a combined-variable equation to predict the dry weight of sycamore trees:

$$\text{Dry bole weight (3-in. top dob)} = -17.67910 + 0.06684 \, dbh^2 H$$

Table 7-9 contains dry-weight values from this equation. When appropriate tree weight tables are available, standing timber can be assessed in pounds or kilograms as readily as in volume units.

TABLE 7-9
DRY-WEIGHT TABLE FOR PLANTATION-GROWN SYCAMORE, IN POUNDS*

dbh, in.	Total height, ft							
	45	50	55	60	65	70	75	80
4	30	36	41	46	52			
5	58	66	74	83	91	99		
6		103	115	127	139	151	163	
7			162	179	195	212	228	
8			218	239	260	282	303	325
9				307	334	361	388	415
10					417	450	484	517

*Equation: $W = -17.67910 + 0.06684 \, dbh^2 H$, where W = dry weight (including bark) to a 3-in. top dob; coefficient of determination = 0.99. From Belanger, 1973.

7-16   Biomass Tables   With increasing emphasis on complete tree utilization and use of wood as a source of energy, tables and equations have been developed to show the weights of total trees and of their components (bolewood, branches, foliage). These tables, commonly referred to as "biomass tables," are generally expressed in terms of dry weight and may include only the aboveground portion, or the entire tree, including roots.

Equations for estimating biomass of individual Engelmann spruce trees were developed by Landis and Mogren (1975). Employing the following model

$$Y = b_0 + b_1 \, \text{dbh}^2$$

where Y is tree component dry weight in kilograms and dbh is tree diameter at breast height in centimeters, they computed the equations shown in Table 7-10. Thus by measuring dbh on standing Engelmann spruce trees, the total aboveground biomass or the dry weight of various tree components can be estimated.

Similar sets of equations have been developed for other species. A commonly employed regression model in several studies of biomass table construction (such as Clark and Schroeder, 1977; Taras and Phillips, 1978; Edwards and McNab, 1979) is

$$\log Y = b_0 + b_1 \log \text{dbh}^2 H$$

where Y = total tree or component weight
 dbh = diameter at breast height (or diameter at ground line for saplings)
  H = total tree height

This model is sometimes referred to as the "logarithmic combined-variable formula."

TABLE 7-10
EQUATIONS FOR ESTIMATING BIOMASS OF INDIVIDUAL ENGELMANN SPRUCE TREES*

Tree component (Y), dry wt, kg	Equation	Coefficient of determination
Total aboveground biomass	$Y = -36.94 + 0.42 \, \text{dbh}^2$	0.97
Stem wood	$Y = -25.26 + 0.25 \, \text{dbh}^2$	0.96
Stem bark	$Y = -2.55 + 0.06 \, \text{dbh}^2$	0.98
Branch wood and bark	$Y = -7.78 + 0.07 \, \text{dbh}^2$	0.90
Foliage	$Y = -0.56 + 0.03 \, \text{dbh}^2$	0.88

*Dbh is in cm. From Landis and Mogren, 1975.

PROBLEMS

7-1 Construct a single-entry tree volume table based on the direct measurement of felled trees.

7-2 Construct a single-entry tree volume table based on a height-dbh relationship for a local species.

7-3 The following data are from 20 loblolly pine trees:

dbh, in.	Total height, ft	Volume, cu ft to 4-in. top dob	dbh, in.	Total height, ft	Volume, cu ft to 4-in. top dob
12	65	20.5	7	50	5.2
8	59	8.4	12	56	20.2
5	42	1.8	9	60	11.8
10	64	13.9	5	35	1.6
11	58	15.5	12	62	21.4
8	64	9.8	11	67	18.8
7	57	6.4	7	71	7.5
5	44	3.3	7	51	5.6
12	64	21.5	11	57	15.6
5	42	2.2	7	61	6.3

a Plot cubic-foot volume versus dbh, dbh^2, total height, and dbh^2 times total height. Do the relationships conform to expected trends?

b Use the given data and the combined-variable formula to compute a cubic-foot volume equation.

c Calculate the coefficient of determination and standard error of estimate for the resulting equation.

d Use your regression equation to predict the cubic-foot volume for a loblolly pine measuring 8 in. dbh and 60 ft total height.

7-4 Use the taper equation for Douglas-fir shown in Sec. 7-9 to solve for the diameter (inside bark) at 1 ft and 17 ft above ground level for a tree 20 in. dbh and 120 ft total height. Compute the cubic-foot volume of this 16-ft. segment, using the predicted diameters and Smalian's formula. Apply mathematical integration techniques to solve for the cubic-foot volume of the same segment and compare the result with that obtained using Smalian's formula.

7-5 Convert an existing tree volume table into a tree weight table.

7-6 Through regression analysis, determine whether the specific gravity for entire tree stems is closely correlated with specific gravity at dbh for a selected species.

REFERENCES

Amateis, R. L., and Burkhart, H. E. 1987. Cubic-foot volume equations for loblolly pine trees in cutover, site-prepared plantations. *So. J. Appl. For.* **11**:190–192.

Amidon, E. L. 1984. A general taper functional form to predict bole volume for five mixed-conifer species in California. *Forest Sci.* **30**:166–171.

Belanger, R. P. 1973. Volume and weight tables for plantation-grown sycamore. *U.S. Forest Serv., Southeast. Forest Expt. Sta. Res. Paper* SE-107. 8 pp.

Bell, J. F., Marshall, D. D., and Johnson, G. P. 1981. Tarif tables for mountain hemlock developed from an equation of total stem cubic-foot volume. *Oregon State Univ. Res. Bull.* 35. 46 pp.

Brister, G. H., and Lauer, D. K. 1985. A tarif system for loblolly pine. *Forest Sci.* **31:**95–108.

————, Clutter, J. L., and Skinner, T. M. 1980. Tree volume and taper functions for site-prepared plantations of slash pine. *So. J. Appl. For.* **4:**139–142.

Burkhart, H. E. 1977. Cubic-foot volume of loblolly pine to any merchantable top limit. *So. J. Appl. For.* **1:**7–9.

Cao, Q. V., Burkhart, H. E., and Max, T. A. 1980. Evaluation of two methods for cubic-volume prediction of loblolly pine to any merchantable limit. *Forest Sci.* **26:**71–80.

Clark, A., III, and Schroeder, J. G. 1977. Biomass of yellow-poplar in natural stands in western North Carolina. *U.S. Forest Serv., Southeast. Forest Expt. Sta. Res. Paper* SE-165. 41 pp.

Clutter, J. L. 1980. Development of taper functions from variable-top merchantable volume equations. *Forest Sci.* **26:**117–120.

Cunia, T. 1964. Weighted least squares method and construction of volume tables. *Forest Sci.* **10:**180–191.

Demaerschalk, J. P. 1972. Converting volume equations to compatible taper equations. *Forest Sci.* **18:**241–245.

Edminster, C. B., Beeson, R. T., and Metcalf, G. E. 1980. Volume tables and point-sampling factors for ponderosa pine in the Front Range of Colorado. *U.S. Forest Serv., Rocky Mt. Forest and Range Expt. Sta. Res. Paper* RM-218. 14 pp.

Edwards, M. B., and McNab, W. H. 1979. Biomass prediction for young southern pines. *J. Forestry* **77:**291–292.

Flewelling, J. W., and Pienaar, L. V. 1981. Multiplicative regression with lognormal errors. *Forest Sci.* **27:**281–289.

Goulding, C. J., and Murray, J. C. 1976. Polynomial taper equations that are compatible with tree volume equations. *N. Z. J. Forestry Sci.* **5:**313–322.

Hann, D. W., and Bare, B. B. 1978. Comprehensive tree volume equations for major species of New Mexico and Arizona: I. Results and methodology. *U.S. Forest Serv., Intermount. Forest and Range Expt. Sta. Res. Paper* INT-209. 43 pp.

Hitchcock, H. C., III. 1978. Aboveground tree weight equations for hardwood seedlings and saplings. *Tappi* **61:**119–120.

Honer, T. G. 1967. Standard volume tables and merchantable conversion factors for the commercial tree species of central and eastern Canada. *Forest Mgt. Res. and Services Inst., Ottawa, Ontario, Information Report* FMR-X-5. 21 pp. plus appendices.

Kozak, A. 1988. A variable-exponent taper equation. *Can. J. For. Res.* **18:**1363–1368.

————, Munro, D. D., and Smith, J. H. G. 1969. Taper functions and their application in forest inventory. *Forestry Chron.* **45:**278–283.

Landis, T. D., and Mogren, E. W. 1975. Tree strata biomass of subalpine spruce-fir stands in southwestern Colorado. *Forest Sci.* **21:**9–12.

Lenhart, J. D., and Hyink, D. M. 1973. Direct estimation of the ovendry weight of plantation-grown loblolly pine trees in the Interior West Gulf Coastal Plain. *For. Prod. J.* **23:**49–50.

Madgwick, H. A. I., Olah, F. D., and Burkhart, H. E. 1977. Biomass of open-grown Virginia pine. *Forest Sci.* **23:**89–91.

Martin, A. J. 1984. Testing volume equation accuracy with water displacement techniques. *Forest Sci.* **30:**41–50.

Max, T. A., and Burkhart, H. E. 1976. Segmented polynomial regression applied to taper equations. *Forest Sci.* **22:**283–289.

Mesavage, C., and Girard, J. W. 1946. Tables for estimating board-foot content of timber. U.S. Forest Service, Washington, D.C. 94 pp.

Perez, D. N., Burkhart, H. E., and Stiff, C. T. 1990. A variable-form taper function for *Pinus oocarpa* Schiede in Central Honduras. *Forest Sci.* **36:**186–191.

Reed, D. D., and Green, E. J. 1984. Compatible stem taper and volume ratio equations. *Forest Sci.* **30:**977–990.

Schumacher, F. X., and Hall, F. 1933. Logarithmic expression of timber-tree volume. *J. Agr. Res.* **47:**719–734.

Scrivani, J. A. 1989. An algorithm for generating "exact" Girard form class volume table values. *No. J. Appl. For.* **6:**140–142.

Solomon, D. S., Droessler, T. D., and Lemin, R. C. 1989. Segmented quadratic taper equations for spruce and fir in the Northeast. *No. J. Appl. For.* **6:**123–126.

Spurr, S. H. 1952. *Forest inventory.* The Ronald Press Company, New York. 476 pp.

Taras, M. A., and Phillips, D. R. 1978. Aboveground biomass of slash pine in a natural sawtimber stand in southern Alabama. *U.S. Forest Serv., Southeast. Forest Expt. Sta. Res. Paper* SE-188. 31 pp.

Turnbull, K. J., Little, G. R., and Hoyer, G. E. 1980. *Comprehensive tree volume tarif tables.* 3d ed. Department of Natural Resources, Olympia, Wash. 132 pp.

Turner, B. J. 1972. Board-foot and cubic-foot volume tables for the commercial forest species of Pennsylvania. *College of Agr., Penn. State Univ.* 68 pp.

Van Deusen, P. C., Sullivan, A. D., and Matney, T. G. 1981. A prediction system for cubic foot volume of loblolly pine applicable through much of its range. *So. J. Appl. For.* **5:**186–189.

Wensel, L. C. 1971. I. Tree volume equations from measurements taken with a Barr and Stroud optical dendrometer. *Hilgardia* **41:**55–64.

———, and Schoenheide, R. L. 1971. II. Young growth gross volume tables for Sierra redwood [*Sequoia gigantea* (Lindl.) Decne]. *Hilgardia* **41:**65–76.

SAMPLING DESIGNS

8-1 Introduction Chapter 2 provides an overview of basic statistical methods. Most statistical methods are based on the assumption that a simple random sample has been drawn. However, samples can be drawn in many possible ways. When the sampling procedure varies, the analysis of the sample data must be varied accordingly. The most appropriate method of selecting the samples, and by extension of analyzing the sample data, depends on the objectives of the survey, the nature of the population to be sampled, and the prior or auxiliary information available about the target population.

The purpose of this chapter is to discuss sampling designs that have been widely applied for inventorying forests. Background information is given on the various sampling procedures, along with the formulas for estimating the population parameters and the reliability of those estimates.

8-2 Sampling versus Complete Enumeration The objective of sample surveys is to gain information about a population. Sometimes the population is relatively small, and so sampling may not be necessary because every unit of the population can be observed. When all individual units of the population are observed, the survey is termed a *complete enumeration.*

Complete enumerations are generally prohibitively expensive and time-consuming to perform. However, there are situations where a complete enumeration is required (that is, the data of interest must be recorded, analyzed, or presented on an individual-unit basis) or is a feasible alternative. For example, a woodlot owner may offer for sale a number of black walnut trees of large size and value. If the owner wishes to sell the trees on an individual-standing-tree basis, then the size and value of each tree must be determined. Alternatively, the owner may sell the trees on a lump-sum basis. The total value of the population of black walnut trees could then be estimated by a complete enumeration or by an appropriately drawn sample.

On small tracts, it is sometimes feasible to measure all trees of interest. The method has the advantage of determining the mean rather than estimating it. However, a combination of high costs and the need for timely information greatly limits application of complete enumerations. For most inventories of forest resources, it is not economically feasible to measure or count 100 percent of the population about which inferences must be made. Furthermore, the time required for complete enumerations of large populations would render the data obsolete by the time they could be amassed, collated, and summarized. Furthermore, foresters confronted with many trees to measure sometimes tend to use imprecise measurement techniques. Sample surveys, with fewer but more precise measurements, are often more accurate than complete enumerations.

Aside from time and cost factors, sampling is also necessary when testing procedures are destructive. All seeds cannot be evaluated in germination tests because there would be none left for sowing. Similarly, all fishes cannot be dissected to study concentrations of chemical elements; otherwise there would be none left for anglers or for human consumption. In business and industry, as well as in forest management, sampling is an accepted means of obtaining information about populations that cannot be subjected to complete census.

The ultimate objective of all sampling is to obtain reliable data from the population sampled and to make certain inferences about that population. How well this objective is met depends on items such as the rule by which the sample is drawn, the care exercised in measurement, and the degree to which bias can be avoided. Of all the techniques described in this book, the concept of sampling is perhaps the most important.

8-3 The Sampling Frame As stated previously, the objective of all sampling is to make some inference about a population from the observations composing the sample. The method of selecting the nonoverlapping sample units to be included in a sample is referred to as the *sampling design,* and a listing of all possible sample units that might be drawn is termed the *sampling frame.*

Establishment of a reliable sampling frame can be a difficult task. For example, if an individual campground visitor is specified as the sample unit, a registration list that includes the occupants of all entering motor vehicles may or may not compose a satisfactory sampling frame. Those persons who enter the campground on foot or horseback would probably be excluded from such a sampling frame; other visitors might arrive at such a time that they somehow avoid the necessity of registration. Fortunately, in most field circumstances, differences between the sampling frame and the population are inconsequential. Otherwise, inferences based on a sample drawn from the frame may be questionable.

After the individual sample unit and the sampling frame have been defined, it is then necessary to decide on the sampling design to be employed. Figure 8-1 illustrates three of the basic designs often used by forest managers: simple random sampling, systematic sampling, and stratified sampling.

COMMON SAMPLING DESIGNS

8-4 Simple Random Sampling Many statistical procedures assume simple random sampling. By this approach, *every possible combination of sample units* has an equal and independent chance of being selected. This is *not* the same as simply requiring that every sample unit in the population have an equal chance of being selected. This latter requirement is met by many forms of restricted randomization and even by some systematic designs.

FIGURE 8-1
Four possible arrangements of 16 samples in a population composed of 256 square plots.

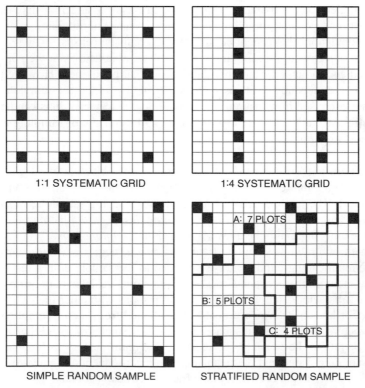

1:1 SYSTEMATIC GRID   1:4 SYSTEMATIC GRID

SIMPLE RANDOM SAMPLE   STRATIFIED RANDOM SAMPLE

Allowing every possible combination of n sample units an equal chance of being selected is easily accomplished. It is only necessary that, at any state of the sampling, the selection of a particular unit be in no way influenced by the other units that have been selected or will be selected. To state it another way, the selection of any given unit should be completely independent of the selection of all other units. One way to do this is to assign every unit in the population a number and to draw n numbers from a table of random digits. A modification of this technique consists of drawing random intersection points in a coordinate system based on column and row numbers designating each plot.

Sample units may be selected with or without replacement. If selection is with replacement, each unit is allowed to appear in the sample as often as it is selected. In sampling without replacement, a particular unit is allowed to appear in the sample only once. Most forest resource sampling is without replacement.

Regardless of whether the selection is with or without replacement, the population mean is estimated by the arithmetic average of the sample observations, i.e.,

$$\bar{x} = \frac{\Sigma x}{n}$$

A first step in computing an estimate of the standard error of the mean is to estimate the variance of individual values of x:

$$s^2 = \frac{\Sigma x^2 - (\Sigma x)^2 / n}{n - 1}$$

In sampling with replacement (or from infinite populations), the standard error is

$$s_{\bar{x}} = \sqrt{\frac{s^2}{n}}$$

The finite population correction should be included if sampling is without replacement from a finite population:

$$s_{\bar{x}} = \sqrt{\frac{s^2}{n}\left(\frac{N - n}{N}\right)}$$

Confidence intervals are established as

$$\bar{x} \pm t s_{\bar{x}}$$

where t (2-tailed) has $n - 1$ degrees of freedom.

8-5   Sampling Intensity   To plan a timber inventory that is statistically and practically efficient, enough sample units should be measured to obtain the desired standard of precision—no more and no less. As an example, one might wish to estimate the mean volume per acre of a timber stand and have a 95 percent probability of being within ± 500 bd ft per acre of the true mean. A formula for computing the required sampling intensity for simple random samples may be derived by transforming the relationship for the confidence limits on the mean (Sec. 2-14). Excluding the finite population correction, the formula may be expressed as

$$n = \left(\frac{ts}{E}\right)^2$$

where E is the desired half-width of the confidence interval and other symbols are as previously described.

Solving this formula requires an estimate of the standard deviation expressed *in the same units* as the desired precision E. This estimate may be obtained by measuring a small preliminary sample of the population or by using the standard deviation obtained from previous sampling of the same or a similar population. The first method is likely to be most reliable, if the expense of a preliminary survey can be accepted. In the example proposed earlier, assume that preliminary measurement of 25 field plots provided the following data:

$$\bar{x} = 4400 \text{ bd ft per acre}$$
$$s = 2000 \text{ bd ft per acre}$$

The original objective was to be within ±500 bd ft per acre, with a confidence probability of 95 percent. For the 95 percent level, t is generally set equal to 2, although initial guesses of the sample size (and thus df and t) can be used in an iterative solution for a more refined estimate of sample size. In most cases, however, the precision of the other required information in the sample-size formula does not justify use of the iterative solution for t. In this example, the *desired* half-width of the confidence interval, $E = 500$, is substituted in the formula, along with the estimated standard deviation and a value of 2 for t:

$$n = \left[\frac{(2)(2000)}{500}\right]^2 = 64 \text{ sample units}$$

When sampling without replacement from finite populations, the sample-size formula is

$$n = \cfrac{1}{\left(\dfrac{E}{ts}\right)^2 + \dfrac{1}{N}}$$

where N represents the number of sampling units in the population.

Formulas for calculating the required sample size for simple random sampling can be written in several ways. If the allowable error is expressed as a percent of the mean and an estimate of the coefficient of variation (CV) in percent is available, the required sample size (for infinite populations) can be calculated from

$$n = \left[\frac{(t)(CV)}{A}\right]^2$$

where A is allowable error, expressed as a percent of the mean, and the other symbols are as previously described.

With an estimated coefficient of variation of 50 percent, for example, one might wish to determine the number of observations needed to estimate a population mean within ±5 percent at a probability level of 0.80. In other words, the desired half-width of the confidence interval is specified as 5 percent of the mean. From the Appendix, the t value (infinite df and probability column of 0.2) is read as 1.282. Therefore, the total number of sample units required to achieve the specified precision is approximately

$$n = \left[\frac{(1.282)(50)}{5}\right]^2 = 164.4, \text{ or } 165, \text{ sample units}$$

Trial substitutions in this formula will demonstrate the fact that sampling intensities are increased *4 times* when (1) the coefficient of variation is doubled, as from 25 to 50 percent, or (2) the specified allowable error is reduced by one-half. These are important facts for consideration in balancing costs and desired precision in resource inventories.

The sample-size formula for sampling without replacement from finite populations (with allowable error expressed as a percent of the mean) is

$$n = \frac{1}{\left(\dfrac{A}{tCV}\right)^2 + \dfrac{1}{N}}$$

As the population size N becomes large, results from this formula approach those from the formula for infinite populations.

8-6   Effect of Plot Size on Variability   At a given scale of measurement, small sample plots usually exhibit more relative variability (i.e., have a larger coefficient of variation) than large plots. The variance in volume per acre on $1/4$-acre plots is usually larger than the variance in volume per acre on $1/2$-acre plots but slightly smaller than that for $1/5$-acre plots. The relation of plot size to variance changes from one population to another. In general, large plots tend to have less relative variability, because they average out the effect of tree clumps and stand openings. In uniform populations (e.g., plantations), changes in plot size have little effect on variance. In nonuniform populations, the relation of plot size to variance depends on how clumps of trees and open areas compare with the sizes of plots.

Although plot sizes have often been chosen on the basis of experience, the objective should be the selection of the most efficient size. Usually, this is the smallest size commensurate with the variability produced. Where the coefficient of variation has been determined for plots of a given size, the coefficient of variation for different-sized plots may be approximated by a formula suggested by Freese (1962):

$$(CV_2)^2 = (CV_1)^2 \sqrt{\frac{P_1}{P_2}}$$

where CV_2 = estimated coefficient of variation for new plot size
CV_1 = known coefficient of variation for plots of previous size
P_1 = previous plot size
P_2 = new plot size

Note that the same relationship can be applied with variances in lieu of coefficients of variation squared. If the coefficient of variation for $1/5$-acre plots is 30 percent, the estimated coefficient of variation for $1/10$-acre plots would be computed as

$$(CV_2)^2 = (30)^2 \sqrt{\frac{0.2}{0.1}} = 900(1.414) = 1272.6$$
$$CV_2 = \sqrt{1272.6} = 36 \text{ percent}$$

The coefficients of variation of 36 percent for $^1/_{10}$-acre plots versus 30 percent for $^1/_5$-acre sample units may now be compared as to relative *numbers* of plots needed. Assume, for example, that the $^1/_5$-acre plots produced a sample mean of 4 cd per plot, which is equivalent to 20 cd per acre. The sample standard deviation is 30 percent of this value, or ±6 cd per acre. The total number of $^1/_5$-acre plots needed to estimate the mean volume per acre within ±2 cd at a probability level of 95 percent (approximating t as 2) is estimated as

$$n = \left[\frac{(2)(6)}{2}\right]^2 = 36 \text{ plots}$$

The reader should note that it is necessary to have the standard deviation and the allowable error in the same units before substitution in the formula. In this instance, these values are ±6 and ±2 cd per acre, respectively. For comparison with the preceding results, the standard deviation for $^1/_{10}$-acre plots, expressed on a per acre basis, would be 0.36(20) = ±7.2 cd per acre. The number of $^1/_{10}$-acre plots required to meet the previous standards of precision would be

$$n = \left[\frac{(2)(7.2)}{2}\right]^2 = 51.84, \text{ or } 52, \text{ plots}$$

The choice between thirty-six $^1/_5$-acre plots versus fifty-two $^1/_{10}$-acre plots is a decision that now rests on the relative time or costs involved. Some aspects of sampling "efficiency" are discussed in Sec. 8-21.

8-7 Systematic Sampling   Under this system, the initial sample unit is randomly selected or arbitrarily established on the ground; thereafter, plots are mechanically spaced at uniform intervals throughout the tract of land. For example, if a 5 percent sample is desired, every twentieth sample unit would be selected.

Systematic sampling has been popular for assessing timber and range conditions because (1) sample units are easy to locate on the ground and (2) they appear to be more "representative," since they are uniformly spaced over the entire population. Although these arguments *may* be true, the drawback is that it is usually difficult—if not impossible—to estimate the variance (or standard error) for one systematic sample.

Rectangular spacings or square grid layouts may yield efficient estimates under certain conditions, but the accuracy can also be poor if there is a periodic or cyclic variation inherent in the population (Fig. 8-2). Furthermore, assessment of the precision presents a formidable problem, since simple random sampling techniques cannot be logically applied to systematic designs. An exception occurs where the elements of the population are in random order. In those rare cases where this situation exists (and can be recognized), a systematic sample

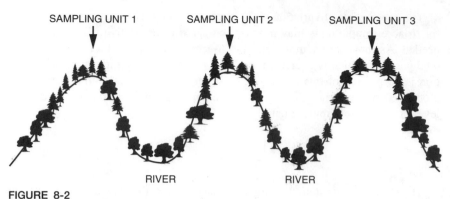

SAMPLING UNIT 1   SAMPLING UNIT 2   SAMPLING UNIT 3

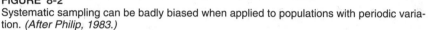

RIVER   RIVER

FIGURE 8-2
Systematic sampling can be badly biased when applied to populations with periodic variation. *(After Philip, 1983.)*

may be analyzed as a simple random sample. Nevertheless, it would be inaccurate to presume that most populations of plant communities are in random order.
Unfortunately, there is no practical alternative to assuming the population is in
random order and using the formulas for simple random sampling when computing estimates from systematic samples. (Note that the standard error would be
computed for sampling *without* replacement.) Although estimates of the mean
are generally satisfactory, estimates of sampling precision based on such manip-
ulations must be regarded, at best, as approximations.

In summary, experience has shown that systematic sampling is generally satisfactory for estimating means in typical forest conditions. When an objective
numerical statement of precision need not be appended to inventory estimates,
systematic sampling may provide more information for the time (or money) expended than simple random sampling. In cases in which estimates of variance
are important or little is known about the basic characteristics of the population
being sampled (e.g., the extent to which it falls into patterns), it is safer to select
a random rather than a systematic sample.

8-8   Stratified Random Sampling   In typical applications of stratified random sampling, a population is divided into subpopulations of known size, and a
simple random sample of at least two units is selected in each subpopulation.
This approach has several advantages. If the strata are constructed so that their
averages are different and their variances are small in relation to the total population variance, the estimate of the population mean will be considerably more
precise than that given by a simple random sample of the same size. Also, it may
be desirable to have separate estimates for each subpopulation (e.g., for different
vegetative types or administrative subunits), and it may be administratively
more efficient to sample by subpopulations.

The first step in estimating the overall population mean from a stratified sample is to compute the sample mean $\bar{x}_h$ for each stratum. Stratum means are computed exactly the same as for simple random samples—i.e., they are the arithmetic averages of the sample observations from an individual stratum. These stratum means are then combined into a weighted overall mean where the weights are equal to the strata sizes. Thus the estimate of the overall population mean $\bar{x}_{st}$ would be

$$\bar{x}_{st} = \frac{\sum\limits_{h=1}^{L} N_h \bar{x}_h}{N}$$

where L = number of strata

N_h = total number of units in stratum h ($h = 1, \ldots, L$)

N = total number of units in all strata ($N = \sum\limits_{h=1}^{L} N_h$)

To determine the standard error of $\bar{x}_{st}$, it is first necessary to compute the variance among individuals within each stratum s_h^2. These variances are computed in the same manner as for simple random sampling. Thus the variance for stratum I would be

$$s_I^2 = \frac{\Sigma x_I^2 - (\Sigma x_I)^2 / n_I}{n_I - 1}$$

where n_I represents the number of sample units observed in stratum I.

From the stratum variances, the standard error of the mean is estimated as

$$s_{\bar{x}_{st}} = \sqrt{\frac{1}{N^2} \sum_{h=1}^{L} \frac{N_h^2 s_h^2}{n_h}}$$

when sampling is with replacement, and

$$s_{\bar{x}_{st}} = \sqrt{\frac{1}{N^2} \sum_{h=1}^{L} \left[\frac{N_h^2 s_h^2}{n_h} \left(\frac{N_h - n_h}{N_h} \right) \right]}$$

when sampling is without replacement. In the standard error formulas, n_h = the number of units observed in stratum h.

Confidence intervals for the mean are computed as

$$\bar{x}_{st} \pm t s_{\bar{x}_{st}}$$

where the number of degrees of freedom for t can be approximated by

$$(n_1 - 1) + (n_2 - 1) + \cdots + (n_L - 1), \left[\text{i.e., by } \sum_{h=1}^{L}(n_h - 1) \right]$$

for moderate to large sample sizes within each stratum.

A stratified random sample may combine the features of aerial and ground estimating, offering a means of obtaining timber volumes with high efficiency. Photographs are used for area determination, for allocation of field samples by volume classes, and for designing the pattern of fieldwork. For each stratum, tree volumes or other data are obtained on the ground by conventional methods. In the example that follows, emphasis is on methods of allocating a fixed number of inventory plots among the various strata recognized.

Assume, for example, that a tract of land containing 300 acres has been subdivided into five distinct timber-volume classes (strata) by interpretation of aerial photographs. Since the tract has been recently inventoried by a systematic sample of 150 field plots, it is possible to compute a preliminary approximation of the standard deviation for each stratum:

Volume class	Stratum area, acres	Std. dev., cords/acre	Area × std. dev.
I	15	20	300
II	45	70	3,150
III	110	35	3,850
IV	60	45	2,700
V	70	25	1,750
Total	300	—	11,750

Assuming that a total of 150 sample units will be measured on the ground, there are two common procedures for distributing the field plots among the five volume classes. These methods are known as *proportional allocation* and *optimum allocation*.

8-9   Proportional Allocation of Field Plots   This approach calls for distribution of the 150 field plots in proportion to the *area* of each type. The general formula is

$$n_h = \left(\frac{N_h}{N} \right) n$$

For the five volume classes, the number of plots in each stratum would be computed as follows:

$$\text{Class I} : \frac{15}{300}(150) = 7 \text{ plots}$$

$$\text{Class II} : \frac{45}{300}(150) = 23 \text{ plots}$$

$$\text{Class III} : \frac{110}{300}(150) = 55 \text{ plots}$$

$$\text{Class IV} : \frac{60}{300}(150) = 30 \text{ plots}$$

$$\text{Class V} : \frac{70}{300}(150) = \underline{35 \text{ plots}}$$
$$\overline{150 \text{ plots}}$$

One disadvantage of proportional allocation is that large areas receive more sample plots than small ones, irrespective of variation in volume per acre. Of course, the same limitation applies to simple random and systematic sampling. Nevertheless, when the various strata can be reliably recognized and their areas determined, proportional allocation will generally be superior to a nonstratified sample of the same intensity.

8-10 Optimum Allocation of Field Plots With this procedure, the 150 sample plots are allocated to the various strata by a plan that results in the smallest standard error possible with a fixed number of observations. Determining the number of plots to be assigned to each stratum requires first a product of the area and standard deviation for each type, as derived earlier. In general terms, for a sample of size n, the number of observations n_h to be made in stratum h is

$$n_h = \left[\frac{N_h s_h}{\sum_{h=1}^{L} N_h s_h} \right] n$$

The number of plots to be allocated to each stratum is computed by expressing each product of "area times standard deviation" as a proportion of the product sum (11,750 in this example). Thus the 150 field plots would be distributed in the following manner:

$$\text{Class I} : \frac{300}{11,750}(150) = 4 \text{ plots}$$

$$\text{Class II} : \frac{3,150}{11,750}(150) = 40 \text{ plots}$$

$$\text{Class III} : \frac{3,850}{11,750}(150) = 49 \text{ plots}$$

$$\text{Class IV} : \frac{2,700}{11,750}(150) = 35 \text{ plots}$$

$$\text{Class V} : \frac{1,750}{11,750}(150) = \underline{22 \text{ plots}}$$
$$\overline{150 \text{ plots}}$$

No matter which method of allocation is used, according to the general theory of stratified random sampling, field plots are located within each stratum by simple random sampling. In practice, systematic sampling is sometimes employed in lieu of simple random sampling. When systematic samples are taken within each stratum, it is commonly assumed that the within-stratum populations are randomly distributed and the standard formulas for estimating the mean, standard error, and confidence intervals are applied.

It will be noted that optimum allocation results in a different distribution of the field plots among the various strata. By comparison with proportional allocation, fewer plots are assigned to classes I, III, and V, i.e., those strata with relatively small standard deviations. On the other hand, more plots are allotted to classes II and IV, which were the strata with relatively large standard deviations. Thus the relative variations within the volume classes more than offset the factor of stratum area in this particular example. When stratum areas and standard deviations can be determined reliably, optimum allocation is usually the preferred method for distributing a fixed number of inventory plots.

8-11 Sample Size for Stratified Sampling   Overall sample size needed to achieve a desired degree of precision at a specified probability level can be computed for stratified random sampling in a similar fashion as was shown for simple random sampling in Sec. 8-5. The exact form of the sample-size formula varies somewhat depending on the method of allocating the sample to the strata. When the sample is to be taken with proportional allocation, the sample-size formula is

$$n = \left(\frac{t}{E}\right)^2 \frac{\sum\limits_{h=1}^{L} N_h s_h^2}{N}$$

if the sampling is with replacement and

$$n = \frac{N \sum\limits_{h=1}^{L} N_h s_h^2}{\dfrac{N^2 E^2}{t^2} + \sum\limits_{h=1}^{L} N_h s_h^2}$$

if the sampling is without replacement. In these equations, n is the total sample size, E is the desired half-width of the confidence interval, and other symbols remain as defined previously for stratified random sampling.

When optimum allocation is applied, the equations for overall sample size are

$$n = \left(\frac{t}{(N)(E)} \right)^2 \left(\sum_{h=1}^{L} N_h s_h \right)^2$$

when sampling is with replacement and

$$n = \frac{\left(\sum_{h=1}^{L} N_h s_h \right)^2}{\dfrac{N^2 E^2}{t^2} + \sum_{h=1}^{L} N_h s_h^2}$$

when sampling is without replacement.

8-12 Regression Estimation Regression estimators, like stratification, were developed to increase the precision or efficiency of sampling by making use of supplementary information about the population being studied (Freese, 1962). In the case of regression estimators, the mean or total of a second variable, which is related to the primary variable we are interested in, is known. The mean or total of the supplementary variable for the population as a whole can be used to improve the precision with which one estimates the population average of the primary variable. In practice, the supplementary variable may be available through a population census (complete enumeration) completed at some time before the sample.

When regression estimation procedures are applied, two related variables (x and y) are measured on each sample unit. Although one is interested in the y variable, because the x variable is chosen such that it is more easily measured and strongly correlated with the y variable, one can increase overall precision by devoting part of the sampling resources to observing the x variable. To use regression estimation one might measure x_i on all N units in a population to determine the population mean μ_x. Then one measures x_i and y_i on a simple random sample of n observations and uses a regression model to combine these data for a precise estimate of the mean of y.

Use of the regression estimator will be illustrated by using data from the *Forestry Handbook* (Wenger, 1984). In this example, a forester wants to estimate the sawtimber volume on a 5-acre tract. The dbh of all 1273 trees on the tract is carefully measured, and the population mean basal area per tree is computed as 1.40 sq ft. A simple random sample of size 20 is then drawn. For the trees in this simple random sample, both the basal area (x_i) and the volume (y_i) are determined. The data from the sample, along with summary statistics, are shown below:

Tree number	dbh, in.	x = basal area, ft^2	y = volume, ft^3	r = ratio, y_i/x_i
1	11.0	0.67	17.5	26.12
2	22.8	2.84	85.7	30.18
3	15.0	1.22	31.8	26.07
4	15.6	1.16	33.9	29.22
5	13.0	0.92	25.0	27.17
6	21.3	2.46	66.8	27.15
7	15.7	1.35	46.4	34.37
8	12.6	0.86	18.6	21.63
9	14.1	1.09	27.1	24.86
10	13.4	0.98	28.2	28.78
11	14.1	1.09	30.0	27.52
12	14.6	1.16	29.3	25.26
13	16.5	1.49	48.2	32.35
14	13.0	0.92	26.8	29.13
15	13.4	0.98	22.9	23.37
16	11.8	0.76	24.3	31.97
17	14.1	1.09	25.0	22.94
18	15.4	1.28	32.1	25.08
19	17.7	1.71	49.6	29.01
20	15.4	1.28	32.9	25.70
Means		$\bar{x}$ = 1.27	$\bar{y}$ = 35.1	$\bar{r}$ = 27.39
Variances		s_x^2 = 0.287	s_y^2 = 284	s_r^2 = 10.78

The linear regression estimate of the population mean of y ($\bar{y}_R$) is

$$\bar{y}_R = \bar{y} + b(\mu_x - \bar{x})$$

where $\bar{y}$ is the mean of y and $\bar{x}$ is the mean of x from the simple random sample, μ_x is the population mean of x, and b is the linear regression slope coefficient (Chap. 2). To apply the linear regression estimator one must assume a straight-line relationship between x and y, where the scatter of the y observations is roughly the same throughout the range of the x observations. Plotting y versus x (Fig. 8-3) shows that the required assumptions are reasonable for these data.

By carrying through with the required computations, the linear regression slope coefficient is estimated as

$$b = \frac{SP_{xy}}{SS_x} = \frac{167.2}{5.455} = 30.65$$

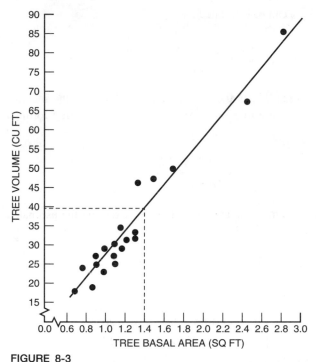

FIGURE 8-3
Relationship between tree basal area and tree volume from a
simple random sample of size 20. *(From Wenger, 1984.)*

where SP_{xy} is the corrected sum of cross products of x and y and SS_x is the corrected sum of squares of x. Substituting the computed slope value and the sample means of y and x into the estimating equation for the population mean of y gives

$$\bar{y}_R = 35.1 + 30.65(1.40 - 1.27)$$
$$= 39.1$$

Note that if there is a positive relationship between x and y and if $\bar{x}$ is below the true mean of x, μ_x, one would expect $\bar{y}$ to be below the true mean of y by an amount $b(\mu_x - \bar{x})$ where b, the slope coefficient, is the number of unit changes in y per unit change in x. Thus, in this example, $\bar{y}$ is adjusted upward; a downward adjustment in $\bar{y}$ would have resulted had the sample mean of x been greater than the population mean.

The standard error of $\bar{y}_R$ can be estimated as

$$S_{\bar{y}_R} = \sqrt{S_{y \cdot x}^2 \left[\frac{1}{n} + \frac{(\mu_x - \bar{x})^2}{SS_x}\right]\left(1 - \frac{n}{N}\right)}$$

In the standard error formula, $s_{y \cdot x}^2$ is the estimate of variability of the individual y values about the regression of y on x (Sec. 2-22); the computing formula is

$$S_{y \cdot x}^2 = \frac{SS_y - (SP_{xy})^2 / SS_x}{n - 2}$$

where SS_y is the corrected sum of squares of y. For this example the estimated variance about regression is

$$S_{y \cdot x}^2 = \frac{5393 - (167.2)^2 / 5.455}{18}$$

$$= 14.90$$

$$S_{\bar{y}_R} = \sqrt{14.90 \left[\frac{1}{20} + \frac{(1.40 - 1.27)^2}{5.455}\right]\left(1 - \frac{20}{1273}\right)}$$

$$= 0.8825 \text{ cu ft}$$

Note that for this small sampling fraction, the finite population correction factor $(1 - n/N)$ could be omitted with no significant impact on the result.

Confidence intervals on the mean can be computed as

$$\bar{y}_R \pm tS_{\bar{y}_R}$$

where t has $n - 2$ degrees of freedom. The 95 percent confidence interval for this example would be

$$39.1 \pm (2.101)(0.8825)$$
$$39.1 \pm 1.854$$
or $$37.246 \text{ to } 40.954$$

8-13 Comparison of Regression Estimation to Simple Random Sampling Suppose that a simple random sample of size 20 were taken in which the y values (cubic volumes) only were measured and that the samples were the same as that shown in Sec. 8-12. How would the estimates compare with those from the regression estimation procedure?

The mean of y would be estimated as

$$\bar{y} = \frac{\sum\limits_{i=1}^{n} y_i}{n} = \frac{702.1}{20} = 35.1$$

as compared with 39.1 for the regression estimate. The estimate of the standard error of the mean for this simple random sample would be

$$S_{\bar{y}} = \sqrt{\frac{S_y^2}{n}\left(1 - \frac{n}{N}\right)}$$

$$= \sqrt{\frac{284}{20}\left(1 - \frac{20}{1273}\right)}$$

$$= 3.7386 \text{ cu ft}$$

as compared with 0.8825 for the regression estimate. Confidence intervals for a simple random sample are computed as

$$\bar{y} \pm tS_{\bar{y}}$$

where t has $n - 1$ degrees of freedom. Substituting in the appropriate values gives

$$35.1 \pm 2.093(3.7386)$$
$$35.1 \pm 7.825$$
or
$$27.275 \text{ to } 42.925$$

The standard error is much bigger and the confidence interval much wider for the simple random sample. In an actual application, the sample size for the simple random sample should be made somewhat larger if part of the sampling resources had not been devoted to measuring x, a supplementary variable. However, impressive gains in precision are possible with regression estimation when the correlation between x and y is near 1, as it is in this case ($r = 0.975$).

8-14 Ratio Estimation The regression estimation procedure presented in Sec. 8-12 is one of a family of related procedures that enables one to increase sampling efficiency by making use of information about a supplementary variable. Two other members of this family of related procedures will be presented: the ratio-of-means estimator and the mean-of-ratios estimator.

The *ratio-of-means* estimator is appropriate when the relationship of y to x is in the form of a straight line passing through the origin and when the variance of y at any given level of x is proportional to the value of x (i.e., the variance of y is not assumed constant as in the regression estimation procedure). The ratio estimate of the mean of y is

$$\bar{y}_{RM} = R\mu_x$$

where $R = \dfrac{\bar{y}}{\bar{x}}$ or $\dfrac{\Sigma y_i}{\Sigma x_i}$

μ_x = known population mean of x

For large samples (generally taken as n greater than 30), the standard error of the ratio-of-means estimator can be approximated by

$$S_{\bar{y}_{RM}} = \sqrt{\left(\frac{S_y^2 + R^2 S_x^2 - 2RS_{xy}}{n}\right)\left(1 - \frac{n}{N}\right)}$$

In practice, the most appropriate estimator should be chosen and applied in any given situation. For purposes of illustration, however, the ratio-of-means estimator will be applied to the same data as were used in the regression estimation example. First, the estimate of R is computed as

$$R = \frac{\bar{y}}{\bar{x}} = \frac{35.1}{1.27} = 27.64$$

and then the population mean of y is estimated as

$$\bar{y}_{RM} = R\mu_x = (27.64)(1.40) = 38.7$$

which is reasonably close to the estimate of 39.1 obtained with the linear regression estimator.

Although the sample size is small ($n = 20$), the formula for large sample sizes will be used to approximate the standard error of the estimate of the mean of y:

$$S_{\bar{y}_{RM}} = \sqrt{\left(\frac{S_y^2 + R^2 S_x^2 - 2RS_{xy}}{n}\right)\left(1 - \frac{n}{N}\right)}$$

$$= \sqrt{\frac{284 + (27.64)^2 (0.287) - 2(27.64)(8.80)}{20}\left(1 - \frac{20}{1273}\right)}$$

$$= 0.9092 \text{ cu ft}$$

Note that, despite having a sample of only 20 trees, this approximation is quite close to the estimate of 0.8825 computed for the linear regression estimator.

The second ratio estimator that is commonly used is called the *mean-of-ratios* estimator. This estimator is appropriate when the relation of y to x is in the form of a straight line passing through the origin and the variance of y at any given level of x is proportional to x^2 (rather than proportional to x as for the ratio-of-means estimator). As the name *mean-of-ratios* implies, the ratio (r_i) of y_i to x_i is computed for each pair of observations. Then the estimate of the mean of y for the population is computed as

$$\bar{y}_{MR} = R\mu_x$$

where $R = \dfrac{\displaystyle\sum_{i=1}^{n} r_i}{n}$

To compute the standard error of this estimate one must first estimate the variability of the individual ratios (r_i) as

$$S_r^2 = \frac{\displaystyle\sum_{i=1}^{n} r_i^2 - \dfrac{\left(\displaystyle\sum_{i=1}^{n} r_i\right)^2}{n}}{n-1}$$

The standard error of the mean-of-ratios estimator ($\bar{y}_R$) is then

$$S_{\bar{y}_{MR}} = \mu_x \sqrt{\frac{S_r^2}{n}\left(1 - \frac{n}{N}\right)}$$

Again, one can use the same data on tree basal area and volume to illustrate the computations with the mean-of-ratios procedure. The individual ratio values are computed from the data as $r_1 = 17.5/0.67 = 26.12$, $r_2 = 85.7/2.84 = 30.18$, ... $r_{20} = 32.9/1.28 = 25.70$. The mean of these r_i values is 27.39. Thus the estimate of the population mean of y would be

$$\begin{aligned}\bar{y}_{MR} &= R\mu_x \\ &= (27.39)(1.40) = 38.3\end{aligned}$$

which is close to the ratio-of-means estimator value of 38.7.

The standard error of the mean for the mean-of-ratios procedure is computed as

$$\begin{aligned}S_{\bar{y}_{MR}} &= \mu_x \sqrt{\frac{S_r^2}{n}\left(1 - \frac{n}{N}\right)} \\ &= 1.4 \sqrt{\frac{10.78}{20}\left(1 - \frac{20}{1273}\right)} \\ &= 1.0197 \text{ cu ft}\end{aligned}$$

In this example, the standard error for the mean-of-ratios procedure is also near the comparable statistic for the ratio-of-means estimator (0.9092).

8-15 Double Sampling Regression and ratio estimators require that the population mean of the supplementary variable x be known. When the population mean is not known, it is sometimes efficient to take a large sample in which x alone is measured. The purpose of this large sample is to develop a good estimate of μ_x, the population mean of x. In a survey designed to make estimates for some other variable y, it may pay to devote part of the resources to this preliminary sample, although this means that the size of the sample in the main survey on y may be reduced. This technique is known as *double sampling* or *two-phase sampling*. Double sampling is applied in cases where the gain in precision from using ratio or regression estimators (or other methods where supplementary information is required) more than offsets the loss in precision due to reduction in the size of the main sample.

Suppose that a large sample is taken in order to obtain a reliable estimate of the mean of x. On a subsample of the units in this large sample, the y values are also measured to provide an estimate of the relationship between y and x. The large sample mean for x is then applied to the fitted relationship to obtain an estimate of the population mean of y. If the relationship between y and x is linear and there appears to be a homogeneous variance of y at all levels of x, then the regression estimator would be appropriate. The computing formula for estimating the mean of y (denoted $\bar{y}_{Rd}$) when the linear regression estimator is applied in a double sample is

$$\bar{y}_{Rd} = \bar{y}_2 + b(\bar{x}_1 - \bar{x}_2)$$

where $\bar{y}_{Rd}$ = estimate of the population mean of y
  $\bar{x}_1$ = mean of x from the large sample
  $\bar{x}_2$ = mean of x from the small sample
  $\bar{y}_2$ = mean of y from the small sample
  b = linear regression slope coefficient computed as SP_{xy}/SS_x from the small sample

The standard error of $\bar{y}_{Rd}$ (denoted $S_{\bar{y}_{Rd}}$) can be estimated as

$$S_{\bar{y}_{Rd}} = \sqrt{S_{y \cdot x}^2 \left(\frac{1}{n_2} + \frac{(\bar{x}_1 - \bar{x}_2)^2}{SS_x} \right) \left(1 - \frac{n_2}{n_1} \right) + \frac{S_y^2}{n_1} \left(1 - \frac{n_1}{N} \right)}$$

where $S_{y \cdot x}^2$ = variance about the fitted regression of y on x from the small sample
  S_y^2 = variance of the y values in the small sample
  SS_x = corrected sum of squares for x from the small sample
  n_1 = number in the large sample
  n_2 = number in the small sample
  N = number in the population

For sampling with replacement or for very small sampling fractions, the term $(1 - n_1/N)$ is omitted.

Confidence intervals, at the 95 percent level, can be approximated as

$$\bar{y}_{Rd} \pm 2S_{\bar{y}_{Rd}}$$

Ratio estimators can also be applied in a double-sampling context. The computing formulas for the ratio-of-means estimator are

$$\bar{y}_{RMd} = R\bar{x}_1$$

where $R = \bar{y}_2/\bar{x}_2$

$$S_{\bar{y}_{RMd}} = \sqrt{\left(1 - \frac{n_2}{n_1}\right)\left(\frac{S_y^2 + R^2 S_x^2 - 2RS_{xy}}{n_2}\right) + \frac{S_y^2}{n_1}\left(1 - \frac{n_1}{N}\right)}$$

where all symbols remain as defined previously.

In the case of the mean-of-ratios estimator, the estimating formulas for the mean of y and the standard error of the mean are

$$\bar{y}_{MRd} = R\bar{x}_1$$

where

$$R = \frac{\displaystyle\sum_{i=1}^{n_2} r_i}{n_2}$$

$$S_{\bar{y}_{MRd}} = \sqrt{\bar{x}_1^2 \frac{S_r^2}{n_2}\left(1 - \frac{n_2}{n_1}\right) + \frac{S_y^2}{n_1}\left(1 - \frac{n_1}{N}\right)}$$

As before, the symbols remain as defined previously.

The total cost (C) of a double sample can be written as

$$C = n_1 C_1 + n_2 C_2$$

where C_1 is the cost per observation in the large sample, n_1 is the large sample size, C_2 is the cost per sample in the small sample, and n_2 is the small sample size. Part of the sampling resources are used to obtain the large sample, but C_1 is generally quite small compared with C_2; thus n_2 is not substantially less than if a simple random sample on y alone had been taken. However, it is worth emphasizing that a double-sampling design is feasible only if the gains due to using regression or ratio estimators (or other procedures that utilize supplementary information about the population) more than offset the smaller sample size for y.

8-16 Double Sampling for Stratification Recall that stratified random sampling requires that the strata sizes be known in advance of sampling. If large gains due to stratification are expected, it may be efficient to use part of the sampling resources to estimate the strata sizes and then to draw a sample and compute population estimates based on the estimated strata sizes. This procedure is another example of double sampling.

In stratified random sampling the N_h values are assumed known. The N_h values can be estimated in a double-sampling design when they are not known in advance of sampling. Consider a case where two timber types are represented on a tract for which an estimate of the overall volume is desired. Past experience indicates that large gains in sampling efficiency will be realized by stratifying by timber type. Unfortunately, the area by timber type is not known. However, aerial photos of the tract are available. A dot grid (Chap. 3) is placed over the tract, which, from a boundary-line survey, has a known total area of 40 acres. The sample size of the random dots is 1000 (that is, $n_1 = 1000$). Of the 1000 dots, 600 fall in the pine type and 400 in the hardwood type. Thus the sizes of the two strata can be estimated as

$$\hat{N}_h = N\left(\frac{n_{1_h}}{n_1}\right)$$

where $\hat{N}_h$ indicates the estimated value for N_h. For this example, the strata sizes are estimated as

$$\text{Pine: } \hat{N}_1 = 40\left(\frac{600}{1000}\right) = 24$$

$$\text{Hardwood: } \hat{N}_2 = 40\left(\frac{400}{1000}\right) = 16$$

Next a small sample ($n_2 = 20$) of ground plots is taken. This is to be a stratified sample with proportional allocation. The estimated strata sizes can be used to allocate the sample as follows:

$$\text{Number in pine type} = n_{21} = \left(\frac{24}{40}\right)20 = 12$$

$$\text{Number in hardwood type} = n_{22} = \left(\frac{16}{40}\right)20 = 8$$

That is,

$$n_{2h} = \left(\frac{\hat{N}_h}{N}\right)n_2$$

After the fieldwork is complete, the estimate of the overall population mean is computed as before, except that estimated weights rather than known weights are applied. Suppose the following strata statistics were computed from the field plots:

$$\text{Pine: } \bar{x}_1 = 5400 \text{ bd ft/acre}$$
$$\text{Hardwood: } \bar{x}_2 = 3800 \text{ bd ft/acre}$$

The estimated overall population mean would be

$$\bar{x}_{\text{std}} = \frac{\sum_{h=1}^{L} \hat{N}_h \bar{x}_h}{N}$$

$$= \frac{24(5400) + 16(3800)}{40}$$

$$= 4760 \text{ bd ft/acre}$$

While the computations thus far are analogous to those where the strata weights are known, the computation of the standard error of the mean is complicated by the fact that the strata sizes are now estimated rather than known. Provided that the strata sizes are precisely estimated, the formula for the standard error of the mean given in Sec. 8-8 can be applied. The $\hat{N}_h$ values must be used in place of N_h, but the formula provides a reasonably good approximation of the standard error of the mean for double samples with reliable estimates of the strata sizes. Confidence intervals at the 95 percent level can then be approximated by setting the t value equal to 2. For additional information about double sampling for stratification and methods for estimating the associated sampling error, readers are referred to the more advanced books (such as Cochran, 1977) cited at the end of this chapter.

8-17   Cluster Sampling   The objective of sample survey design is to obtain a specified amount of information about a population parameter at minimum cost. In some circumstances, cluster sampling gives more information per unit cost than do simple random or other alternative sampling methods. A *cluster sample* is a sample in which each sampling unit is a collection, or cluster, of elements (Scheaffer et al., 1990). There are two primary reasons for applying cluster sampling:

1 A list of the elements of the population may not be available from which to select a random sample, but it may be feasible to develop a list of all clusters. A forester may wish to estimate the mean height of all seedlings in a nursery. No list of seedlings exists, so it is not feasible to select individual seedlings at random. However, a sampling frame of all possible rows in the seedling beds can be constructed. Then a random sample of rows can be drawn, and heights of seedlings in the rows can be measured. The seedlings are the elements, and the rows are the clusters.

2 Even when a list of elements is available, it may be more economical to select clusters randomly rather than individual elements. In the case of estimating mean seedling height, even if a list of all seedlings were available, it may be more efficient to measure seedlings in rows rather than individual seedlings scattered over all nursery beds.

In summary, cluster sampling is attractive if the cost of obtaining a frame that lists all population elements is high or if the cost of obtaining observations increases as the distance separating the elements increases.

The first task in cluster sampling is to specify appropriate clusters. For maximum precision in cluster sampling, clusters should be formed so that the individuals within a cluster vary as much as possible. In practice, elements within a cluster are generally physically close together and, hence, tend to have similar characteristics. Consequently, the amount of information about the population parameters may not be increased substantially as the cluster sizes increase. As a general rule, the number of elements within a cluster should be small relative to the population size, and the number of clusters in the sample should be reasonably large.

The following notation, from Scheaffer et al. (1990), will be used for cluster sampling:

N = number of *clusters* in the population

n = number of clusters selected by simple random sampling

m_i = number of elements in cluster i ($i = 1, . . . , N$)

$$\overline{m} = \frac{\sum\limits_{i=1}^{n} m_i}{n} = \text{average cluster size for the sample}$$

$$M = \sum\limits_{i=1}^{N} m_i = \text{number of elements in the population}$$

$$\overline{M} = \frac{M}{N} = \text{average cluster size for the population}$$

y_i = total of all observations in the ith cluster

The estimator for the population mean of y is the sample mean $\overline{y}_c$, which is computed as

$$\overline{y}_c = \frac{\sum\limits_{i=1}^{n} y_i}{\sum\limits_{i=1}^{n} m_i}$$

The standard error of $\overline{y}_c$ can be estimated as

$$S_{\overline{y}_c} = \sqrt{\left(\frac{N - n}{Nn\,\overline{M}^2}\right) \frac{\sum\limits_{i=1}^{n}(y_i - \overline{y}_c m_i)^2}{n - 1}}$$

where $\overline{M}$ can be estimated by $\overline{m}$ if M is unknown and where the 95 percent confidence interval can be approximated as

$$\overline{y}_c \pm 2S_{\overline{y}_c}$$

To illustrate cluster sampling, suppose that a forester wishes to estimate the average height of trees in a plantation. The plantation is divided into 400 row-segments, each containing an equal number of original planting spaces (but not an equal number of surviving trees due to differential mortality rates). A simple random sample of 20 row-segments is selected from the 400 segments that compose the plantation. All trees in the sampled segments are measured and the following results obtained:

Number of trees, m_i	Sum of tree heights, y_i	Average height, ft	Number of trees, m_i	Sum of tree heights, y_i	Average height, ft
4	144	36	5	210	42
3	120	40	3	111	37
5	175	35	4	132	33
2	82	41	7	294	42
4	156	39	4	176	44
6	264	44	2	78	39
3	120	40	6	276	46
3	117	39	4	164	41
5	230	46	5	165	33
4	164	41	5	185	37
$\sum_{i=1}^{20} m_i = 84$			$\sum_{i=1}^{20} y_i = 3363$		

The estimate of the population mean is

$$\bar{y}_c = \frac{\sum\limits_{i=1}^{n} y_i}{\sum\limits_{i=1}^{n} m_i} = \frac{3363}{84} = 40.04 \text{ ft}$$

Since $\overline{M}$ is unknown, it must be estimated with $\overline{m}$ when computing the standard error of the mean; for this example

$$\overline{m} = \frac{\sum\limits_{i=1}^{n} m_i}{n} = \frac{84}{20} = 4.2$$

Thus, the estimate of the standard error of the mean is

$$S_{\bar{y}_c} = \sqrt{\left(\frac{400-20}{(400)(20)(4.2)^2}\right)\left(\frac{[144-(40.04)(4)]^2 + \cdots + [185-(40.04)(5)]^2}{19}\right)}$$

$$= 1.678 \text{ ft}$$

While worthwhile gains in precision are theoretically possible with appropriate clustering, in practice this is generally not the case since clustering is typically based on physical proximity, which tends to ensure that the elements of a cluster are similar rather than dissimilar. Cluster sampling is applied for reasons of convenience and in an effort to achieve a given precision at a lower cost than is possible with alternative sampling designs. A cluster sample will be more precise than a simple random sample of the same size if the elements within clusters vary more on the average than do the elements in the population as a whole.

8-18 Two-Stage Sampling Clusters often contain too many elements to obtain a measurement on each, or the elements are similar and measurement of only a few elements provides information on the entire cluster (Scheaffer et al., 1990). When either situation occurs, one can select a simple random sample of clusters and then take a simple random sample of elements within each cluster. This procedure is called a *two-stage sample.*

To illustrate two-stage sampling, one might consider the problem of estimating the mean volume per acre on 640 acres of timberland which is divided into square blocks of 40 acres each. Assume that the 40-acre blocks are plantations of various ages and that the 40-acre plantations can in turn be subdivided into squares of 100 original planting spaces (the number of trees now present in each 100-space square will vary). Sample units will be the square plots consisting of 100 original planting spaces. Since the 40-acre blocks are scattered, the travel time between blocks is large; however, it is obviously not possible to measure all elements (plots) in the blocks selected for sampling. Conversely, once a block is located, it would not be efficient to measure only one plot. Hence the designer of the inventory decides to select six blocks by simple random sampling and then to select three sample plots at random from each block selected in the first stage. In the sampling literature, the 40-acre blocks would be termed *primary sampling units* (primaries) and the plots *secondary sampling units* (secondaries). There will be n blocks (in our case, $n = 6$) selected with m plots (in this example, $m = 3$) selected within each of the selected blocks.

If y_{ij} designates the volume of the jth sample plot ($j = 1 . . . m$) on the ith sample block ($i = 1 . . . n$), the estimated mean volume per plot is

$$\bar{y}_{ts} = \frac{\sum_{i=1}^{n}\sum_{j=1}^{m} y_{ij}}{mn}$$

The standard error of the estimated mean for this case in which the primaries are equal in size and the number of elements per cluster is equal is given by

(Freese, 1962):

$$S_{\bar{y}_{ts}} = \sqrt{\frac{1}{mn}\left[S_B^2\left(1-\frac{n}{N}\right) + \frac{nS_W^2}{N}\left(1-\frac{m}{M}\right)\right]}$$

where n = number of primaries sampled

N = total number of primaries in the population

m = number of secondaries sampled in each of the primaries selected for sampling

M = total number of secondaries in each primary

S_B^2 = sample variance between primaries when sampled by m secondaries per primary

S_W^2 = sample variance among secondaries within primaries

The variances S_B^2 and S_W^2 are computed from the equations

$$S_B^2 = \frac{\dfrac{\sum_{i=1}^{n}\left(\sum_{j=1}^{m} y_{ij}\right)^2}{m} - \dfrac{\left(\sum_{i=1}^{n}\sum_{j=1}^{m} y_{ij}\right)^2}{mn}}{n-1}$$

$$S_W^2 = \frac{\sum_{i=1}^{n}\sum_{j=1}^{m} y_{ij}^2 - \dfrac{\sum_{i=1}^{n}\left(\sum_{j=1}^{m} y_{ij}\right)^2}{m}}{n(m-1)}$$

If the number of primary units sampled (n) is a small fraction of the total number of primary units (N), the standard error formula simplifies to

$$S_{\bar{y}_{ts}} = \sqrt{\frac{S_B^2}{mn}}$$

When n/N is fairly large but the number of secondaries (m) sampled in each selected primary is only a small fraction of the total number of secondaries (M) in each primary, the standard error formula can be expressed as

$$S_{\bar{y}_{ts}} = \sqrt{\frac{1}{mn}\left[S_B^2\left(1-\frac{n}{N}\right) + \frac{nS_W^2}{N}\right]}$$

The timber-volume estimation example will be carried through to illustrate the application of these formulas. Suppose the six primaries with three secondaries within each primary were selected with the following result:

Block, primary	Plot, secondary	Volume, ft³	Block, primary	Plot, secondary	Volume ft³
1	1	500	4	1	210
	2	650		2	185
	3	610		3	170
2	1	490	5	1	450
	2	475		2	300
	3	505		3	500
3	1	940	6	1	960
	2	825		2	975
	3	915		3	890

Substituting into the computing formula for the estimated mean plot gives

$$\bar{y}_{ts} = \frac{\sum\limits_{i=1}^{n}\sum\limits_{j=1}^{m} y_{ij}}{mn} = \frac{(500 + 650 + \cdots + 890)}{3(6)} = \frac{10,550}{18} = 586.1 \text{ ft}^3 \text{ per plot}$$

In order to compute the standard error of the mean, the quantities S_B^2 and S_W^2 are first computed as

$$S_B^2 = \frac{\dfrac{(1760^2 + \cdots + 2825^2)}{3} - \dfrac{(10,550)^2}{(3)(6)}}{6-1}$$

$$= 250,188.9$$

$$S_W^2 = \frac{\left(500^2 + \cdots + 890^2\right) - \left(\dfrac{1760^2 + \cdots + 2825^2}{3}\right)}{6(3-1)}$$

$$= 3869.4$$

Since the total number of 40-acre blocks in the 640-acre tract is $N = 16$ and the total number of plots in each 40-acre block is $M = 160$, the standard error of the mean, on a plot basis, is

$$S_{\bar{y}_{ts}} = \sqrt{\frac{1}{(3)(6)}\left[250,188.9\left(1 - \frac{6}{16}\right) + \frac{(6)(3869.4)}{16}\left(1 - \frac{3}{160}\right)\right]}$$

$$= 93.628 \text{ ft}^3 \text{ per plot}$$

The 95 percent confidence interval, on a per plot basis, is approximated as

$$586.1 \pm (2)(93.628)$$
$$398.844 \text{ to } 773.356 \text{ ft}^3$$

or

For a fixed number of sample observations, two-stage sampling is usually less precise than simple random sampling. Two-stage sampling may, however, permit us to obtain the desired precision at a lower cost by reducing the cost per observation. Usually the precision and cost both increase as the number of primaries is increased and the number of secondaries per sampled primary is decreased. The total cost of the survey can generally be reduced by taking fewer primaries and more secondaries per primary, but precision usually suffers. References listed at the end of this chapter (Freese, 1962; Cochran, 1977) contain formulas for estimating the optimum number of secondaries per primary from the standpoint of giving the greatest precision for a given expenditure.

In the preceding examples, equal weight was given to all primaries. While equal weight is logical for the illustration in which the tract was divided into equal-sized blocks of 40 acres each, it may not be the most efficient selection scheme if the primaries vary greatly in size. If, for example, stands are used as the primary and the forest population of interest is composed of stands that vary widely in size, a two-stage sampling scheme that takes primary size into account may be desired. Extensions of two-stage sampling, including selecting primaries with probability proportional to size, are contained in numerous textbooks on sample survey techniques.

8-19   Sampling for Discrete Variables   The formulas for estimates, standard deviations, confidence limits, etc., that were discussed in the previous section apply to data that are on a continuous or nearly continuous quantitative scale of measurement. These methods may not be applicable if each unit observed is classified according to a qualitative attribute, such as alive or dead, deciduous or evergreen, forest or nonforest. Data such as counts in two mutually exclusive classes follow what is known as the binomial distribution, and slightly different statistical formulas are required.

Assume that a forester wishes to estimate the proportion of surviving seedlings in a young plantation. Because the plantation was planted at exact regular spacing, it is possible to choose individual seedlings for observation by randomly selecting pairs of random numbers and letting the first number stand for a row and the second number designate the seedling within the row. The forester can go to each randomly chosen spot and observe the seedling, recording its status as live or dead. A survey in which 50 seedlings were randomly chosen showed that 40 of these were alive. The estimate of the proportion alive ($\overline{P}$) is

$$\overline{P} = \frac{\text{number alive}}{\text{total number observed}} = \frac{40}{50} = 0.80$$

The standard error of $\bar{P}$ (Freese, 1962) is

$$S_{\bar{p}} = \sqrt{\frac{\bar{P}(1 - \bar{P})}{n - 1}\left(1 - \frac{n}{N}\right)}$$

where n = number of units (seedlings in this example) observed. Given that the total number of trees in the plantation N is 5000, one can substitute into the formula for the standard error of $\bar{P}$ to obtain

$$S_{\bar{p}} = \sqrt{\frac{(0.80)(1 - 0.80)}{50 - 1}\left(1 - \frac{50}{5000}\right)}$$
$$= 0.05686$$

For large samples, the 95 percent confidence interval can be computed as

$$\bar{P} \pm \left[2\,S_{\bar{p}} + \frac{1}{2n}\right]$$

In this example the confidence interval, approximated with the formula for large samples, is

$$0.80 \pm \left[(2)(0.05686) + \frac{1}{2(50)}\right]$$
$$0.80 \pm 0.12372$$
or $\qquad$ 0.67628 to 0.92372

8-20 Cluster Sampling for Attributes In attribute sampling the cost of selecting and locating an individual is usually very high relative to the cost of determining whether or not the individual has a certain characteristic (Freese, 1962). Hence, some form of cluster sampling is usually preferred over simple random sampling. In cluster sampling, a group of individuals becomes the unit of observation, and the unit value is the proportion of the individuals in the group having the specified attribute. When clusters are reasonably large and all of the same size, the procedures for computing estimates of means and standard errors are similar to those for data measured on a continuous scale.

Returning to the example of estimating the proportion of surviving seedlings in a plantation, it would likely be more efficient to select plots of a given size and to determine the proportion of surviving trees in each plot. The plots, in this example, can be constructed such that there is an equal number of seedlings in each. Suppose that plots with 20 seedlings each are constructed; thus there are 250 possible plots in the population (that is, $N = 250$). A simple random sample of size 10 gives the following results:

Plot no.	1	2	3	4	5	6	7	8	9	10
Proportion alive	0.75	0.80	0.80	0.85	0.70	0.90	0.70	0.75	0.80	0.65

If P_i is the proportion alive in the ith plot, the mean proportion alive would be estimated by

$$\bar{P} = \frac{\sum_{i=1}^{n} P_i}{n} = \frac{(0.75 + \cdots + 0.65)}{10} = \frac{7.7}{10} = 0.77$$

The variance of P can be computed as

$$S_p^2 = \frac{\sum_{i=1}^{n} P_i^2 - \frac{\left(\sum_{i=1}^{n} P_i\right)^2}{n}}{n-1}$$

$$= \frac{\left(0.75^2 + \cdots + 0.65^2\right) - \frac{(7.7)^2}{10}}{9}$$

$$= 0.005667$$

Substituting into the formula for the standard error of the mean gives

$$S_{\bar{p}} = \sqrt{\frac{S_p^2}{n}\left(1 - \frac{n}{N}\right)} = \sqrt{\frac{0.005667}{10}\left(1 - \frac{10}{250}\right)} = 0.02332$$

As in simple random sampling with data measured on a continuous scale, a confidence interval for the mean can be computed as

$$\bar{P} \pm t\left(S_{\bar{p}}\right)$$

where $t =$ the value of t with $n - 1$ degrees of freedom. For this example, the 95 percent confidence interval would be

$$0.77 \pm 2.262\,(0.02332)$$
$$0.77 \pm 0.05275$$

or $\qquad\qquad 0.71725$ to 0.82275

8-21   Relative Efficiencies of Sampling Plans   In controlling the intensity of various sampling plans, one must fix (1) the sample size, (2) the sampling variance, or (3) the cost. The best sampling design for a given estimation problem is one which provides the desired precision (in terms of confidence limits on the estimate) for the lowest cost. Or if the cost itself is fixed in advance of sampling, the objective is to obtain an estimate of the greatest precision for the funds available.

The relative efficiency of alternative procedures or various sampling plans may, therefore, be calculated from the elements of cost and precision. If the costs required to achieve the *same* level of precision (sample variance) are known for plans A and B, then the relative efficiency of the two plans may be computed as a ratio of the two expenditures. For example, if plan A cost $800 and plan B achieved the same level of precision for $600, then the relative efficiency would be calculated as $800 \div 600 = 1.33$; i.e., plan B is 1.33 times as efficient as plan A.

For those situations where alternative plans do not result in the same level of precision, an index of efficiency may be computed as the product of the squared standard error and the cruising time (or expenditure) required. To illustrate, the following results of five inventories of the same tract may be considered:

Inventory plan	Standard error of mean, bd ft	Time, hr	(Standard error)^2 × time	Efficiency (rank)
A	215.6	10	464,834	3
B	192.5	11	407,619	2
C	224.0	8	401,408	1
D	316.8	6	602,173	5
E	267.8	8	573,735	4

In this example, inventory plan C proved most efficient (rank 1) because of the lowest product of squared standard error and time. Although cruise B was more precise, it ranked second in overall efficiency because of the larger time factor. If this method of ranking timber cruises appears to penalize estimates with large standard errors, it should be remembered that halving a standard error requires not merely twice as many sampling units but *4 times as many*. On the other hand, time is regarded as a linear variable, because reducing time by one-half is expected to lower cruising costs proportionately.

PROBLEMS

8-1 Following is a map of a forested tract. Each number represents the volume in cords per acre on a 1-acre plot.

	Column									
Row	1	2	3	4	5	6	7	8	9	10
1	10	15	14	34	41	42	39	31	47	46
2	15	17	22	27	46	19	48	42	39	60
3	10	9	17	29	34	28	36	41	35	40
4	11	14	13	42	48	39	41	34	39	39
5	13	19	20	48	40	49	44	31	47	25
6	15	20	23	35	37	27	39	47	19	49
7	14	19	18	31	45	38	40	44	54	65
8	10	18	22	38	31	42	48	33	36	57
9	9	17	21	41	25	36	37	25	19	59
10	13	15	18	29	43	41	40	32	15	35

Stratum 1	Stratum 2	Stratum 3
Pine plantation	Mixed pine-hardwood	Bottomland hardwood

a Use the random numbers table in the Appendix to obtain a simple random sample (without replacement) of size 10 from the 100-acre tract. Estimate the mean volume per acre and the standard error of the mean, and compute 95 percent confidence intervals for the mean.

b For a total sample of size 10 ($n = 10$), compute the appropriate sample sizes by strata for stratified random sampling by using proportional allocation. Draw the stratified random sample without replacement, estimate the overall population mean volume per acre, compute an estimate of the standard error of the mean, and place 95 percent confidence intervals on the mean.

c Past experience indicated that the standard deviation for 1-acre plots for these three strata should be approximately as follows: $S_1 = 5$, $S_2 = 7$, $S_3 = 14$. Using this prior information on standard deviations, compute the number of samples to be drawn from each stratum for a total sample size of 10. Select the sample without replacement, estimate the overall mean volume per acre, estimate the standard error of the mean, and compute 95 percent confidence intervals for the mean.

8-2 A forest scientist wishes to estimate the average leaf area on a plant. The relation between leaf area and leaf weight can be used to increase precision of the area estimate. Since the weight of individual leaves and the total weight and number of leaves on the plant can be determined quickly and inexpensively, this suggests use of regression estimates. On a plant the *true mean* (μ_x) weight per leaf was determined to be 2.5 g. Following is a sample of leaves which were weighed and for which leaf area was determined:

x, weight, g	y, area, cm^2
2.0	161
2.5	193
1.7	129
2.1	174
3.0	232

    a Use a linear regression estimator to obtain an estimate of the mean area per leaf.

    b Calculate the standard error of the estimated mean area per leaf. (Total number of leaves on the plant = 50; that is, $N = 50$.)

8-3 Given the following sample data

x	y
5	20
6	18
8	21
9	30
10	38
12	45

and given μ_x (true mean of x) = 8.5,

    a Estimate the population mean of y using the ratio-of-means approach.

    b Estimate the population mean of y using the mean-of-ratios approach.

8-4 An estimate of the mean volume (with associated measures of precision) is desired for a large timbered tract. Aerial photographs of the area are available. It is inexpensive to measure volume per acre on aerial photo plots but expensive to measure volume per acre by locating field plots on the ground. Because we expect a high correlation between volume measured on aerial photo plots and volume measured at the same point on the ground, a double sample is suggested. A large sample of aerial photo volume plots is established. The large sample data follow:

$$n_1 = \text{number of observations in large sample} = 250$$
$$N = \text{number of sample units in the population} = 2000$$
$$\bar{x}_1 = \text{large sample mean of } x = 1800 \text{ ft}^3/\text{acre}$$

From the large sample photo points, 12 points are selected at random. Each of these 12 points is visited on the ground, and the ground volume per acre is observed. The small-sample data are shown below:

x, photo estimate, ft^3/acre	y, ground measured, ft^3/acre
700	900
1700	2400
1000	1450
3400	3700
1300	1900
1400	1500
3000	3700
3000	3200
2200	2300
1700	1900
2500	2800
2700	3300

a Plot the 12 pairs of observations. Does the variability of y appear to be approximately the same at all values of x? Does the relationship of y to x appear to be linear?

b Apply the linear regression estimate in a double-sampling context to estimate the mean volume per acre.

c Estimate the standard error of the mean from part b. Place 95 percent confidence bounds on the mean.

d Assume that the values for y were obtained in a simple random sample of size 12. Estimate the mean volume per acre, estimate the standard error of the mean, and compute the 95 percent confidence interval for the mean. Compare these estimates (realizing that if part of the sampling resources were not devoted to the aerial photo plot measurements, the simple random sample size could be increased slightly) with those from the double sample.

8-5 A nursery manager wants to estimate the average height of seedlings in the nursery beds. Because the beds were sowed at uniform spacing and because the germination percentage of seeds was high, it is possible to divide the beds into 500 equal-sized plots of 50 seedlings each. From these 500 plots (primaries), 10 are selected at random, and the height is measured on 5 randomly selected seedlings (secondaries) in each plot. The data follow:

Plot	Height of seedlings, in.
1	12, 11, 12, 10, 13
2	10, 9, 7, 9, 8
3	6, 5, 7, 5, 6
4	7, 8, 7, 7, 6
5	10, 11, 13, 12, 11
6	14, 15, 13, 12, 13
7	8, 6, 7, 8, 7
8	9, 10, 8, 9, 9
9	7, 10, 8, 9, 9
10	12, 11, 12, 13, 12

a Estimate the average height of seedlings in the nursery beds.

b Compute the standard error of the mean from part a, and place 95 percent confidence bounds on the mean.

8-6 Draw a simple random sample of at least 30 observations from a population in your field of interest. Then, (a) place 95 percent confidence limits on the sample mean, and (b) compute the total number of sample units that would be required to estimate the population mean within ±5 percent at a confidence probability of 90 percent. (*Note:* Remember to apply the finite population correction, if applicable.)

8-7 Select a local tract of land that can be subdivided into three or more strata according to vegetation, soil types, or timber volumes. Then design ground inventories to estimate some population parameter by (a) systematic sampling, (b) simple random sampling, and (c) an optimum allocation of field plots based on stratified random sampling. Use the *same number and size* of field plots for all three sampling designs. Compare results and relative efficiencies of the three systems.

REFERENCES

Barrett, J. P., and Nutt, M. E. 1979. *Survey sampling in the environmental sciences: A computer approach.* COM-Press, Inc., Wentworth, N.H. 319 pp.

Cochran, W. G. 1977. *Sampling techniques.* 3d ed. John Wiley & Sons, New York. 428 pp.

Freese, F. 1962. Elementary forest sampling. *U.S. Dept. Agr. Handbook* 232, Government Printing Office, Washington, D.C. 91 pp.

Hamilton, D. A., Jr. 1979. Setting precision for resource inventories: The manager and the mensurationist. *J. Forestry* **77**:667–670.

MacLean, C. D. 1972. Improving inventory volume estimates by double sampling on aerial photographs. *J. Forestry* **70**:748–749.

Mesavage, C., and Grosenbaugh, L. R. 1956. Efficiency of several cruising designs on small tracts in north Arkansas. *J. Forestry* **54**:569–576.

Philip, M. S. 1983. *Measuring trees and forests.* Aberdeen University Press, Aberdeen, Scotland. 338 pp.

Scheaffer, R. L., Mendenhall, W., and Ott, L. 1990. *Elementary survey sampling.* 4th ed. PWS-Kent Publishing Company, Boston. 390 pp.

Stuart, A. 1984. *The ideas of sampling.* Statistical Monograph No. 4. Charles Griffin and Company, Ltd., High Wycombe, England. 91 pp.

Vries, P. G. de. 1986. *Sampling theory for forest inventory.* Springer-Verlag, New York. 399 pp.

Wenger, K. F. (ed.). 1984. *Forestry handbook.* 2d ed. John Wiley & Sons, New York. 1335 pp.

Yandle, D. O., and Wiant, H. V., Jr. 1981. Comparison of fixed-radius circular plot sampling with simple random sampling. *Forest Sci.* **27**:245–252.

Zeide, B. 1980. Plot size optimization. *Forest Sci.* **26**:251–257.

FOREST INVENTORY

9-1 Introduction The usual purpose of a timber inventory is to determine, as precisely as available time and money will permit, the volume (or value) of standing trees in a given area. To attain this objective requires (1) a reliable estimate of the forest area and (2) measurement of all or an unbiased sample of trees within this area. No reliable timber inventory can be planned until the forester knows the location of all tract corners and boundary lines; recent aerial photographs and maps are therefore genuine assets for working in unfamiliar terrain.

The choice of a particular inventory system, often made at the forester's discretion, is governed by relative costs, size and density of timber, area to be covered, precision desired, number of people available for fieldwork, and length of time allowed for the estimate. Other things being equal, the intensity of sampling tends to increase as the size of the tract decreases and as the value of the timber increases.

9-2 Classes of Timber Surveys The organization, intensity, and precision required in a timber inventory are logically based on the planned use of information collected. Therefore, no work should be initiated until inventory objectives have been clearly outlined and the exact format of summary forms to be compiled is known.

Depending on primary objectives, timber surveys may be conveniently classified as (1) land acquisition inventories, (2) inventories for logging or timber sales, (3) management plan inventories or continuous forest inventory systems, and (4) special surveys designed for evaluating conditions such as stand improvement needs, plantable areas, insect and disease infestations, or timber trespass.

For land acquisition or timber sales, the principal information desired is net volume and value of merchantable trees growing in operable areas. In simple terms, a stand is usually classed as "operable" when merchantable trees can be logged at a profit. Notations on timber quality, by species, are commonly required. For land acquisition surveys, additional information is needed on soil or site quality, presence of nonmerchantable growing stock, and proximity of the tract to mills or primary markets. Where timber values are relatively high, acquisition or sale inventories should be of an intensity that will produce estimates of mean volume within ±10 percent or less.

Management plan cruises, designed for providing information on timber growth, yield, and allowable cut, are no longer considered essential in all regions. In many instances, such cruises have been replaced by systems that make use of permanent sample plots. As a rule, either type of cruise is of low intensity, and the information collected is primarily intended for broad-based management decisions and long-range planning. As a result, inventory data are summarized by large administrative units rather than by cutting compartments or logging units.

Special surveys are so diversified that few general rules can be stipulated. For locating spot insect or disease infestations, a survey might merely consist of an accurate forest type map with "trouble areas" located visually from aerial observations. Similarly, understocked stands in need of planting might be mapped from existing aerial photographs. In other instances, a 100 percent tree tally might be made for determining the number and volume of trees suitable for poles, piling, or veneer logs. Special surveys are also required in timber trespass cases. The estimation of timber volumes removed from cutover areas is discussed in a later section of this chapter.

9-3   Inventory Planning   Regardless of the kind of inventory being undertaken, a carefully developed plan is needed to execute the inventory efficiently. Inventory planning may be informally developed and executed by the forester, in the case of many timber cruises, or thoroughly documented in a formal plan, in the case of regional or national assessments.

The following checklist includes the major items that should be considered in planning a forest inventory. All items do not apply to all types of inventories, and the checklist is applicable to inventory of timber. If other resources are to be assessed at the same time as the timber assessment, the checklist should be expanded appropriately.

1 Purpose of the inventory
   a Why the inventory is required
   b How the resulting information will be used

2 Background information
  a Past surveys, reports, maps, photographs
  b Personnel, funding, or time constraints
3 Description of area to be inventoried
  a Location
  b Size
  c Terrain, accessibility, and transportation factors
4 Information required from inventory
  a Tables and graphs
  b Maps
  c Outline of narrative report
5 Inventory design
  a Identification of the sampling unit
  b Construction of the sampling frame
  c Selection of a sampling technique
  d Determination of the sample size
6 Measurement procedures
  a Location of sampling units
  b Establishment of sampling units
  c Measurements of sampling units
    i Instruments and instructions for use
    ii Tree and other plot measurements
  d Recording of field data
  e Supervision and quality control
7 Compilation and calculation procedures
  a Data editing
  b Relationships to convert observed or measured attributes to quantities of
    interest (e.g., tree-volume tables)
  c Formulas for estimating means, totals, and corresponding standard errors
  d Description of calculation methods
8 Reporting of results, maintenance and storage of records
  a Number and distribution of copies
  b Storage and retrieval of data
  c Plans for updating inventory

9-4 Forest Inventory and Analysis Forest Inventory and Analysis (FIA), formerly called the U.S. Forest Survey, is a periodic assessment of the nation's forest and range resources. State-by-state inventories, originally authorized by the McSweeney-McNary Forest Research Act of 1928, provide the basis for these periodic assessments. The initial forest inventories were begun around 1930. Reinventories, normally planned at intervals of 8 to 15 years, have been completed on much of the 731 million acres (296 million hectares) of forest land in the United States. The Renewable Resources Planning Act of 1974, the National Forest Management Act of 1976, and the Forest and Rangeland

Renewable Resources Research Act of 1978 amended and broadened the assessment provision of the original act to include 820 million acres (332 million hectares) of range lands.

FIA is essentially a continuous inventory system based on permanent sample units that are sometimes supplemented by temporary sample units. FIA work units located at regional experiment stations of the Forest Service, USDA, have primary responsibility for the fieldwork, data analysis, and published results. At the conclusion of fieldwork in each state, data are collated and summarized by electronic data processing equipment; results are publicized by FIA reports describing the current supply of and demand for renewable resources in that state. Statistical and analytical reports for the various states are available from USFS experiment stations.

The primary objectives of FIA are to provide the resource data needed for national assessments of this nation's forest and range resources. Every 10 years, FIA information from the individual states is summarized and incorporated into a national assessment. Each assessment includes a determination of the present and potential productivity of the land, and of such other facts as may be necessary and useful in the determination of ways and means to balance the demand for and supply of these renewable resources. Assessments encompass all ownerships of forest and range lands and all benefits derived from uses of these lands. In turn, each assessment serves as the basis for development of a program for the nation's renewable resources. Each program is intended to implement those actions deemed necessary to assure an adequate supply of forest and range resources in the future while maintaining the integrity and quality of the environment.

SPECIAL INVENTORY CONSIDERATIONS

9-5 Tree Tallies In accordance with the tree volume or tree weight tables to be used, standing trees may be tallied by simple counts, by diameter at breast height (dbh) and species only, or by various combinations of dbh, species, merchantable height, total height, form, individual tree-quality classes, and so on. Merchantable height limitations, and tree form expressions have been described in Chap. 6.

Neophyte foresters should be particularly careful in estimating tree heights; upper limits of stem merchantability often change from one species or locality to another. When ocular estimates of tree heights are permitted, the conscientious forester will nevertheless "check his eye" by *measuring* every tenth or twentieth tree. Proficiency and consistency in inventory work are dependent on constant checks of estimation techniques.

It is essential that neat, concise, and accurate records be maintained when trees are measured and tallied in the field. Foresters have generally adopted the dot-dash system for indicating the number of trees tallied. The first four tallies are made by forming a small square with four pencil dots; the next four tallies are indicated by drawing successive lines between the dots to make a completed square; and the ninth and tenth tallies are denoted by diagonals placed within the square. A simplified tally form illustrating this technique is shown in Fig. 9-1.

Portable electronic data recorders are now commonly used. However, when paper tally sheets are used, data should be recorded in pencil, because inked recordings tend to smear and become illegible when record sheets get wet. Erasures can be avoided by circling erroneous tallies; partial erasures often result in confusion and lead to later errors in office computations, particularly when several different persons are required to decipher field tabulations. Tally sheets, including pertinent locational headings, should be filled out completely *in the field*—not several hours later back at headquarters. Organization is just as important in field record keeping as it is in office bookkeeping.

9-6 Tree-Defect Estimation The ability to make proper allowance for defective trees encountered on timber inventories requires experience that can be gained only by (1) getting repeated practice in estimating standing tree defects and (2) observing the sawing and utilization of defective logs at various mills.

When entire trees are classed as culls, they are either omitted from the field tally or recorded by species and dbh in a separate column of the tally sheet. For merchantable trees with evidence of interior defects, deductions for unsound portions of the stem may be handled by one of the following techniques:

FIGURE 9-1
Sample tally of standing trees by dbh and height classes.

dbh (INCHES)	TREE HEIGHT CLASSES (FEET)					TREE TOTALS
	20	40	60	80	100	
10	• • •					3
12	• •	• • • •				6
14	•	• • •	⊓ •	• •		13
16		• •	⊠ • • •	⊠	• •	27
18		•	• • • •	• • • •	⊡	17
20			•	⊓ • •	⊿	16
TREE TOTALS	6	10	24	23	19	82

DOT-DASH TALLY METHOD

1 For *visible defects,* dimensions of tallied trees are reduced in proportion to the estimated amount of defect. Thus a 22-in.-dbh three-log tree might be recorded as an 18-in.-dbh tree with three logs or possibly as a 22-in.-dbh tree with $2^1/_2$ logs. Refinements may be made in this technique by applying the defect-deduction formulas for log volumes as suggested by Grosenbaugh (Sec. 5-13).

2 For *hidden defects,* all trees are tallied in the field as sound. After gross volumes have been computed, a percentage is deducted in proportion to the total amount of timber presumed to be defective. Although this method will usually produce more consistent results than individual tree allowances, the drawback is the difficulty of deciding on the amount of the deduction to be applied to various species.

A promising technique has been developed by Aho (1974). From a detailed stem analysis of more than 1000 felled trees, defect-deduction percentages have been derived through multiple regression analysis for grand fir. These regression estimators are based on tree dbh, age, exterior indicators of decay, and infection courts. It is probable that similar regression equations can be developed for other timber species of commercial importance.

9-7 The Complete Tree Tally Under limited circumstances when scattered, high-value trees occur on small tracts, a complete or 100 percent tree tally may be feasible. Every tree of the desired species and size class may be measured, or the tally may consist of a 100 percent *count* of all stems plus a subsample (every *n*th tree) of actual measurements. The choice of methods depends on the stumpage value of trees inventoried, allowable costs, and desired precision.

Advantages of the complete tree tally are as follows:

1 More accurate estimates of total volume are possible, because every tree can be tallied by species, dbh, height, and quality class.

2 Deductions for defect can be accurately assessed, because cull percentages can be applied to individual trees as they are tallied.

3 It is not necessary to determine the exact area of the tract. Once boundaries have been located, the total estimate can be made without regard to area.

Disadvantages of the complete tree tally are:

1 High costs. Because of expense and time required, the 100 percent inventory is usually limited to small tracts or to individual trees of extra-high stumpage value.

2 Trees must often be marked as they are recorded to avoid omissions or duplications in the field tally. This requires additional time and/or added personnel.

9-8 Organizing the Complete Tree Tally For dense stands of timber with large numbers of trees, it is desirable to have three persons in the field party. Two carry tree calipers for quick dbh measurements, while the third serves as recorder. If the area exceeds 10 acres in size, it is helpful to first lay out rectangles of about 4 × 10 chains by using stout cord or twine. Then, depending on topography and underbrush, parallel strips 1 to 2 chains wide can be traversed through each rectangle.

Fieldwork in dense stands proceeds most efficiently when it is feasible to merely count merchantable trees and restrict actual measurements to every tenth or twentieth stem. For pure stands that require little or no cull deductions, an alternative procedure might employ caliper measurements of dbh only for all stems, with volumes derived from single-entry volume tables. To ensure that no trees are overlooked or tallied twice, each stem should be marked at eye level on the side facing the unmeasured portion of the stand. In deciduous forests, complete stem tallies are preferably made during the dormant season when trees are leafless.

9-9 Timber Inventory as a Sampling Process Except for those circumstances where a complete tree tally is justified, the conduct of a timber inventory, or "cruise," is a sampling process. Among the considerations involved in developing an efficient sampling scheme are sample size, plot size and shape, and the sampling design, e.g., whether systematic, simple random, stratified random, etc. Many of these considerations were discussed in Chap. 8.

Regardless of inventory objectives, the method of selecting sample trees for measurement is based on the concept of sampling probability. The methods of concern are (1) probability proportional to frequency (Chap. 10), (2) probability proportional to size, or point sampling (Chap. 11), and (3) probability proportional to prediction, or 3P sampling (Chap. 12).

SUMMARIES OF CRUISE DATA

9-10 Stand and Stock Tables Although total stand volume is a major objective of most forest inventories, such information is most useful when it is summarized by tree sizes and species groups. It is important to know that a given stand contains 1 million bd ft of timber, but it is more valuable to know how this volume is distributed among various species groups and diameter classes. Thus the compilation of stand and stock tables is often a prime requisite in summarizing cruising results.

A *stand table* is a tabulation of the total *number* of stems (or average number of stems per acre) in a stand or compartment, by dbh classes and species. A *stock table* lists the total *volume* of stems (or average volume per acre) in a stand, by dbh classes and species. As stock tables are derived from stand tables, they are sometimes combined into the same tabulation as shown by Table 9-1. This summary for mixed hardwood species includes oaks, gums, yellow-poplar, ash, and sycamore on a tract of 107.2 acres. Total numbers of trees and volumes are shown as well as per acre averages.

TABLE 9-1
COMBINED STAND AND STOCK TABLE FOR MIXED HARDWOODS
ON A TRACT OF 107.2 ACRES IN GREENE COUNTY, GEORGIA

dbh, in.	No. of trees		Cubic-foot volumes*	
	Tract total	Per acre	Tract total	Per acre
5	1195.5	11.2	1,730.4	16.1
6	1432.6	13.4	3,767.6	35.1
7	1455.1	13.6	5,667.8	52.9
8	1150.6	10.7	6,294.4	58.7
9	1128.0	10.5	8,379.0	78.2
10	1082.9	10.1	10,622.4	99.1
11	823.4	7.7	11,142.7	103.9
Total	8268.1	77.2	47,604.3	444.0

*Cubic-foot volumes are inside bark of the merchantable stem to a variable top diameter not smaller than 3 in.

9-11   Timber Volumes from Stump Diameters   In timber trespass cases, it may be necessary to determine the volume of trees illegally removed during a clandestine logging operation. Since stem diameters cannot be measured at breast height, they must be estimated, by species or species groups, from available stump diameters. The conversion may be estimated by use of regression equations, or ratios may be derived "on location" from sample measurements of trees left standing.

As an example, correlations between stump diameter (inside bark) and dbh (outside bark) have been established for Black Hills ponderosa pine by measurements of 503 felled trees (Van Deusen, 1975). The linear regression is

$$\text{dbh} = 0.8018 + 0.9267 \text{ (stump dib)}$$

The coefficient of determination (Sec. 2-22) is 0.99 for this relationship, and the standard error of estimate is ±0.85 in.

Once each dbh has been ascertained, volumes may be determined from single-entry volume tables. Or if tree tops have not been scattered, lengths of removed merchantable stems may be obtained by measuring distances between paired stumps and tops. With this additional information, volumes can be derived from multiple-entry tables. When the cutover area is too large for a 100 percent stump tally, partial estimates based on sample plots may be used as in conventional inventories. The final volume summary should be accompanied by an appraisal of the stumpage value of timber removed, along with notes and photographs documenting damage to real property or to residual standing trees.

SALES OF STANDING TIMBER

9-12 Stumpage Value   The sale value of standing timber is known as its stumpage value. For a given species, size (volume), and quality of timber, stumpage prices are highest when trees are accessible (easily logged) and located near concentration yards or primary markets. If a forest owner participates in harvesting, his income from the enterprise is increased in accordance with the value added by cutting, loading, and hauling to wood dealers or directly to mills. To determine *which* trees to sell and when to sell them, foresters must be intimately acquainted with local markets and prevailing prices, and they should learn to anticipate seasonal or periodic fluctuations in demands for various types of stumpage. Only by becoming thoroughly familiar with specifications for various wood products can the forest manager expect to realize consistently high returns from timber sales.

9-13 Methods of Selling Standing Timber   For handling sales of stumpage, trees to be cut (or those to be left standing) should be marked, or a cruise should be made by species and product designations. High-value trees, such as those to be utilized for veneer logs or poles, should be logged first. This cutting may be followed by removal of sawlogs, specialty bolts and billets, and tie logs. Finally, residual tops and smaller trees may then be converted into pulpwood, round mine timbers, fence posts, or fuel wood. Relative values of these products are dictated by local markets, and failure to observe rational priorities in cutting operations may severely penalize the seller of standing timber.

If the forest owner has made a reliable inventory or if timber is to be removed by clearcutting, sales may be made on a lump-sum basis. As with all business transactions, this is a reasonable approach provided both buyer and seller are well informed as to market values and volume of wood involved. In other instances, sale prices may be based on marked trees, on minimum diameter or merchantability limits, or on log scale as determined after trees are cut and skidded to a landing.

9-14 Timber-Sale Contracts   For most sales of standing timber, even for small tracts of low-value species, it is desirable to draw up a simple written agreement to protect both buyer and seller and to avoid unnecessary misunderstandings in the transaction. A timber-sale contract contains sections on (1) location and description of timber, (2) prices and terms of payment, (3) utilization standards and related conditions of timber removal, and (4) procedures for settling disputes. The sample contract that follows, intended primarily for farmers and small woodlot owners, is reproduced from a bulletin of the USDA.

SAMPLE TIMBER-SALE AGREEMENT

_____, of _____, _____,
(I or we) (Name of Purchaser)   (Post Office)   (State)
hereinafter called the purchaser, agree to purchase from _____ of
(Seller's Name)
_____, _____, hereinafter called the seller, the designated
(Post Office)   (State)
trees from the area described below.

I DESCRIPTION OF SALE AREA:
(Describe by legal subdivisions, if surveyed)

II   TREES DESIGNATED FOR CUTTING: (Cross out A or B—use only one clause)

A   All _____ trees marked by the seller, or his agent, with paint spots
(species)
below stump height; also dead trees of the same species which are
merchantable for _____.
(Kind of forest products)

B   All _____ trees merchantable for _____
(species)   (Kind of forest products)
which measure _____ inches or more outside the bark at a point
not less than 6 inches above the ground, also other _____ trees
(species)
marked with paint spots below stump height by the seller or his agent.

III CONDITIONS OF SALE:

A The purchaser agrees to the following:

1 To pay the seller the sum of $_____ for the above-described
trees and to make payments in advance of cutting in amounts of at
least $_____ each.

2 To waive all claim to the above-described trees unless they are cut
and removed on or before _____.
(Date)

3 To do all in his power to prevent and suppress forest fires on or
threatening the Sale Area.

4 To protect from unnecessary injury young growth and other trees
not designated for cutting.

5 To pay the seller for undesignated trees cut or injured through
carelessness at the rate of $_____ each for trees measuring
10 to _____ inches in diameter at stump height and $_____
each for trees _____ inches or over in diameter.

      6 To repair damage caused by logging to ditches, fences, bridges, roads, trails or other improvements damaged beyond ordinary wear and tear.

      7 Not to assign this agreement in whole or in part without the written consent of the seller.

  B The seller agrees to the following:

      1 To guarantee title to the forest products covered by this agreement and to defend it against all claims at his expense.

      2 To allow the purchaser to use unmerchantable material from tops of trees cut or from trees of _____ species for necessary logging improvements free of charge, provided such improvements are left in place by the purchaser.

      3 To grant the freedom of entry and right-of-way to the purchaser and his employees on and across the area covered by this agreement and also other privileges usually extended to purchasers of stumpage which are not specifically covered, provided they do not conflict with specific provisions of this agreement.

  C In case of dispute over the terms of this agreement we agree to accept the decision of an arbitration board of three selected persons as final. Each of the contracting parties will select one person and the two selected will select a third to form this board.

Signed in duplicate this _____ day of _____ 19__

_____	_____
(Witness)	(Purchaser)
_____	_____
(Witness)	(Seller)

PROBLEMS

9-1 Prepare a written report on a compartment inventory system used by a local forest industry.

9-2 Present an oral report on permanent plot remeasurement and field-tally procedures for an industrial CFI system.

9-3 For oral presentation, prepare a 20-minute review of the most recent U.S. Forest Service report that deals with the timber resources of your state.

9-4 The following data were collected by measuring dbh and stump diameter (inside bark) on 10 red oak trees:

dbh, in.	Stump, dib, in.	dbh, in.	Stump, dib, in.
14.1	17.2	7.9	9.1
10.5	11.3	10.4	12.3
9.0	10.0	12.8	14.2
16.4	18.6	17.0	20.0
7.1	7.5	13.2	16.0

a Compute a simple linear regression, $Y = b_0 + b_1 X$, where $Y = $ dbh in inches and $X = $ stump dib in inches for estimating tree dbh from measured stump diameter.

b Compute the coefficient of determination (r^2) and the standard error of estimate for the regression equation fitted in part a.

c Do the measures of goodness of fit computed in part b indicate that dbh can be reliably estimated from measured stump diameter?

9-5 The following tree tally was recorded on 20 $^1/_5$-acre plots established in a 50-acre stand of yellow-poplar:

dbh, in.	Number of trees on 20 plots	dbh, in.	Number of trees on 20 plots
8	50	12	130
9	95	13	100
10	105	14	70
11	110	15	40

Using the following local volume equation for yellow-poplar (Sec. 7-11): $V = -8.4166 + 0.2679$ dbh^2, where V is cubic-foot volume (outside bark) to a 4-in. top dob and dbh is in inches, compile a combined stand and stock table in the same format as that of Table 9-1.

REFERENCES

Aho, P. E. 1974. Defect estimation for grand fir in the Blue Mountains of Oregon and Washington. *U.S. Forest Serv., Pacific Northwest Forest and Range Expt. Sta., Res. Paper* PNW-175. 12 pp.

Bylin, C. V. 1982. Estimating dbh from stump diameter for 15 southern species. *U.S. Forest Serv., Southern Forest Expt. Sta., Res. Note* SO-286. 3 pp.

Hair, D. 1973. The nature and use of comprehensive timber appraisals. *J. Forestry* **71:**565–567.

Hamilton, R. 1977. Timber sale contracts. Cooperative Extension Serv., Univ. of Nebraska, Lincoln, Neb. 4 pp.

Husch, B. 1971. *Planning a forest inventory.* F.A.O., Rome, Italy. 120 pp.

Johnson, E. W. 1976. Information needed and the design of an inventory. *U.S. Forest Serv., N.E. Area, State and Private Forestry, Res. Inventory Note* 6. 4 pp.

McClure, J. P., Cost, N. D., and Knight, H. A. 1979. Multiresource inventories—A new concept for forest survey. *U.S. Forest Serv., Southeast. Forest Expt. Sta. Res. Paper* SE-191. 68 pp.

Rennolls, K. 1989. Design of the census of woodlands and trees 1979–82. *Forestry Commission, Occasional Paper* 18, Edinburgh, Scotland. 36 pp.

Schlieter, J. A. 1986. Estimation of diameter at breast height from stump diameter for lodgepole pine. *U.S. Forest Serv., Intermountain Research Sta., Res. Note* INT-359. 4 pp.

Stage, A. R., and Alley, J. R. 1972. An inventory design using stand examinations for planning and programming timber management. *U.S. Forest Serv., Intermountain Forest Expt. Sta., Res. Paper* INT-126. 17 pp.

U.S. Department of Agriculture. 1958. Measuring and marketing farm timber. *Farmers' Bull.* 1210, U.S. Forest Service, Washington, D.C. 33 pp.

————. 1982. Forest Service resource inventory: An overview. *U.S. Forest Serv., Forest Resources Economics Research Staff,* Washington, D.C. 22 pp.

Van Deusen, J. L. 1975. Estimating breast height diameters from stump diameters for Black Hills ponderosa pine. *U.S. Forest Serv., Rocky Mt. Forest and Range Expt. Sta., Res. Note* RM-283. 3 pp.

Warton, E. H. 1984. Predicting diameter at breast height from stump diameters for northeastern tree species. *U.S. Forest Serv., Northeast. Forest Expt. Sta., Res. Note* NE-322. 4 pp.

CHAPTER 10

INVENTORIES WITH SAMPLE
STRIPS OR PLOTS

10-1 Fixed-Area Sample Units Many forest inventories are carried out using fixed-area sample units. These fixed-area sample units are called *strips* or *plots,* depending on their dimensions. Sample plots can be any shape (square, rectangular, circular, or triangular); however, square- and circular-plot shapes are most commonly employed. A strip can be thought of as a rectangular plot whose length is many times its width.

When employing sample plots or strips, the likelihood of selecting trees of a given size for measurement is dependent on the *frequency* with which that tree size occurs in the stand. That is, strip and plot inventories are methods of selecting sample trees with *probability proportional to frequency.* Within the sample area defined by the strips or plots, individual trees are tallied in terms of the characteristics to be assessed, such as species, dbh, and height. Then the sample-area tallies are expanded to a per-unit-area basis by applying an appropriate expansion factor.

STRIP SYSTEM OF CRUISING

10-2 Strip-Cruise Layout With this system, sample areas take the form of continuous strips of uniform width which are established through the forest at equally spaced intervals, such as 5, 10, or 20 chains. The sample strip itself is usually 1 chain wide, although it may be narrowed to $1/2$ chain in dense stands of young timber or increased to 2 chains and wider in scattered, old-growth saw-timber. Strips are commonly run straight through the tract in a north-south or

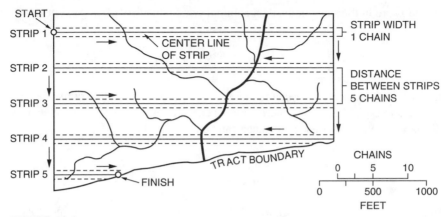

FIGURE 10-1
Diagrammatic plan for a 20 percent systematic strip cruise. Sample strips 1 chain wide are spaced at regular intervals of 5 chains.

east-west direction, preferably oriented to cross topography and drainage at right angles (Fig. 10-1). By this technique, all soil types and timber conditions from ridge top to valley floor are theoretically intersected to provide a representative sample tally.

Strip cruises are usually organized to sample a predetermined percentage of the forest area. One-chain sample strips spaced 10 chains apart provide a nominal 10 percent estimate, and $1/2$-chain strips at 20-chain intervals produce a nominal $2^{1}/_{2}$ percent cruise (Table 10-1). The conversion factor to expand sample volume to total volume may be derived by (1) dividing the cruising percentage into 100 or (2) dividing the total tract acreage by the number of acres sampled.

The computation of cruise intensity and expansion factor can be expressed in formula form. If W = strip width, D = distance between strip centerlines, and

TABLE 10-1
EXAMPLES OF CRUISING INTENSITIES FOR
1-CHAIN SAMPLE STRIP WIDTHS

Distance between strip centerlines,		No. of strips per "forty"	Nominal cruise percent
ft	chains		
1320	20	1	5
660	10	2	10
330	5	4	20
165	$2^{1}/_{2}$	8	40

W and D are in the same units, then nominal cruise intensity (I) in percent equals

$$I = \frac{W}{D}(100)$$

It is important to remember that nominal cruise percent and actual cruise percent are seldom equal because timbered tracts generally are not perfectly rectangular in shape. The actual cruise percentage can be calculated as

$$\left(\frac{\text{Area in sample}}{\text{Total tract area}}\right)100$$

In this calculation, area in sample and total tract area must be in the same units.

To convert sample volume to total tract volume, one computes the expansion, or blow-up, factor (EF) as

$$\text{EF} = \frac{100}{\text{cruise percent}}$$

In the computation of the expansion factor, the cruise percent should be the actual, not the nominal, percent. Alternatively, the expansion factor can be computed as total tract area/area in sample. The estimate of total volume for the entire tract is obtained by multiplying volume tallied in all the sample strips times the expansion factor.

10-3 Computing Tract Acreage from Sample Strips   If the boundaries of a tract are well established, but the total area is unknown, a fixed cruising percentage may be decided upon, and the tract area can be estimated from the total chainage of strips composing the sample. A 5 percent cruise utilizing strips 1 chain wide spaced at 20-chain intervals provides a good example. The centerline of the first sample strip is offset 10 chains from one corner of the tract (i.e., one-half the planned interval between lines), and parallel strips are alternately run 20 chains apart until the entire area has been traversed by a pattern similar to that shown in Fig. 10-1. If 132 lineal chains of sample strips are required, the area sampled is $(132 \times 1)/10 = 13.2$ acres. Because the strips were spaced for a 5 percent estimate, the total tract area is approximately 20×13.2, or 264 acres. The expansion factor of 20 is also used to convert the sampled timber volume to total tract volume.

When trees are tallied according to forest types and acreages are desired for each type encountered, the preceding technique may also be used to develop these breakdowns. If the 132 lineal chains of strip were made up of 90 chains intersecting a coniferous type and 42 chains intersecting a hardwood type, sampled areas would be 9 and 4.2 acres, respectively. Applying the expansion factor of 20 would result in estimated areas of 180 acres for conifers and 84 acres for hardwoods. Although this procedure does not necessarily provide exact values, it generally gives a reasonably good indication of the relative proportions by types.

10-4   Field Procedure for Strip Cruising   Accurate determination of strip lengths and centerlines on the ground requires that distances be chained rather than paced; thus a two-person crew is needed for reliable fieldwork. One person locates the centerline with a hand or staff compass and also serves as head chainman; the other cruises the timber on the sample strip and acts as rear chainman. Either person may handle the tree tally, depending on underbrush and density of the timber. The width of the sample strip is ordinarily checked by occasional pacing from the 2-chain tape being dragged along as a moving centerline. Trailer tapes may be used where slope corrections are necessary. When offsetting between strip centerlines, it is important that the distance be carefully measured *perpendicular* to the orientation of the strips. Because many timbered tracts have irregular borders, it is also important to "square off" the ends of strips in order that the strip area can be computed easily as a rectangle.

In an efficient cruising party, the compassman is always 1 to 1½ chains ahead of the cruiser, and the sampling progresses in a smooth, continuous fashion. Experienced cruisers learn to "size up" tree heights well ahead, because there is a tendency toward underestimation when standing directly under a tree. At the end of each cruise line, the strip chainage should be recorded to the nearest link. Strip cruising can be speeded up appreciably by tallying tree diameters only and determining timber volumes from single-entry volume tables.

When timber type maps are prepared as cruising progresses, strips are preferably spaced no more than 10 chains apart. There are few forest stands where the cruiser can map more than 5 chains to either side of the centerline without having to make frequent side checks to verify the trends of type boundaries, streams, trails, or fence lines. The preferred technique for mapping is to sketch cruise lines directly on a recent aerial photograph; approximate type lines and drainage can also be interpreted in advance of fieldwork. Then, during the conduct of fieldwork, type lines can be verified and cover types correctly identified with the photographs in hand.

10-5   Pros and Cons of Strip Cruising   The strip system of cruising is not as popular as in previous years. Its loss of favor is probably due to the fact that two-person crews are needed and volume estimates are difficult to analyze statistically unless the tally is separated every few chains (resulting in a series of contiguous rectangular plots). In addition to items cited previously, the principal advantages claimed for strip cruising are:

1 Sampling is continuous, and less time is wasted in traveling between strips than would be the case for a plot cruise of equal intensity.

2 In comparison with a plot cruise of the same intensity, strips have fewer borderline trees, because the total perimeter of the sample is usually smaller.

3 With two persons working together, there is less risk to personnel in remote or hazardous regions.

Disadvantages of strip cruising are as follows:

1 Errors are easily incurred through inaccurate estimation of strip width. Since the cruiser is constantly walking while tallying, there is little incentive to leave the centerline of the strip to check borderline trees.

2 Unless tree heights are checked at a considerable distance from the bases of trees, they may be easily underestimated.

3 Brush and windfalls are more of a hindrance to the strip cruiser than to the plot cruiser.

4 It is difficult to make spot checks of the cruise results, because the strip centerline is rarely marked on the ground.

LINE-PLOT SYSTEM OF CRUISING

10-6 The Traditional Approach As the name implies, line-plot cruising consists of a systematic tally of timber on a series of plots that are arranged in a rectangular or square grid pattern. Compass lines are established at uniform spacings, and plots of equal area are located at predetermined intervals along these lines. Plots are usually circular in shape, but they may also take the form of squares, rectangles, or triangles. In the United States, $1/4$- and $1/5$-acre circular plots are most commonly employed for sawtimber tallies; smaller plots are preferred for cruising poletimber or sapling stands. For inventories where a wide variety of timber sizes will be encountered, it is often efficient to use concentric circular plots with each centered at the same point. As an example, $1/5$-acre plots might be used to tally sawtimber trees, $1/10$-acre plots for pulpwood trees, and $1/1000$-acre plots for regeneration counts. Radii for circular plots frequently used in timber inventory are given in Table 10-2.

As with the strip method, systematic line-plot inventories are often planned on a percent cruise basis. In Fig. 10-2, for example, $1/5$-acre plots are spaced at intervals of 4 chains on cruise lines that are 5 chains apart. As each plot "represents" an area of 20 square chains, the nominal cruising percentage is computed as

$$\frac{\text{Plot size in acres}}{\text{Acres represented}} \times 100 = \frac{0.2 \text{ acres}}{2 \text{ acres}} \times 100 = 10 \text{ percent}$$

By the same token, 10 percent estimates may also be accomplished by spacing the same $1/5$-acre plots at intervals of $2^1/2 \times 8$ chains, 2×10 chains, and so

TABLE 10-2
RADII FOR SEVERAL SIZES OF CIRCULAR SAMPLE PLOTS

Plot size, ac	Plot radius, ft	Plot size, ha	Plot radius, m
1	117.8	1	56.42
$1/2$	83.3	$1/2$	39.89
$1/4$	58.9	$1/4$	28.21
$1/5$	52.7	$1/5$	25.23
$1/10$	37.2	$1/10$	17.84
$1/20$	26.3	$1/20$	12.62
$1/25$	23.5	$1/25$	11.28
$1/40$	18.6	$1/40$	8.92
$1/50$	16.7	$1/50$	7.98
$1/100$	11.8	$1/100$	5.64
$1/500$	5.3	$1/500$	2.52
$1/1000$	3.7	$1/1000$	1.78

FIGURE 10-2
Diagrammatic plan for a 10 percent systematic line-plot cruise utilizing $1/5$-acre circular sample units.

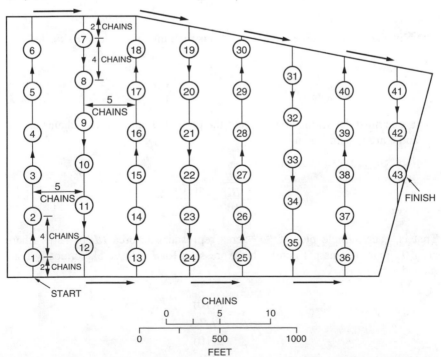

on. If a 1:1 square grid arrangement is desired, the intervals between both plot centers and compass lines would be calculated as $\sqrt{20}$ square chains, or 4.47 × 4.47 chains. Similar computations can be made for other plot sizes and cruising intensities. Cruise expansion factors are calculated by the same methods described for strip cruises.

The number and spacing of plots for line-plot cruises can be expressed in formula form; in order to express the desired relationships algebraically, the following symbols (after Burkhart et al., 1984) are defined

A = total tract area,
A_p = area of all plots,
$P = A_p/A$ = intensity of cruise as a decimal,
a = area of one plot,
n = number of plots,
L = distance between lines,
B = distance between plots on a line,

where A, A_p, and a are all in square units of L and B. When line plot cruises are designed on a percentage basis, P is specified. Assuming that the total area of the tract being inventoried is known, the area of the sample is

$$A_p = AP$$

The plot size (a) is specified in advance, thus the number of plots (n) needed is

$$n = \frac{A_p}{a}$$

Next, one must determine how to space the plots. If B and L are in chains and A, A_p, and a are in acres, then

$$P = \frac{\dfrac{a}{BL}}{10} = \frac{a(10)}{BL}$$

That is, each sample plot of "a" acres represents an area $BL/10$ acres (there are 10 square chains per acre). The expression for P can be algebraically rearranged as:

$$B = \frac{a(10)}{LP}$$

or

$$L = \frac{a(10)}{BP}$$

Thus, if either B or L (as well as P) is specified, the other can be computed readily. For square spacing (that is, $B = L$), we have

$$P = \frac{a(10)}{B^2}$$

and

$$B = \sqrt{\frac{a(10)}{P}}$$

To design a 10 percent line-plot cruise for an 86-acre tract using $\frac{1}{5}$-acre sample plots, the sample area is computed as

$$A_p = AP = 86(0.1) = 8.6 \text{ acres}$$

Next, the number of plots needed is computed as

$$n = \frac{A_p}{a} = \frac{8.6}{0.2} = 43$$

If the distance between lines (L) is specified to be 5 chains, then the distance between plots on lines will be

$$B = \frac{a(10)}{LP} = \frac{0.2(10)}{5(0.1)} = 4 \text{ chains}$$

Figure 10-2 shows a line-plot cruise utilizing $\frac{1}{5}$-acre plots spaced 5 by 4 chains.

10-7 Sampling Intensity and Design   The intensity of plot sampling is governed by the variability of the stand, allowable inventory costs, and desired standards of precision. The coefficient of variation in volume per unit area should first be estimated, either on the basis of existing stand records or by measuring a preliminary field sample of, perhaps, 10 to 30 plots. Then the proper sampling intensity can be calculated by the procedures outlined in Chap. 8.

The trend is away from the concept of fixed cruising percentages, for it is not the sampling fraction that is important; it is the number of sample units (of a specified kind) needed to produce estimates with a specified precision. In the final analysis, the best endorsement for a given plot size and sampling intensity is an unbiased estimate of stand volume that is bracketed by acceptable confidence limits.

In addition to determining the sampling intensity, it is necessary to decide on the *sampling design*, i.e., the method of selecting the nonoverlapping plots for field measurement. When sample plots are employed, the *sampling frame* is defined as a listing of all possible plots that may be drawn from the specified (finite) population or tract of land. The sample plots to be visited on the ground can be drawn from such a listing by use of a table of random numbers.

In spite of the statistical difficulties associated with systematic sampling designs, such cruises are still employed frequently. Where estimates of sampling precision are regarded as unnecessary, systematic sampling may provide a useful alternative to random sampling methods.

10-8   Cruising Techniques   Circular-plot inventories are often handled by one person, but two or three persons can be used efficiently when square or rectangular plots are employed. Field directions are established with a hand or staff compass, and intervals between sample plots may be either taped or paced. The exact location of plot centers is unimportant, provided that the centers are established in an unbiased manner. When "check cruises" are to be made, plot centers or corners should be marked with stakes, with cairns, or by reference to scribed trees.

With square or rectangular plots, the four corner stakes make it a simple matter to determine which trees are inside the plot boundaries. However, with circular plots, inaccurate estimation of the plot radii is a common source of error. As a minimum, four radii should be measured to establish the sample perimeter. If an ordinary chaining pin is carried to denote plot centers, a steel tape can be tied to the pin for one-person checks of plot radii. When trees appear to be borderline, the center of the stem (pith) determines whether they are "in" or "out." For plots on sloping ground, one must be careful to measure horizontal (not slope) distances when checking plot boundaries.

Inaccurate estimation of plot radii is one of the greatest sources of error in using circular samples. The gravity of such errors is exemplified by a $2^{1}/_{2}$ percent cruise; every stem erroneously tallied or ignored has its volume expanded 40 times. Thus the failure to include one tree having a volume of 300 bd ft will result in a final estimate that is 12,000 bd ft too low.

Separate tally sheets are recommended for each plot location and species; descriptive plot data can be handwritten or designated by special numerical codes. It is usually most efficient to begin the tally at a natural stand opening (or due north) and record trees in a clockwise sweep around the plot. When the tally is completed, a quick stem count made from the opposite direction provides a valuable check on the number of trees sampled.

10-9   Boundary Overlap   A problem arises when the last plot on a line does not lie wholly within the area being sampled. This problem, commonly referred to as *edge-effect bias* or *boundary overlap,* can introduce a bias in the plot cruise statistics if it is not treated properly. When large areas are cruised with small circular plots, the bias due to boundary overlap is usually negligible. However, for small areas, especially long, narrow tracts with a high proportion of edge trees, appropriate precautions should be taken to guard against bias caused by boundary overlap.

One method of dealing with the boundary-overlap problem is to move plot centers (back on the line of travel in the case of line-plot cruises) until the entire plot lies in the area being sampled. This method is generally satisfactory if the timber in the edge is similar to that in the remainder of the tract, but it is not likely to be suitable for small woodlots that have edges strikingly different from the tract interior. Adjustment of the plot-center location may introduce bias because the trees in the edge zone may be undersampled.

In a cruise of small tracts with a high proportion of "edge," a procedure for dealing with boundary overlap should be adopted. The mirage method developed by Schmid in 1969 and described by Beers (1977) and others in the American forestry literature is a simple and, for most situations, easily applied technique. When the plot center falls near the stand boundary so that the plot is not completely within the tract being sampled, the cruiser measures the distance D from plot center to the boundary. A correction-plot center is then established by going this distance D beyond the boundary. All trees in the overlap of the original plot and the correction plot are tallied twice (Fig. 10-3).

Similar boundary-overlap problems arise when volume estimates are being summarized by different types and a sample plot happens to fall at a transition line that divides two types. If the cruise estimate is to be summarized by types and expansion factors for each type (including nonforest areas) are determined, the plot should be moved until it falls *entirely* within the type indicated by its original center location, or a boundary-overlap correction, such as the mirage method, should be applied.

In contrast to the foregoing, plots should not be shifted if a single area expansion factor is to be used for deriving total tract volumes. Under these conditions, edge effects, type transition zones, and stand openings are typically part of the population; therefore, a representative sample would be *expected* to result

FIGURE 10-3
The mirage method for correction of boundary-overlap bias when circular plots are used. Trees in the shaded area are tallied twice.

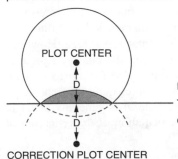

PLOT CENTER

D

FORESTED AREA OF CONCERN

TRACT BOUNDARY

OUTSIDE CRUISE AREA

D

CORRECTION PLOT CENTER

in occasional plots that are part sawtimber and part seedlings—or half-timbered and half-cutover land. To arbitrarily move these plot locations would result in a biased sample.

10-10   Merits of the Plot System   The principal advantages claimed for line-plot cruising over the strip system are as follows:

1 The system is suitable for one-person cruising.

2 Cruisers are not hindered by brush and windfalls as in strip cruising, for they do not have to tally trees while following a compass line.

3 A pause at each plot center allows the cruiser more time for checking stem dimensions, borderline trees, and defective timber.

4 The tree tally is separated for each plot, thus permitting quick summaries of data by timber types, stand sizes, or area condition classes.

USE OF PERMANENT SAMPLE PLOTS

10-11   Criteria for Inventory Plots   The periodic remeasurement of permanent sample plots is statistically superior to successive independent inventories for evaluating *changes* in forest conditions. When independent surveys are repeated, the sampling errors of both inventories must be considered in assessing stand differences or changes over time. But when identical sample plots are remeasured, sampling errors relating to such differences are apt to be lower; i.e., the precision of "change estimates" is improved. In addition, trees initially sampled but absent at a later remeasurement can be classified as to the cause of removal, e.g., harvested yield, natural mortality, and so on.

Regardless of whether temporary or permanent sample units are employed for an inventory, two basic criteria must be met: the field plots must be *representative* of the forest area for which inferences are made, and they must be *subjected to the same treatments* as the nonsampled portion of the forest. If these conditions are not fully achieved, inferences drawn from such sample units will be of questionable utility.

One attempt to ensure that sample units are representative of equal forest areas is illustrated by some rigid continuous forest inventory (CFI) procedures whereby field plots are systematically arranged on a square grid basis; thus each plot represents a fixed and equal proportion of the total forest area. However, such sampling designs tend to be inflexible in meeting the changing requirements of management, and therefore are not recommended for most forest inventories. Even though systematic samples are sometimes quite efficient, especially from the viewpoint of reducing field travel time, it is generally better to use other methods of sampling that will permit calculation of the reliability of sample estimates.

10-12 Sample Units: Size, Shape, and Number Circular sample plots of ⅕ acre have been widely used for CFI systems in the past. Nevertheless, square or rectangular plots may be more efficient because the establishment of four corner stakes, however inconspicuously, improves the chances for plot relocation at a later date. Depending on the size and variability of timber stands, an ideal plot size for second-growth forests will generally fall in the range of ½ to ¼ acre.

As outlined previously, the *number* of permanent sample plots to be established and measured is dependent on the variability of the quantity being assessed and the desired sampling precision. For tracts of 50,000 to 100,000 acres, sampling errors of ±10 to 20 percent might be desired for current volume, with ±20 to 30 percent being accepted for growth (probability level of 0.95). If this precision is maintained on parcels of 50,000 to 100,000 acres, the overall precision for an entire forest holding of 1 to 3 million acres should be approximately ±2 to 3 percent for current volume and ±5 percent for growth.

10-13 Field-Plot Establishment Recent aerial photographs and topographic maps are invaluable aids for the initial location, establishment, and relocation of permanent sample plots. All pertinent data relative to bearings of approach lines, distances, and reference points or monuments should be recorded on a "plot-location sheet" *and* on the back of the appropriate aerial photograph. It is essential that such information be complete and coherent because subsequent plot relocations are often made by entirely different field crews.

Plot centers or corner stakes are preferably inconspicuous and are referenced by using a permanent landmark at least 100 to 300 ft distant and by recording bearings and distances to two or more scribed or tagged "witness trees" that are nearer (but not within) the plot. There is some disagreement as to whether permanent plots should be marked (1) conspicuously, so that they can be easily relocated, or (2) inconspicuously, to ensure that they are accorded the same treatment as nonsampled portions of the forest (Fig. 10-4). The trend is toward essentially "hidden plots," for it is mandatory that they be subjected to *exactly* the same conditions or treatments as the surrounding forest, whether this be stand improvement, harvesting, fires, floods, or insect and disease infestations. Only under these conditions can it be assumed that the sample plots are representative.

Small sections of welding rods, projecting perhaps 6 to 12 in. above ground level, are useful for plot corner stakes. Where it becomes feasible to use more massive iron stakes, it may be possible to find them again with a "dip needle" or other magnetic detection devices. If individual trees on the plot are marked at all, the preferable method is to nail numbered metal tags into the stumps near ground level so that they will not be noticeable to timber markers and other forest workers. As an alternative to tagging the sample trees, individual stem locations may be numbered and mapped by coordinate positions on a plot-diagram sheet.

FIGURE 10-4
A permanent sample plot with trees conspicuously marked at breast height. *(U.S. Forest Service photograph.)*

10-14   Field-Plot Measurements   The inventory forester in charge of the permanent plot system should assume the responsibility for training field crews and for deciding how measurements should be taken on each sample plot. Standardized field procedures are emphasized because *consistency in measurement techniques* is as important as precision for evaluating changes over time.

To avoid problems arising from periodic variations in tree merchantability standards, field measurements should be planned so that tree volumes are expressed in terms of cubic measure (inside bark) for the entire stem, including stump and top. It may also be necessary to estimate the volume of branch wood on some operations. Techniques for predicting merchantable volumes for various portions of trees are given in Chap. 7.

The field information collected for each sample unit is recorded under one of two categories: plot description data and individual tree data. The exact measurements required will differ for each inventory system; thus the following listings merely include *examples* of the data that may be required:

Plot data	Individual tree data
Plot number and location	Tree number
Date of measurement	Species
Forest cover type	dbh
Stand size and condition	Total height
Stand age	Merchantable stem lengths
Stocking or density class	Form or upper-stem diameters
Site index	Crown class
Slope or topography	Tree-quality class
Soil classification	Vigor
Understory vegetation	Diameter growth
Treatments needed	Mortality (and cause)

All field data are numerically coded and recorded on tally forms or directly onto a machine-readable medium for computer processing. Plot inventories are preferably made immediately after a growing season and prior to heavy snowfall. For tracts smaller than 100,000 acres, it may be possible to establish all plots in a single season and remeasure them within similar time limitations. On larger areas, fieldwork may be conducted each fall on a rotation system that reinventories about one-fifth of the forest each year.

10-15 Periodic Reinventories   Permanent sample plots are commonly remeasured at intervals of 3 to 10 years, depending on timber growth rates, expected changes in stand conditions, and the intensity of management. The interval must be long enough to permit a measurable degree of change, but short enough so that a fair proportion of the trees originally measured will be present for remeasurement. At each reinventory, trees that have attained the minimum diameter during the measurement interval are tallied as ingrowth. Also, felling records are kept to correct yields for those plots cut during the measurement interval. This information, along with mortality estimates, is essential for the prediction of future stand yields.

The data needed to calculate volume growth include stand tables prepared from two consecutive inventories, felling records, mortality estimates, and a volume table (or volume-prediction equation) that is applicable to the previous and present stands. First, the stand tables for the two inventories are converted to corresponding stock tables; then, the difference in volume, after accounting for harvested yields and mortality, represents the growth of the plot.

One of the problems facing field crews who must remeasure permanent sample units is that of *finding the plots*. When plots are inconspicuously marked, relocation time can make up a sizable proportion of the total time allotted for reinventories. A study conducted by Nyssonen (1967) in Norway revealed that, after a 7-year interval, 4 to 8 percent of the permanent sample plots could not be found again. Where plots *could* be relocated, the time required for transportation, relocation, and measurement was distributed as follows:

Activity	Percent of total time
Transport by a vehicle	20.6
Walking to, between, and from the plots	22.6
Searching for the plots	12.9
Sample plot measurement	35.7
Pauses	8.2
Total	100.0

Even though time factors will obviously differ for every inventory system, the foregoing tabulation serves to illustrate some of the nonproductive aspects that should be recognized in the application of permanent plot-inventory systems.

REGENERATION SURVEYS WITH SAMPLE PLOTS

10-16   Need for Regeneration Surveys   Evaluations of forest regeneration efforts are of critical importance in on-the-ground forest management. Regeneration may be evaluated at different times in the forest production cycle, and various methods may be used. To devise the most suitable regeneration evaluation system, one must first identify survey objectives.

The primary needs for regeneration information are (Stein, 1984a) (1) to determine regeneration status or potential, (2) to demonstrate compliance with conservation laws, (3) to determine the effectiveness of the regeneration method employed, (4) to identify the needs for additional cultural treatments (e.g., thinning, release from competing vegetation), and (5) to collect data for predicting future yields.

The two main methods used to conduct regeneration surveys are distance sampling methods and fixed-area sample-plot methods. Distance sampling methods will not be discussed here. Rather, attention will be focused on sampling techniques based on sample plots. Fixed-area plots can serve three distinctive regeneration evaluation objectives (Stein, 1984b): (1) determine the presence or absence of trees, (2) obtain a quantitative estimate of the number of trees per unit area, and (3) measure changes in numbers, size, or composition of trees that occur with the passage of time. The methodology that serves each objective is known, respectively, as the stocked-quadrat method, plot-count method, and staked-point method.

10-17   Stocked-Quadrat Method   The stocked-quadrat method is based on the presence or absence of a tree on the plot. It was developed to place evaluation emphasis on tree distribution. The basic concept of the stock-quadrat method is that if a given area is divided into squares of such a size that one established seedling per square will fully stock the square at maturity, then the percentage of units stocked, regardless of the total number of seedlings per acre, indicates the proportion of land being utilized for tree growth. Developers of the stocked-quadrat method have identified its two key features as (1) it automatically compares actual stocking against a defined fully stocked stand and (2) the size of the sample plot used must have a logical relation to full stocking.

If, for example, 250 well-distributed stems per acre at maturity were defined as "full stocking," then the appropriate plot size for applying the stocked-quadrat method would be 1/250 acre. One would proceed by locating 1/250-acre plots (randomly or perhaps systematically) through the area of interest. Field application of the stocked-quadrat method is simple and fast. Each sample plot is classified as stocked if at least one acceptable tree (to be acceptable the tree must meet the species, size, and competitive position criteria set for the survey) is found and as nonstocked if no acceptable tree is found. The tally obtained by locating a sufficient number of plots of the correct size provides a direct estimate of stocking—that is, the percentage of the area occupied by trees. Stocking percent is computed by dividing the number of plots stocked by the total number established. The stocking percent so obtained provides an estimate of the area occupied by well-spaced trees; however, it does not reveal the pattern of stocking on the area. Stocking pattern can be ascertained by plotting the location of stocked and nonstocked plots on a map of the surveyed area. While a single plot at each sample location is recommended (Stein, 1984b), cluster samples—large quadrats divided into four quadrants—are sometimes used to increase data collected relative to the time spent in travel between plots.

10-18   Plot-Count Method   This method is applied when the objective of the regeneration survey is to estimate the number of trees per acre. The plot-count method is simple in concept and straightforward in application. Plots are located, randomly or systematically, throughout the area of interest. Sample plots of uniform size and shape are searched for acceptable trees. The average number of trees per plot is determined, and this average is expanded to a per-acre basis. No specific stocking goal need be specified prior to sampling to apply the plot-count method.

Circular plots are often used for regeneration plot counts. The most appropriate plot size will vary depending on the density (numbers per unit area) of the regeneration and the amount of competing vegetation. When adopting a plot size, several practical matters must be taken into consideration. Because the entire area of each plot must be searched for trees, search time increases with increasing plot size. Furthermore, large plots are more difficult to search thoroughly than small plots, and some trees may be missed on large plots.

Computing the average number of trees per acre from plot-count data is a straightforward procedure. However, the average number of acceptable seedlings does not provide useful information on the pattern of tree distribution in the area. Plotting tree-count data by sample-plot location can provide useful information on distribution. Plot-count data can also be analyzed statistically to aid in interpreting the uniformity of distribution (for example, a small coefficient of variation would indicate relatively uniform distribution, whereas a large coefficient of variation would indicate nonuniform distribution of trees per acre).

10-19 Staked-Point Method This method is used when the survey objective is to estimate changes in tree survival or growth over time. Permanent plots are required for the staked-point method, whereas temporary plots are usually used in the stocked-quadrat and plot-count methods.

Extra costs are associated with the staked-point method, because the plots and trees must be marked so that they can be relocated. However, repeated examination of a representative sample is required for obtaining reliable data for certain objectives.

As with other regeneration survey methods, different plot sizes and shapes can be employed in the staked-plot method. The most appropriate approach to analyzing the data will depend on the sample design used and the measurement data obtained.

PROBLEMS

10-1 Compute the nominal cruising percents and expansion factors for the following systematic samples:
  a Strips $^1/_2$ chain wide spaced 20 chains apart
  b Four 1-chain strips run through a quarter section of land
  c Plots of $^1/_{10}$ acre spaced at $2^1/_2 \times 5$ chains
  d Plots of $^1/_5$ acre spaced at 5×15 chains
10-2 **a** If you space 1-chain strips at 10-chain intervals through a square section of land and tally 350 MBF on the sample, what would be the total-volume estimate for the entire tract?
  b If you space $^1/_4$-acre circular plots at 5×10 chains through a 240-acre tract, and the volume tallied on the sample is 68.4 MBF, what would be the total-volume estimate for the entire tract?
  c How many lineal chains of sample strips 1 chain wide would be run through a township to obtain a 2 percent cruising intensity?
  d If you made a 0.05 percent inventory of the total land area in a state consisting of 30 million acres, how many $^1/_4$-acre circular plots would be required? For a square grid arrangement of samples, what would be the distance (in chains) between plots?
10-3 The coefficient of variation for $^1/_{10}$-acre circular plots was estimated to be 90 percent for a timbered tract of 50 acres. If one wishes to estimate the mean volume per acre of this tract within ±15 percent unless a 1-in-20 chance occurs,
  a Compute the number of plots to be measured assuming simple random sampling without replacement.

    b Calculate the distance between plot centers in chains assuming the plots will be systematically established on a square grid.

10-4 For the same plot sizes shown in Table 10-2, compile a similar tabulation for *square* sample plots. In lieu of plot radii, show the length of one side of the squares in feet and in meters.

10-5 Design and conduct a field study to compare the relative efficiencies of circular, square, and rectangular sample plots in your locality.

10-6 Assume that desirable stocking for mature timber of species of interest is 150 trees per acre.

    a When conducting a stocked-quadrat survey of regeneration for this species, what plot size should be used?

    b Suppose that 50 plots were established. Acceptable trees were found on 42 plots. What is the stocking percent?

10-7 A plot-count regeneration survey was conducted using $1/100$-acre plots located randomly over the tract of interest. The tree count per plot follows:

Plot	Count	Plot	Count
1	5	6	1
2	2	7	2
3	4	8	2
4	0	9	6
5	3	10	3

    a Estimate the mean number of trees per acre.

    b Compute the coefficient of variation for numbers of trees per acre.

    c Compute the 90 percent confidence interval for the mean.

REFERENCES

Avery, T. E., and Newton, R. 1965. Plot sizes for timber cruising in Georgia. *J. Forestry* **63:**930–932.

Beers, T. W. 1977. Practical correction of boundary overlap. *So. J. App. For.* **1:**16–18.

Brand, D. G. 1988. A systematic approach to assess forest regeneration. *Forestry Chron.* **64:**414–420.

Burkhart, H. E., Barrett, J. P., and Lund, H. G. 1984. Timber inventory. Pp. 361–411 in *Forestry handbook,* K. F. Wenger (ed.), John Wiley & Sons, New York.

Fowler, G. W., and Arvanitis, L. G. 1979. Aspects of statistical bias due to the forest edge: Fixed-area circular plots. *Can. J. Forest Res.* **9:**383–389.

Johnson, F. A., and Hixon, H. J. 1952. The most efficient size and shape of plot to use for cruising in old-growth Douglas-fir timber. *J. Forestry* **50:**17–20.

Kendall, R. H., and Sayn-Wittgenstein, L. 1960. A rapid method of laying out circular plots. *Forestry Chron.* **36:**230–233.

Nyssonen, A. 1967. *Remeasured sample plots in forest inventory.* Norwegian Forest Research Inst., Vollebekk, Norway, 25 pp.

Schmid, P. 1969. Sichproben am Waldrand. *Mitt. Schweiz. Anst. Forstl. Versuchswes* **45:**234–303.

Stein, W. I. 1984a. Regeneration surveys: An overview. Pp. 111–116 in *New forests for a changing world,* Proceedings of the 1983 Society of American Foresters National Convention, Portland, Oreg.

————. 1984b. Fixed-plot methods for evaluating forest regeneration. Pp. 129–135 in *New forests for a changing world,* Proceedings of the 1983 Society of American Foresters National Convention, Portland, Oreg.

Wiant, H. V., and Yandle, D. O. 1980. Optimum plot size for cruising sawtimber in Eastern forests. *J. Forestry* **78:**642–643.

Zeide, B. 1980. Plot size optimization. *Forest Sci.* **26:**251–257.

CHAPTER 11

INVENTORIES WITH POINT SAMPLES

11-1 The Concept of Point Sampling Point sampling is a method of selecting trees to be tallied on the basis of their *sizes* rather than by their frequency of occurrence. Sample points, somewhat analogous to plot centers, are located within a forested tract, and a simple prism or angle gauge that subtends a fixed angle of view is used to "sight in" each tree diameter at breast height (dbh). Tree boles close enough to the observation point to completely fill the fixed sighting angle are tallied; stems too small or too far away are ignored. The resulting tree tally may be used to compute basal areas, volumes, or numbers of trees per unit area.

The probability of tallying a given tree depends on its cross-sectional area and the sighting angle used. The smaller the angle, the more stems will be included in the sample.

Point sampling does not require direct measurements of either plot areas or tree diameters. A predetermined basal-area factor (BAF) is established in advance of sampling, and resulting tree tallies can be easily converted to basal area per unit area. And the relationship between basal area and tree volume makes it feasible to use point sampling for obtaining conventional timber inventory data when "counted" trees are recorded by merchantable or total height classes. Point sampling was developed in 1948 by Walter Bitterlich, a forester of Salzburg, Austria. The introduction and adoption of the method in North America were largely due to the efforts of Lewis R. Grosenbaugh.

11-2 Selecting a Sighting Angle BA conversion factors are dependent on the sighting angle (or "critical angle") arbitrarily selected. The sighting angle chosen, in turn, is largely based on the average size and distribution of trees to be sampled. Furthermore, from the standpoint of subsequent volume computations, it is desirable to select a sighting angle having a BAF that can be expressed as a whole number rather than as a fractional number.

In eastern United States, a predetermined sighting angle of 104.18 min (BAF of 10 sq ft per acre) is commonly used in second-growth sawtimber or dense poletimber stands. Critical angles of 73.66 min (BAF 5) and 147.34 min (BAF 20) are often employed for light-density pole stands and for large, old-growth sawtimber, respectively. With small, scattered stems, the sighting angle is narrowed so that it will extend farther out for trees of minimum diameter; conversely, where large tree diameters are common, the angle is enlarged to reduce excessively heavy field tallies.

Depending on the region, average tree size, and amount of underbrush restricting line-of-sight visibility, the BAF is usually chosen to provide an average tally of 5 to 12 trees per sample point. In western United States where larger timber predominates, a BAF of 20 to 60 is in common use. For "West Side" Douglas-fir, a BAF of 40 might be regarded as typical, but an instrument with a BAF of 20 would be more frequently encountered in sampling stands of "East Side" ponderosa pine.

11-3 Plot Radius Factor To illustrate the meaning of BA conversions listed in Table 11-1, a sighting angle of 104.18 min (BAF 10) may be presumed.

TABLE 11-1
COMMON BASAL-AREA FACTORS AND ANGLE SIZES USED IN POINT SAMPLING

Basal-area factor	Angle size, min	Angle size, diopters	Ratio (tree diameter to plot radius)	Plot radius factor
1	32.94	0.96	1/104.4	8.696
2	46.59	1.36	1/73.8	6.149
3	57.06	1.66	1/60.2	5.021
4	65.89	1.92	1/52.2	4.348
5	73.66	2.14	1/46.7	3.889
10	104.18	3.03	1/33.0	2.750
15	127.59	3.71	1/26.9	2.245
20	147.34	4.29	1/23.3	1.944
25	164.73	4.79	1/20.9	1.739
30	180.46	5.25	1/19.0	1.588
35	194.92	5.67	1/17.6	1.470
40	208.38	6.07	1/16.5	1.375
50	232.99	6.79	1/14.8	1.230
60	255.23	7.44	1/13.5	1.123

Source: Hovind and Rieck, 1970.

As this angle can also be defined by placing a 1-in. horizontal intercept on a sighting base of 33 in. (column 4 of Table 11-1), it follows that all trees located no farther than 33 times their diameter from the sample point will be tallied. Accordingly, a 1-in.-dbh tree must be within 33 in. of the point, a 12-in.-dbh tree will be tallied up to 396 in. (33 ft) away, and a 24-in.- or 2-ft- dbh tree will be recorded up to a distance of 66 ft (Fig. 11-1). This 1:33 ratio of tree diameter to plot radius, a constant for the specified angle of 104.18 min, has a value of 2.75 ft (33 ÷ 12) when expressed as a "plot radius factor." Thus for each full inch added to stem diameter, a tree can be 2.75 ft farther from the sample point and still be tallied.

HOW POINT SAMPLING WORKS

11-4 Imaginary Tree Zones As the plot radius factor for BAF 10 has been developed in the preceding section, all subsequent explanations of point-sampling theory and tree volume conversions in this chapter will presume a sighting angle of 104.18 min and a BAF of 10 sq ft per acre. Nevertheless, the underlying principles discussed may be applied to any other angle or BAF.

Because each tree "sighted in" must be within 33 times its diameter of the sample point to be tallied, it is convenient to presume that all trees are encircled

FIGURE 11-1
Tree sizes and limiting distances for a 1:33 angle gauge.

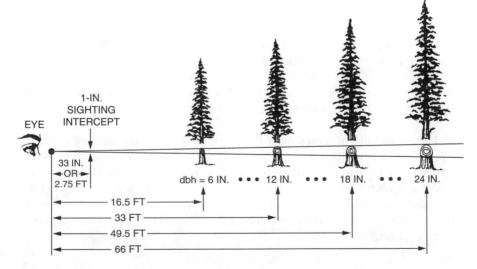

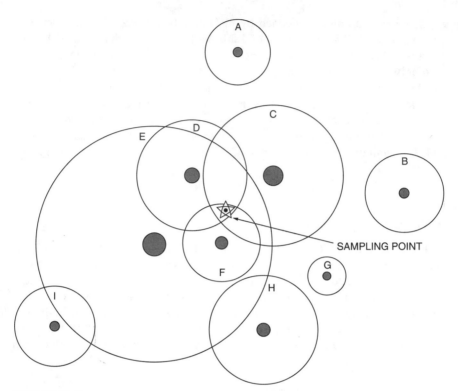

FIGURE 11-2
Imaginary zones proportional to stem basal area and encircling each tree determine which trees will be tallied at a given point. *(Adapted from Hovind and Rieck, 1970.)*

with imaginary zones whose radii are exactly 33 times the diameter of each tree stem. All these imaginary circles that encompass a given sampling point on the ground represent trees to be tallied (Fig. 11-2). Thus the probability of tallying any given tree is proportional to its stem BA. A 12-in.-dbh stem has 4 times the probability of being counted as a 6-in.-dbh stem.

11-5   Equality of Trees on a Per-Acre Basis   For the sighting angle of 104.18 min, each tallied tree (regardless of its size or relative position to the sampling point) represents 10 sq ft of BA on a *per-acre basis*. The reason for this is that each stem and its imaginary zone "represent" a definite part of an acre and a specific number of trees per acre, depending on its size. The derivation of values for 6-in.- and 12-in.-dbh trees in Table 11-2 provides an explanation, or "proof," of this theory.

TABLE 11-2
DERIVATION OF THE BASAL-AREA FACTOR OF 10 SQ FT PER ACRE FOR
POINT SAMPLING

Tree dbh, in. (1)	Imaginary plot radius, ft (2)	Imaginary plot size, acres (3)	Trees per acre,* no. of stems (4)	Basal area per tree, sq ft (5)	Basal area per acre, sq ft (6)
4	11.00	0.0087	114.94	0.087	10
6	16.50	0.0196	51.02	0.196	10
8	22.00	0.0349	28.65	0.349	10
10	27.50	0.0545	18.35	0.545	10
12	33.00	0.0785	12.74	0.785	10
14	38.50	0.1069	9.35	1.069	10
16	44.00	0.1396	7.16	1.396	10
18	49.50	0.1767	5.66	1.767	10
20	55.00	0.2182	4.58	2.182	10
22	60.50	0.2640	3.79	2.640	10
24	66.00	0.3142	3.18	3.142	10
26	71.50	0.3687	2.71	3.687	10
28	77.00	0.4276	2.34	4.276	10
30	82.50	0.4909	2.04	4.909	10
32	88.00	0.5585	1.79	5.585	10
34	93.50	0.6305	1.59	6.305	10
36	99.00	0.7069	1.41	7.069	10
Method of calculation:	dbh $\times$ 2.75	$\dfrac{\pi\, r^2}{43,560}$	$\dfrac{1}{\text{plot size}}$	$0.005454 \times \text{dbh}^2$	$(4) \times (5)$

*Exact value for number of trees per acre may vary slightly, depending upon number of decimal places expressed for imaginary plot size.

Considering the 6-in. dbh first, its imaginary "plot" radius is read from Table 11-2 as 16.50 ft. This hypothetical zone represents an imaginary plot of 0.0196 acre around each 6-in.-dbh stem (column 3 of Table 11-2). By dividing 0.0196 into 1 acre, it can be seen from column 4 that there can be 51.02 such areas fitted into a single acre. Thus when one 6-in.-dbh tree is tallied, it is tacitly assumed that there are 51.02 such stems per acre. Accordingly, the BA of a 6-in.-dbh tree (0.196 sq ft from column 5), multiplied by 51.02 trees per acre, yields the "constant" BAF of 10 sq ft *per acre* (column 6).

For 12-in.-dbh stems, the imaginary plot radius is 33 ft, and the implied plot size is 0.0785 acre. Only 12.74 trees per acre are assumed—one-fourth the number of 6-in.-dbh trees expected. However, 12-in.-dbh trees have 4 times the BA of 6-in.-dbh stems, and this value (0.785 sq ft) from column 5, multiplied by 12.74 trees per acre, again produces a BA of 10 sq ft per acre. The same result applies to all other tree sizes encountered when sampling with a BAF 10 angle gauge.

INSTRUMENTS FOR POINT SAMPLING

11-6   The Stick-Type Angle Gauge   This simple, horizontal angle gauge often consists of a wooden rod with a peep sight at one end and a metal intercept at the other. To establish a sighting angle of 104.18 min (BAF 10), an intercept 1 in. wide on a 33-in. sighting base can be easily improvised. Gauges for other factors can be constructed according to ratios provided in Table 11-1. Regardless of the ratio desired, the sighting base should be at least 24 in. long; otherwise, it is difficult to keep both the intercept and the tree in focus simultaneously.

When the stick gauge is used, all tree diameters larger than the defined angle are counted; those smaller are ignored. Trees that appear to be exactly the same size as the intercept should be checked by measuring their exact dbh and the distance from the sampling point to the tree center. The product of dbh and the appropriate plot radius factor (2.75 for BAF 10) determines whether the tree is "in" or "out."

With a stick gauge, the observer's eye represents the vertex of the sighting angle; hence the stick must be pivoted or revolved about this exact point for a correct tree tally. When properly calibrated for use by a particular individual, the stick gauge may be just as accurate as other more expensive point-sampling devices. In dense sapling or pole stands and where heavy underbrush is encountered, the stick gauge is often easier to use than more sophisticated relascopes or prisms.

11-7   The Spiegel Relascope   This is a versatile, hand-held instrument developed for point sampling by Walter Bitterlich (Fig. 11-3). It is a compact and rugged device that may be used for determining BA per acre, upper-stem diameters, tree heights, horizontal distances of 66 and 99 ft with correction for slope, and measurement of slope on percent, degree, and topographic scales. Sighting angles are provided for factors of 5, 10, 20, or 40, and the instrument automatically corrects each angle for slope. The base has a tripod socket for use when especially precise measurements are desired.

Establishment of sighting angles with the Spiegel relascope is somewhat analogous to measuring distances with transit and stadia rod; the principal difference is that the relascope subtends a horizontal angle, and the transit and stadia system is based on a vertically projected angle. The Spiegel relascope is complex in design but relatively simple to use. Its principal disadvantages are that it is relatively expensive and lacks the optical qualities for good sighting visibility on dark and rainy days.

11-8   The Wedge Prism   A properly ground and calibrated prism is merely a tapered wedge of glass that bends or deflects light rays at a specific offset angle. When a tree stem is viewed through such a wedge, the bole appears to be displaced, as if seen through a camera rangefinder. The amount of offset or displacement is controlled by the prism strength, measured in diopters. As one prism diopter is equal to a right angle displacement of one unit per 100 units of

FIGURE 11-3
The Spiegel relascope can be used for doing point sampling and for determining upper-stem diameters or tree heights. *(U.S. Forest Service photograph.)*

distance, a 3.03-diopter prism will produce a displacement of one unit per 33 units of distance, i.e., a critical angle of 104.18 min. Other prism-strength relationships are given in Table 11-1.

Field use of the prism requires that it be held precisely over the sampling point at all times, for this point and *not the observer's eye* is the pivot from which the stand is "swept" by a 360° circle. All tree stems not completely offset when viewed through the wedge are counted; others are not tallied (Fig. 11-4). Trees that appear to be borderline should be measured and checked with the appropriate plot radius factor.

The prism may be held at any convenient distance from the eye, provided it is always positioned directly over the sampling point. Proper orientation also requires that the prism be held in a vertical position and at right angles to the line of sight; otherwise, large errors in the tree tally may result (Fig. 11-5).

The wedge prism is simple, relatively inexpensive, portable, and as accurate as other angle gauges when properly calibrated and used. Some sighting difficulties are found in dense stands where displaced bole sections offset into one another, and special corrections must be applied when slopes of 15 percent and greater are encountered. However, the latter disadvantage may be cited for all point-sampling devices except the Spiegel relascope.

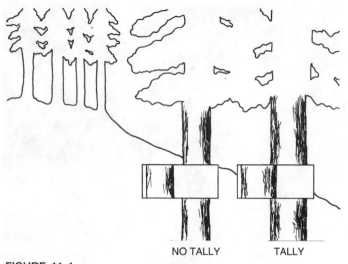

NO TALLY   **TALLY**

FIGURE 11-4
Use of the wedge prism for point sampling.

11-9 Calibration of Prisms or Angle Gauges Precision-tested prisms and angle gauges should be used whenever feasible, because a BAF of *exactly* 5, 10, or 40 is conducive to faster computations than such values as 4.9, 9.8, or 39.5. Prisms ground to within ±1 min of a specified angle are desired, for such deviation will usually result in a maximum error of about 2 percent for a BAF 10 prism.

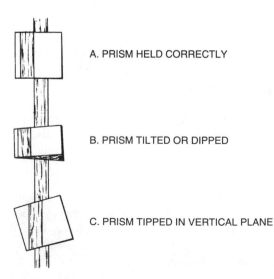

A. PRISM HELD CORRECTLY

B. PRISM TILTED OR DIPPED

C. PRISM TIPPED IN VERTICAL PLANE

FIGURE 11-5
Correct and incorrect methods of holding the wedge prism.
(Adapted from Hovind and Rieck, 1970.)

Where inexpensive prisms or angle gauges are employed and the exact BAF is unknown, such devices should be carefully calibrated. As individual eyesights may vary appreciably, it may even be desirable to calibrate *all instruments* (regardless of price or supposed precision) for each cruiser's own sighting habits or peculiarities.

To calibrate a prism or angle gauge, a target of known width (for example, 1 ft) is set up against a contrasting background. With the angle gauge or prism in proper orientation, the observer backs away from the target until the target exactly fills the sighting angle. The exact distance from target to instrument is then measured and the BAF computed by

$$BAF = 10,890 \left(\frac{W}{D}\right)^2$$

where W is the target width in feet and D is the distance to target in feet.

The foregoing formula is not exact when a flat target is used in calibration but only when the target is a circular cross section with diameter W. With small critical angles, however, this simple approximation is usually satisfactory, because of the near equality of sine and tangent functions for narrow angles. For reliable results, the instrument should be mounted on a plane table, tripod, or vise and several readings made of the horizontal distance to the target. When a fractional BAF is derived, conversions similar to those in Table 11-2 should be prepared to simplify subsequent inventory computations.

11-10   Corrections for Slope   Unless the Spiegel relascope is used to establish sighting angles, corrections must be made in point sampling when slope is 15 percent or greater, i.e., a 15-ft rise or drop in elevation per 100 horizontal feet (Stage, 1959). In general, the sighting angle must be reduced so that, when sighting along the slope, trees will appear appropriately "in" or "out" for a corresponding horizontal distance from the sampling point. The reduction in sighting angle is usually accomplished through instrument adjustments.

When the wedge prism is used, an approximate on-the-ground compensation for slope can be made by tilting the top edge of the prism through the estimated slope angle—at right angles to the line of slope. If a flat-based prism is used and an Abney hand level or Suunto clinometer is employed to determine slope, an efficient means of accomplishing the prism rotation is:

1 Measure the slope with the Abney level or Suunto clinometer.
2 Set the prism on the clinometer and rotate the combined unit through an angle equivalent to the slope.
3 Sight the tree in question at breast height.

For stick-type angle gauges, the sighting angle can be reduced by making the intercept narrower or the sighting base longer. One practical solution is to construct an angle gauge with a sliding intercept so that the sighting base can be made longer for slope correction. If the sighting base for zero slope is denoted by L, then the appropriate sighting base for any given amount of slope is

$$L_S = \frac{L}{\cos S}$$

where S is the slope angle in degrees. Note that $\cos S$ will be less than 1 and, thus, L_s will be greater than L. A stick of sufficient length can be used as the base and a sliding intercept mounted on the stick. Slope percents are marked directly on the stick, thus enabling the observer to make a rapid adjustment of the target setting. In the field, slope percent from the observer's eye at the sampling point to the dbh of trees that appear just "out" is determined. The intercept is appropriately adjusted on the stick, and the tree is sighted again. Trees that appear "in" or "borderline" on moderate to steep slopes are tallied, and no additional checking is needed. Most trees that are "out" will be obviously so, and the checking procedure just described will not be needed.

In lieu of making instrument adjustments, one can compensate for slope by checking all doubtful trees by measuring dbh and horizontal distance from the sampling point to the tree center at dbh. Doubtful trees are those that appear slightly "out" along the slope distance. This procedure of measuring doubtful trees is satisfactory in situations where slope corrections are infrequently needed.

11-11   Doubtful Trees, Limiting Distances, and Bias   Most cruisers possess some degree of observer bias when "sighting in" doubtful trees. In a strict sense, questionable trees occur only when the distance from the sampling point to the stem center is precisely equal to tree dbh times plot radius factor. Therefore, if doubtful trees are regularly checked by careful measurement, the "borderline" tree is effectively eliminated.

To encourage regular field checks of doubtful trees and to speed up the sampling process, it is helpful to compile tables of limiting distances in advance of fieldwork. These are easily prepared in feet or in metric units by multiplying various tree diameters by the appropriate plot radius factor (Table 11-3).

For trees that lean to the left or right of the observer, the angle gauge should be rotated so that the vertical axis of the gauge parallels the axis of the leaning tree.

Trees that lean severely toward or away from the observer and appear questionable should be checked by measuring the tree dbh and horizontal distance to the tree center. When checking the status of such leaning trees, the tree center is commonly assumed to be a point vertically above the center of the tree cross section at the groundline.

Precautions are necessary to avoid missing or double-counting trees. After each obvious tree is tallied, the cruiser can sway from side to side to detect trees behind others. When brush or other obstructions make it necessary to move from

TABLE 11-3
HORIZONTAL LIMITING DISTANCES IN FEET FOR BAF 10 POINT-SAMPLING INSTRUMENTS

dbh, in.	0.0	0.1	0.2	0.3	0.4	0.5	0.6	0.7	0.8	0.9
5	13.75	14.02	14.30	14.57	14.85	15.12	15.40	15.67	15.95	16.22
6	16.50	16.77	17.05	17.32	17.60	17.87	18.15	18.42	18.70	18.97
7	19.25	19.52	19.80	20.07	20.35	20.62	20.90	21.17	21.45	21.72
8	22.00	22.27	22.55	22.82	23.10	23.37	23.65	23.92	24.20	24.47
9	24.75	25.02	25.30	25.57	25.85	26.12	26.40	26.67	26.95	27.22
10	27.50	27.77	28.05	28.32	28.60	28.87	29.15	29.42	29.70	29.97
11	30.25	30.52	30.80	31.07	31.35	31.62	31.90	32.17	32.45	32.72
12	33.00	33.27	33.55	33.82	34.10	34.37	34.65	34.92	35.20	35.47
13	35.75	36.02	36.30	36.57	36.85	37.12	37.40	37.67	37.95	38.22
14	38.50	38.77	39.05	39.32	39.60	39.87	40.15	40.42	40.70	40.97
15	41.25	41.52	41.80	42.07	42.35	42.62	42.90	43.17	43.65	43.72
16	44.00	44.27	44.55	44.82	45.10	45.37	45.65	45.92	46.20	46.47
17	46.75	47.02	47.30	47.57	47.85	48.12	48.40	48.67	48.95	49.22
18	49.50	49.77	50.05	50.32	50.60	50.87	51.15	51.42	51.70	51.97
19	52.25	52.52	52.80	53.07	53.35	53.62	53.90	54.17	54.45	54.72
20	55.00	55.27	55.55	55.82	56.10	56.37	56.65	56.92	57.20	57.47
21	57.75	58.02	58.30	58.57	58.85	59.12	59.40	59.67	59.95	60.22
22	60.50	60.77	61.05	61.32	61.60	61.87	62.15	62.42	62.70	62.97
23	63.25	63.52	63.80	64.07	64.35	64.62	64.90	65.17	65.45	65.72
24	66.00	66.27	66.55	66.82	67.10	67.37	67.65	67.92	68.20	68.47
25	68.75	69.02	69.30	69.57	69.85	70.12	70.40	70.67	70.95	71.22
26	71.50	71.77	72.05	72.32	72.60	72.87	73.15	73.42	73.70	73.97
27	74.25	74.52	74.80	75.07	75.35	75.62	75.90	76.17	76.45	76.72
28	77.00	77.27	77.55	77.82	78.10	78.37	78.65	78.92	79.20	79.47
29	79.75	80.02	80.30	80.57	80.85	81.12	81.40	81.67	81.95	82.22
30	82.50	82.77	83.05	83.32	83.60	83.87	84.15	84.42	84.70	84.97

the sampling point to view certain stems, special care must be exercised to maintain the correct distances from obscured trees. Failure to maintain proper distance relationships can result in sizable errors in the tally, especially when using a large BAF.

When dbh is obscured by limbs or underbrush, the cruiser can sight the tree at a visible point higher on the stem. Trees that qualify for inclusion at some point above dbh also qualify at dbh, unless the tree leans toward the sampling point and the distance is critical.

11-12 Boundary Overlap When part of a tree's imaginary "plot" extends beyond the boundary of the forest tract being sampled, boundary overlap occurs. In these situations, the probability of the sample point falling within the tree's imaginary plot zone is less than that for a tree of the same size which is not near the boundary. The bias created by boundary overlap is negligible on large forest areas having a small proportion of edge or where the forest outside the boundary is similar to that within the boundary. However, the bias can be considerable on small areas where the proportion of edge is great. If compensation is not made, especially on long, narrow tracts where the proportion of edge is large, sizable bias can result.

The mirage method of boundary-overlap correction (described for fixed-area plots in Sec. 10-9) is recommended when point-sample cruising is employed. If a point falls sufficiently close to the boundary such that boundary overlap *might* occur, one determines the distance D from the sampling point to the forest boundary. The cruiser then proceeds the distance D out from the boundary to establish a correction point, sights back toward the forested area with the angle gauge, and tallies all trees that appear "in" from the correction point. If overlap is present, certain trees will appear "in"; thus they are tallied twice—at the original point and again at the correction point. If there is no overlap, no trees will appear "in" from the correction point.

11-13 Choice of Instruments In summary, the selection of a point-sampling sighting gauge is largely a matter of balancing such factors as costs, efficiency, and personal preferences. All the devices described here will provide a reliable tree tally if they are properly calibrated and carefully used. Accordingly, the following generalizations will be primarily useful to the newer advocates of point sampling:

1 When steep slopes are regularly encountered, the Spiegel relascope is preferred.

2 For relatively flat topography, either the wedge prism or the stick gauge may be used. The prism is particularly desirable for persons who wear eyeglasses, because the vertex of the sighting angle occurs at the prism rather than at the observer's eye. However, the prism is difficult to use in dense stands due to displacement of stem sections into one another.

3 The simple stick gauge, though largely supplanted by the prism, is preferable in dense stands—especially if the cruiser does not wear eyeglasses. Cruisers who use point sampling only occasionally will find the stick gauge more reliable, because there are fewer ways for errors to result with this device than with the wedge prism.

VOLUME CALCULATIONS

11-14 Example of Computational Procedures It may be assumed that a point-sample cruise was performed using a BAF 10 instrument at 12 points on a 40-acre tract. The objective of the inventory was to estimate the BA, number of trees, and board-foot volume for trees 10-in. dbh and larger. A summary of the tree tally is given in Table 11-4.

11-15 Basal Area per Acre As previously described, each tree tallied in point sampling, *regardless of its size,* represents the same amount of BA on a per-acre basis. Thus an estimate of BA per acre for any tract may be computed by

$$\text{BA per acre} = \frac{\text{total trees tallied}}{\text{no. of points}} \times \text{BAF}$$

With a BAF 10 and 96 trees tallied at 12 points, the estimated BA per acre of trees 10-in. dbh and larger is $(96/12) \times 10 = 80$ sq ft per acre.

11-16 Trees per Acre Because each diameter class has a different imaginary plot zone, the per-acre conversion factor varies from class to class. Consequently, it is necessary to compute the per-acre conversion factor for each dbh

TABLE 11-4
FREQUENCY OF STEMS TALLIED BY DBH AND
HEIGHT CLASSES

	Height (no. of logs)			
dbh, in.	1	2	3	Total
10	12	7	. . .	19
12	8	19	7	34
14	. . .	16	15	31
16	. . .	4	8	12
Total	20	46	30	96

class, convert the tree tally in each class to a per-acre basis, and then summarize for an overall estimate of trees per acre. In formula form, the number of trees per acre for any given diameter class is

$$\text{Trees per acre} = \frac{(\text{no. trees tallied})(\text{per-acre conversion factor})}{\text{total no. of points}}$$

where the per-acre conversion factor for BAF 10 (Table 11-2) is

$$\frac{43,560}{\pi(\text{dbh} \times 2.75)^2} \quad \text{or} \quad \frac{\text{BAF}}{\text{BA per tree}}$$

Computing values for the cruise data given in Table 11-4, one obtains:

$$10\text{-in. class} = \frac{19(18.35)}{12} = 29 \text{ trees per acre}$$

$$12\text{-in. class} = \frac{34(12.74)}{12} = 36 \text{ trees per acre}$$

$$14\text{-in. class} = \frac{31(9.35)}{12} = 24 \text{ trees per acre}$$

$$16\text{-in. class} = \frac{12(7.16)}{12} = \underline{7} \text{ trees per acre}$$

Total $= 96$ trees per acre in the 10-in.-dbh class and larger

11-17   Volume per Acre by the Stand-Table Method   Prior to conducting a timber inventory, one must select an appropriate volume table. The Mesavage-Girard form-class volume table shown in Chap. 7 was selected for this cruise example. The relevant volume-table entries for the cruise data are

dbh, in.	Board-foot volume by 16-ft logs, International 1/4-in., form class 80		
	1	2	3
10	39	63	. . .
12	59	98	127
14	. . .	141	186
16	. . .	190	256

Volume per acre can be computed readily by applying the per-acre conversion factors to data from each dbh-height class and by then summing class volumes for an overall estimate. Thus volume per acre for any given diameter-height class can be derived as

Volume per acre

$$= \frac{(\text{no. of trees})(\text{volume per tree})(\text{per-acre conversion factor})}{\text{total no. of sample points}}$$

Applying the foregoing relationship to the example cruise data gives:

$$10\text{-in., 1 log} = \frac{(12)(39)(18.35)}{12} = \quad 716 \text{ bd ft per acre}$$

$$10\text{-in., 2 log} = \frac{(7)(63)(18.35)}{12} = \quad 674 \text{ bd ft per acre}$$

$$12\text{-in., 1 log} = \frac{(8)(59)(12.74)}{12} = \quad 501 \text{ bd ft per acre}$$

$$12\text{-in., 2 log} = \frac{(19)(98)(12.74)}{12} = 1,977 \text{ bd ft per acre}$$

$$12\text{-in., 3 log} = \frac{(7)(127)(12.74)}{12} = \quad 944 \text{ bd ft per acre}$$

$$14\text{-in., 2 log} = \frac{(16)(141)(9.35)}{12} = 1,758 \text{ bd ft per acre}$$

$$14\text{-in., 3 log} = \frac{(15)(186)(9.35)}{12} = 2,174 \text{ bd ft per acre}$$

$$16\text{-in., 2 log} = \frac{(4)(190)(7.16)}{12} = \quad 453 \text{ bd ft per acre}$$

$$16\text{-in., 3 log} = \frac{(8)(256)(7.16)}{12} = \underline{1,222} \text{ bd ft per acre}$$

Total $= 10,419$ bd ft per acre

11-18   Volume per Acre by the Volume/Basal-Area Ratios Approach
An alternative approach to using the per-acre conversion factors for computing volume-per-acre estimates (as shown in the previous section) involves the use of volume/BA ratios. As an initial step, one calculates the volume per square foot of BA for the volume table or equation being applied. For example, a 10-in., 1-log tree for the Mesavage-Girard International $\frac{1}{4}$-in., form class 80 table has 39 bd ft and the BA is $0.005454(10)^2 = 0.545$ sq ft. Thus the volume/BA ratio is $39/0.545 = 72$ bd ft per sq ft of BA. Dividing all entries of the form-

class volume table by the corresponding BA gives the following ratios:

dbh, in.	Board-foot volume per sq ft of basal area by 16-ft logs		
	1	2	3
10	72	116	. . .
12	75	125	162
14	. . .	132	174
16	. . .	136	183

The volume per acre can then be estimated as the average volume/BA ratio times the BA per acre; that is

$$\text{Volume per acre} = \frac{\text{sum of ratios}}{\text{no. of trees}} \times \text{BA per acre}$$

In this example, the sum of the volume/BA ratios would be

$$12(72) + 7(116) + 8(75) + 19(125) + 7(162) + 16(132)$$
$$+ 15(174) + 4(136) + 8(183) = 12{,}515$$

It will be recalled that 96 trees were tallied on 12 points, thus giving an estimated BA per acre of 80 sq ft. Substituting this information into the volume-per-acre formula gives

$$\frac{12{,}515}{96} \times 80 = 10{,}429 \text{ bd ft per acre}$$

The discrepancy between this volume-per-acre estimate and that obtained through the stand-table approach (10,419 bd ft per acre) is due solely to rounding off errors.

An alternative formula to volume computation when the ratios approach is used is

$$\text{Volume per acre} = \frac{\text{sum of ratios}}{\text{no. of points}} \times \text{BAF}$$

Substituting values for this example gives

$$\frac{12{,}515}{12} \times 10 = 10{,}429 \text{ bd ft per acre}$$

11-19 Field Tally by Height Class Examination of the volume/basal-area ratios computed in the previous section shows that they do not vary greatly within height classes (indeed, the ratios are constant within height classes for some volume tables). Thus dbh tallies can be omitted and trees can be recorded by height classes only without much loss of accuracy when estimating overall volume by point-sampling techniques.

Tree volume can be expressed by the constant form factor equation

$$V = b_1(\text{dbh})^2 H$$

where H represents some measure of tree height. Since BA is equal to a constant times dbh squared, volume per square foot of BA can be expressed as

$$\frac{V}{\text{BA}} = \frac{b_1(\text{dbh})^2 H}{c(\text{dbh})^2} = kH$$

and the variable of dbh is eliminated. In the foregoing expression, k represents the volume per square foot of BA per unit of height. Hence $k \times$ BAF is equal to the volume per acre represented by each unit of height, and a cumulative height tally of "in" trees is all that is needed to get a quick estimate of volume. The average number of height units per point times the volume per acre represented by each height unit equals average volume per acre.

One might, for example, use the following board-foot volume equation:

$$V = 0.30624(\text{dbh})^2 H$$

where dbh is in inches and H is the number of 16-ft sawlogs. The volume per square foot of basal area is

$$\frac{V}{\text{BA}} = \frac{0.30624(\text{dbh})^2 H}{0.005454(\text{dbh})^2} = 56.1H$$

Assuming a BAF of 10, each 16-ft sawlog represents 561 (56.1×10) bd ft per acre.

If a cruiser tallied eighty-six 16-ft sawlogs on 20 points by using a BAF 10 instrument, then the estimated volume per acre is

$$\frac{86}{20} \times 561 = 2412 \text{ bd ft per acre}$$

To obtain an estimate of BA per acre, it would be necessary to know how many trees were tallied on the 20 points. Further, per-acre conversion factors, and thus number of trees per acre, cannot be computed unless "in" trees are tallied by dbh.

11-20   Point Sampling in a Double-Sampling Context   Point sampling can be efficiently applied in a double-sampling design (Sec. 8-15). Basal area per acre is easily and quickly determined with point-sampling methodology, because only a tree count is needed. Determining volume per acre, however, requires that the "in" trees be tallied by dbh and/or height classes. Thus the volume points are more time-consuming and expensive to establish. Consequently part of the sampling resources might be devoted to establishing a large number of points in which basal area per acre only is determined. Basal area per acre is highly correlated with volume per acre, the variable of ultimate interest. Hence, double sampling with a regression or ratio estimator is suggested.

As an example, suppose that 20 BAF 10 points were randomly located on a forested tract. On 10 of the points, called *basal-area* points, only a tree count was taken. Trees were tallied by dbh and total height on the other 10 points, called *volume points*. A summary of the 20 points follows.

N_1	N_2	
Basal-area points	Volume points	
Basal area, sq ft per acre	Basal area, sq ft per acre	Volume, cu ft per acre
30	20	378
40	10	284
10	60	1,239
60	90	2,132
80	10	257
70	70	1,484
40	50	1,070
20	80	1,762
90	30	690
50	100	2,173
Total 490	520	11,469
x_1	x_2	

Plotting the volume per acre versus basal area per acre shows a strong linear relationship with a homogeneous variance. Thus, the linear regression estimator was chosen.

The estimate of the mean for a linear regression estimator in a double sample is computed as

$$\bar{y}_{Rd} = \bar{y}_2 + b(\bar{x}_1 - \bar{x}_2)$$

where $\bar{y}_{Rd}$ = estimate of the population mean volume per acre
$\quad\quad \bar{y}_2$ = mean volume per acre for the small sample
$\quad\quad b$ = linear regression slope coefficient (computed from the small sample)
$\quad\quad \bar{x}_1$ = mean basal area per acre from the large sample
$\quad\quad \bar{x}_2$ = mean basal area from the small sample

For the data in this example, with a large sample of size 20 ($n_1 = 20$) and a small sample of size 10 ($n_2 = 10$), the numerical values are

$$\bar{x}_1 = \frac{490 + 520}{20} = 50.5 \text{ sq ft per acre}$$

$$\bar{x}_2 = \frac{520}{10} = 52.0 \text{ sq ft per acre}$$

$$\bar{y}_2 = \frac{11{,}469}{10} = 1146.9 \text{ cu ft per acre}$$

$$b = \frac{SP_{xy}}{SS_x} = \frac{219{,}142}{9960} = 22.0$$

$$\bar{y}_{Rd} = 1146.9 + 22.0\,(50.5 - 52.0)$$
$$= 1113.9 \text{ cu ft per acre}$$

The standard error of the mean can be estimated as

$$S_{\bar{y}Rd} = \sqrt{ S_{y\cdot x}^2 \left[\frac{1}{n_2} + \frac{(\bar{x}_1 - \bar{x}_2)^2}{SS_x} \right] \left(1 - \frac{n_2}{n_1} \right) + \frac{S_y^2}{n_1} }$$

where

$$S_{y\cdot x}^2 = \frac{SS_y - (SP_{xy})^2 / SS_x}{n_2 - 2}$$

$$S_y^2 = \frac{SS_y}{n_2 - 1}$$

Following through with this numerical example gives

$$S_{\bar{y}Rd} = \sqrt{ 5630.4 \left[\frac{1}{10} + \frac{(50.5 - 52.0)^2}{9960} \right] \left(1 - \frac{10}{20} \right) + \frac{540{,}685}{20} }$$

$$= 165.3 \text{ cu ft per acre}$$

The 95 percent confidence limits for the mean can be approximated as

$$\bar{y}_{Rd} \pm 2 S_{\bar{y}Rd}$$

Using numerical values from this example gives

$$1113.9 \pm 2(165.3)$$
$$1113.9 \pm 330.6$$

or   783.3 to 1444.5 cu ft per acre

In practice, foresters generally establish four or five basal-area points for each volume point, and the ratio-of-means estimator (Sec. 8-15) is often applied. However, this numerical example illustrates the general concept and utility of applying point sampling in a double-sampling design.

POINT-SAMPLE CRUISING INTENSITY

11-21   Comparisons with Conventional Plots   There is no fixed plot size when using point sampling; hence it is difficult to compute cruise intensity on a conventional area-sample basis. Each tree has its own imaginary plot radius (depending on the BAF used), and the exact plot size cannot be easily determined, even after the tally has been made. However, approximations can be made on the basis of the *average* stem diameter encountered at a given point.

Assuming an even-aged plantation with a single dbh class of 6 in. and a critical angle of 104.18 min, the area sampled would have a radius of 6 × 2.75, or 16.5 ft—equivalent to about $1/50$ acre. If the dbh class were doubled to 12 in., the effective sample area would quadruple to about $1/12$ acre. To sample a full $1/5$ acre, average dbh would have to be about 19 in.

From the foregoing, it follows that use of BAF 10 sample points in lieu of the same number of $1/5$- or $1/4$-acre plots will usually result in a tally of fewer trees. From a statistical standpoint, however, the selection of trees according to size rather than frequency may more than offset this reduction of sample size—and with an additional saving in time. Conversely, it must be remembered that smaller samples of any kind require larger expansion, or blow-up, factors. Thus when point sampling is adopted, the so-called borderline trees must always be closely checked, for the erroneous addition or omission of a single stem can greatly reduce accuracy.

11-22   Number of Sampling Points Needed   The only accurate method of determining how many sample points should be measured is to determine the standard deviation (or coefficient of variation) of BA or volume per acre from a preliminary field sample. When this has been done, sampling intensity may be derived by formulas described in Chap. 8. If the statistical approach is not feasible, the following rules of thumb will often provide acceptable results:

1 If the BAF is selected according to tree size so that an average of 5 to 12 trees are counted at each point, use the same number of points as $1/5$-acre plots.

2 With a BAF 10 angle gauge and timber that averages 12 to 15 in. in diameter, use the same number of points as $1/10$-acre plots.

3 For reliable estimates, never use fewer than 30 points in natural timber stands or less than 20 points in even-aged plantations.

11-23   Point Samples versus Plots   Of the numerous field comparisons of point sampling and plot cruising, one of the more extensive evaluations was made by the U.S. Forest Survey in southeast Texas (Grosenbaugh and Stover,

1957). In this test, BAF 10 point samples were measured from the centers of 655 circular ¼-acre plots that were distributed throughout 12 counties. Per-acre comparisons were made for BA, cubic-foot volumes, and board-foot volumes.

Differences in mean volumes by the two sampling methods were not significant at the 5 percent level. Coefficients of variation for point sampling were only 7 to 12 percent larger than for the ¼-acre plots, and standard errors were within 0.5 percent of each other. It was estimated that 20 percent more point samples would be needed to provide the same precision in cubic volume as derived from the plots; however, even with these additional samples the points could be measured in considerably less field time.

11-24 Attributes and Limitations In summary, the principal advantages of point sampling over plot cruising are:

1 It is not necessary to establish a fixed plot boundary; thus greater cruising speed is possible.

2 Large high-value trees are sampled in greater proportions than smaller stems.

3 BA and volume per acre may be derived without direct measurement of stem diameters.

4 When volume-per-acre conversions are developed in advance of fieldwork, efficient volume determinations can be made in a minimum of time. Thus the method is particularly suited to quick, reconnaissance-type cruises.

The main drawbacks to point sampling are:

1 Heavy underbrush reduces sighting visibility and cruising efficiency.

2 Because of the relatively small size of sampling units, carelessness and errors in the tally (when expanded to tract totals) are likely to be more serious than in plot cruising.

3 Slope compensation causes difficulties that may result in large errors unless special care is exercised. Similar difficulties are encountered in strip and line-plot cruising, of course.

4 Unless taken into account, problems can arise in edge-effect bias when sampling very small tracts or long, narrow tracts.

PROBLEMS

11-1 Prepare a compilation similar to Table 11-2 for dbh values 4 through 10 and BAF 20.

11-2 Prepare a table of limiting distances (similar to Table 11-3) in feet and links (or in meters) for BAF 20.

11-3 Construct a calibrated angle gauge for an appropriate BAF. Then establish 10 or more sample points in a forested tract, and design a simple inventory to compare relative efficiencies of the stick-type angle gauge, the Spiegel relascope, and the wedge prism.

11-4 Derive an appropriate set of cubic-foot or cubic-meter volume conversions for point sampling in your locality.

11-5 Establish 30 to 50 randomly selected points in a forest area. Make independent point-sample and circular-plot inventories based on the same center points. Compare results as to mean volumes, confidence limits on the sample means, average number of trees tallied per sample unit, and inventory time per sample unit.

11-6 Suppose that the tree tally shown in Table 11-4 was distributed as follows for the 12 sample points:

Point no.	Tree tally	Point no.	Tree tally
1	3 10-in., 1-log 2 12-in., 2-log	7	6 14-in., 3-log 2 16-in., 2-log 3 16-in., 3-log
2	1 12-in., 1-log 6 14-in., 2-log	8	4 12-in., 2-log
3	3 10-in., 1-log 1 10-in., 2-log 4 12-in., 2-log	9	6 10-in., 1-log 3 10-in., 2-log
4	No trees tallied	10	7 12-in., 1-log 5 14-in., 2-log
5	5 12-in., 2-log	11	3 10-in., 2-log 4 12-in., 2-log 4 12-in., 3-log
6	3 12-in., 3-log 5 14-in., 2-log 2 14-in., 3-log 3 16-in., 3-log	12	7 14-in., 3-log 2 16-in., 2-log 2 16-in., 3-log

Assuming that a simple random sample was performed,
a Estimate the mean volume per acre.
b Compute the standard error of the mean.
c Compute 90 percent confidence limits for the mean.

11-7 By using a BAF 20 instrument, 180 trees were tallied at 25 points. The sum of the volume (cu ft)/basal-area ratios for all tallied trees is 5340.
a Estimate the average BA per acre.
b Estimate the average cubic-foot volume per acre.

11-8 The following constant form factor volume equation predicts volume V in cords from dbh in inches and the number of 5-ft pulpwood sticks H in loblolly pine trees:

$$V = 0.00022(\text{dbh})^2 H$$

By using a BAF 5 instrument, a total of 980 pulpwood sticks was tallied at 20 points. Estimate the mean cordwood volume per acre.

11-9   Apply the ratio-of-means technique (Sec. 8-15) to the data in Sec. 11-20 to estimate the population mean volume per acre and the standard error of the mean. Compute approximate 95 percent confidence limits for the mean. Compare the results with those shown in Sec. 11-20 for the linear regression estimator.

11-10   Assume that the 10 volume-per-acre observations shown in Sec. 11-20 were obtained using simple random sampling. Estimate the mean volume per acre and the standard error of the mean, and place 95 percent confidence limits on the mean. Compare the results with those shown in Sec. 11-20 for the linear regression estimator.

REFERENCES

Barrett, J. P. 1969. Estimating averages from point-sample data. *J. Forestry* **67:**185.

———, and Nevers, H. P. 1967. Slope correction when point-sampling. *J. Forestry* **65:**206–207.

Beers, T. W., and Miller, C. I. 1964. Point sampling: Research results, theory, and applications. *Purdue Univ. Agr. Expt. Sta. Res. Bull.* 786. 56 pp.

Bell, J. F., and Dilworth, J. R. 1990. *Log scaling and timber cruising.* O.S.U. Bookstores, Inc., Corvallis, Oreg. 394 pp.

Bitterlich, W. 1948. Die Winkelzahlprobe. *Allgem. Forest-u, Holzw. Ztg.* **59**(1/2):4–5.

———. 1984. *The relascope idea.* Commonwealth Agricultural Bureau, Slough, England. 242 pp.

Bruce, D. 1955. A new way to look at trees. *J. Forestry* **53:**163–167.

Burkhart, H. E., Barrett, J. P., and Lund, H. G. 1984. Timber inventory. Pp. 361–411 in *Forestry handbook,* K. F. Wenger (ed.), John Wiley & Sons, New York.

Gregoire, T. G. 1982. The unbiasedness of the mirage correction procedure for boundary overlap. *Forest Sci.* **28:**504–508.

Grosenbaugh, L. R. 1952. Plotless timber estimates—New, fast, easy. *J. Forestry* **50:**32–37.

———, and Stover, W. S. 1957. Point-sampling compared with plot-sampling in southeast Texas. *Forest Sci.* **3:**2–14.

Hovind, H. J., and Rieck, C. E. 1970. Basal area and point-sampling: Interpretation and application. *Wisconsin Conservation Dept. Tech. Bull.* 23. 52 pp. (Revised.)

Hunt, E. V., and Baker, R. D. 1967. Practical point-sampling. *SFA State College, Bull. 14.* Nacogdoches, Texas. 43 pp.

Myers, C. A. 1963. Point-sampling factors for southwestern ponderosa pine. *U.S. Forest Serv., Rocky Mt. Forest and Range Expt. Sta., Res. Paper* RM-3. 15 pp.

Oderwald, R. G. 1981. Point and plot sampling—The relationship. *J. Forestry* **79:**377–378.

———, and Jones, E. 1992. Sample sizes for point, double sampling. *Can. J. For. Res.* **22:**980–983.

Robinson, D. W. 1969. The Oklahoma State angle gauge. *J. Forestry* **67:**234–236.

Stage, A. R. 1959. A cruising computer for variable plots, tree heights, and slope correction. *J. Forestry* **57:**835–836.

Wensel, L. C., Levitan, J., and Barber, K. 1980. Selection of basal area factor in point sampling. *J. Forestry* **78:**83–84.

Wiant, H. V., Jr., and Maxey, W. R. 1979. Board-foot factors for point sampling. *J. Forestry* **77:**29.

CHAPTER 12

INVENTORIES WITH
3P SAMPLING

12-1 Introduction The 3P (probability proportional to prediction) system of timber inventory was designed for situations where a highly precise estimate of the volume or value of standing trees is required. Although 3P sampling is not a feasible alternative to conventional fixed-area and point-sampling methods (Chaps. 10 and 11) for many operational cruises, it has found use in special circumstances. An example is a lump-sum timber sale made on the basis of cruising standing timber. If a list of individual trees in the timber sale were available prior to sampling, sample trees could be selected from the list to estimate the overall sale volume. In most forest inventory work, a list is not available, nor is it feasible to compile a list of the population prior to sampling. The 3P system is a variation of list sampling that does not require a list of the population prior to performing the inventory.

The 3P scheme exploits the ability of timber cruisers to estimate volume or some related numerical value in trees. By visiting every tree in a timber sale and estimating its volume, for example, the cruiser would have an estimate of total volume in the sale. This estimate will be either higher or lower than the actual volume. To correct the total estimated volume, a few sample trees are selected and measured. The average ratio of the measured volume to the estimated volume in the sample trees is computed and used to adjust the total estimated volume.

12-2 Components of 3P Inventory The 3P method of timber inventory, as conceived and developed by L. R. Grosenbaugh, consists of three components:

(1) a rule (3P) for selecting sample trees, (2) a method for observing the variables of interest (such as volume determined by using an optical dendrometer or scaled volume from felled trees) on the sample trees selected, and (3) use of computer programs to transform the sample observations to estimates for the whole forest (for example, to compute the volume of sample trees from dimensions observed with an optical dendrometer). These three components—selection, measurement, and computation—are interrelated in any forest inventory and all must be considered. The purpose of this chapter is to present the basic concepts and principles involved in the design and execution of a 3P timber inventory. Details on the use of various instruments for measurement of standing trees and on the use of computer programs developed by Grosenbaugh to convert these measurements to whole tract estimates can be found in the references listed at the end of the chapter.

The 3P procedure involves measurement of volume or a related numerical value for a sample of trees in the population of interest. Because direct measurement of tree volumes (and similar values) is time-consuming and expensive, only a small number can be chosen. Thus an efficient, unequal-probability selection rule is needed to determine which trees to measure (the 3P selection rule). In 3P sampling, the predicted tree attribute (e.g., volume) is paired with a random number from a specially constructed list. Trees with predicted values greater than or equal to the matched random number are measured—thus the probability of selection is proportional to prediction. With this selection scheme, the larger the tree's numerical value, the more chance it has of being selected for measurement.

To illustrate how this unequal-probability selection rule works, suppose that a numerical value is predicted for each tree in a specific population, and the predicted number will always be an integer from 1 through 10. After the value is estimated, it is compared with a random number drawn from the integers 1 through 10; each integer is equally likely to occur. If the predicted value for the tree is greater than or equal to the random number drawn, the tree is selected as a part of the sample. Thus if the tree's predicted value is 2, it has 2 chances in 10 of being selected as part of the sample (i.e., it will be selected only if the random integer is 1 or 2; the probability of drawing a 1 or a 2 is 0.1 plus 0.1, or 0.2). If the tree's predicted value is 8, for example, the chances of its being selected as part of the sample are 8 in 10. Hence, with this selection rule, the probability of inclusion in the sample is proportional to the predicted numerical value of the tree.

Tree volume tables (Chap. 7) are appropriate and highly useful in many timber inventory situations. However, the sample trees used to construct the volume table may have been selected from a population of trees that is different from the one being inventoried. Consequently, some bias is likely when volume tables are applied. In cases where accurate and precise estimates of volume are needed, and when this volume must be determined in standing trees (i.e., there is no opportunity to measure the cut products directly), measurements can be made on a sample of trees in the population being inventoried. This approach thus avoids the application of volume tables and the bias that is likely to be incurred. Direct measurement of tree volume or some related value constitutes the measurement aspect of a 3P inventory.

Measurement of sample trees creates large quantities of data which are best handled through computer programs to produce estimates of total tract volume and associated sampling errors.

HOW 3P IS APPLIED

12-3   Timber-Sale Example   Inventories with 3P sampling were originally applied to timber sales where each tree in the population (all of the marked trees) was visited and a volume, value, or related attribute predicted. A sample of this marked-tree population was selected for detailed measurement. The steps in conducting a 3P sample for the purpose of estimating the total volume of timber marked for sale will serve to illustrate the application of 3P sampling. For this example, the "3P variable" is volume—i.e., the predicted variable for each tree is its volume.

12-4   Preliminary Steps   Before the actual 3P sample is conducted, some preliminary information is needed. The steps involved in obtaining this preliminary data are:

1 Determine the number of trees n_e to be precisely measured for volume (this is termed the expected sample size in 3P inventories) with the following formula, previously described in Chap. 8:

$$n_e = \left[\frac{(t)(CV)}{A}\right]^2$$

where n_e = number of sample trees needed to achieve precision of A, with probability level determined by t

t = quantity from t distribution (generally taken as 1 for 67 percent and 2 for 95 percent probability levels, respectively)

CV = coefficient of variation, percent, of y_i = measured-volume/estimated-volume values

A = allowable error, percent

The variation in 3P sampling is related to the ratio of measured to estimated volume, and most cruisers obtain a coefficient of variation (CV) of 15 to 20 percent. In typical 3P sampling situations, the allowable sampling error is set at 1.5 to 2 percent, which might then require that around 100 to 200 trees be measured

(67 percent probability level). The CV for the ratio of measured to estimated volume depends greatly on the skill and experience of the cruiser, of course. As a general guide, with trained personnel, 100 or so sample trees are usually sufficient for an allowable error of 1.5 percent; beginners require about 200 sample trees for this precision.

In addition to the computation of the expected number of sample trees, some preliminary information about the area to be inventoried is required. This information can be obtained from previous cruises of similar forest types or from a reconnaissance cruise through the area of interest. Specifically, the cruiser must obtain the following information.

2 Estimate the sum of the volumes $\hat{T}_x$ of the N trees in the population of interest

$$\hat{T}_x = \sum_{i=1}^{N} X_i$$

where X_i is the cruiser's estimate of individual-tree volume. Note that the actual sum of estimated volumes is known only after the inventory is completed.

3 Estimate the maximum individual-tree volume expected K. That is,

$$K = \text{maximum } X_i$$

The maximum tree volume expected is obtained at the time the total volume $\hat{T}_x$ is estimated.

Besides obtaining an estimated total volume and maximum tree volume, it is also necessary to obtain the following information.

4 Estimate the number N of trees in the population of interest.

With information on the total volume and maximum tree volume expected, it is now possible to perform step 5.

5 Generate a population-specific set of random numbers from 1 to $K + Z$. The variable Z is used to control the expected number of sample trees in the actual 3P cruise. An equal number of each integer from 1 to K is generated; numbers greater than K are assigned a rejection symbol to facilitate the use of the selection rule in the field (-0 will be used for the rejection symbol). In 3P sampling the probability P of a given tree's being selected for measurement is

$$P = \frac{X_i}{K + Z}$$

where X_i represents the tree's estimated volume. This implies that the number of sample trees that will be measured in any given 3P cruise will be approximately $\hat{T}_x/(K + Z)$. We want the actual number to approximate the expected number as computed by the formula shown in step 1. Thus we can set n_e equal to $\hat{T}_x/(K + Z)$ and compute the value for Z as

$$Z = \frac{\hat{T}_x}{n_e} - K$$

Computer programs are available for generation of these tailor-made random numbers, the quantity of which must equal or exceed the number N of trees in the population of interest.

12-5   Field Procedure   The preliminary work just outlined (steps 1 to 5) can commonly be achieved in about one day. After completing these initial steps, the cruiser is ready to proceed with the 3P cruise. In the field the cruiser visits each of the N trees in the population. The area being cruised is often divided into strips 1 to 2 chains wide to ensure that each tree is visited but that the same tree is not tallied twice. The edges of these strips can be marked with kite string, paint, or ribbons. At each tree the cruiser must:

1 Estimate tree volume.
2 Record the estimate.
3 Draw a random number from the set of integers from 1 through K.
   a If the random number is less than or equal to the volume estimate X_i, precisely measure the tree volume.
   b If a rejection symbol ("null") is drawn or if the integer is greater than the estimated volume, move on to the next tree.

In field practice, two-person field crews are commonly used. The cruiser estimates the volume and calls this estimate to the assistant who carries the random number list and makes the comparison. This practice avoids the possibility of bias, since the cruiser has no notion of what the next random number might be.

Because ocular volume estimates are difficult for inexperienced cruisers, they sometimes measure diameter at breast height (dbh) with a Biltmore stick and enter a single-entry volume table to obtain the "estimated" volume. This "local" volume table value is compared with the appropriate random integer to determine whether the tree should be precisely measured for volume.

In some cases, trees are found that are larger than K, the estimated largest tree. When this happens, the tree is cruised and its volume is kept separate and later added to the 3P estimated volume. (Trees with volumes larger than K are commonly termed *sure-to-be-measured*.) It is sometimes desirable to set K slightly less than the largest tree value expected, thus ensuring that all the largest, most valuable trees in the population will be measured.

The sample trees can be measured immediately, or they can be marked and located on a map so that the cruiser can return to obtain the measurements after all volume estimates are completed.

12-6 Sample-Tree Measurement Various methods of sample-tree measurement, including conventional tree measurements, have been used with 3P selection. A description of the two primary methods that have been applied for obtaining detailed sample-tree measurements in 3P inventories follows.

1 Optical dendrometers have been used to measure upper-stem diameters which are then converted to volumes of standing trees. A number of instruments for measuring upper-stem diameters without resorting to climbing or felling trees are available commercially. One instrument that has been widely applied for dendrometry of sample trees in 3P samples is the Barr and Stroud dendrometer. This sophisticated dendrometer, which is a short-base rangefinder with magnifying optics, permits the measurement of diameter and the height to that diameter for points on the visible portion of tree boles. These measurements are converted to "measured volume" for the 3P sample trees.

Although the Barr and Stroud instrument provides excellent measurement data, it is an expensive instrument that requires skilled operators to use it effectively. Also, computer programs are needed to convert efficiently the instrument readings to tree volumes. Studies have shown (e.g., Yocom and Bower, 1975) that, for many purposes, satisfactory tree volume determinations can be made with less sophisticated instruments such as the Spiegel relascope.

2 Falling, bucking, and scaling of sample trees has been applied. In situations where timber is defective and breakage is likely for felled trees (as with mature timber in the western United States), measurement of net volume in standing trees is subject to considerable error. In these circumstances, the 3P sample trees may be felled, bucked, and scaled according to local utilization standards for their volume, defect, and grade. Sample trees felled as part of the inventory are sent to the mill when the area is logged. This system is sometimes referred to as *fall, buck, and scale cruising* (Johnson and Hartman, 1972).

12-7 3P Computations After the cruise has been completed and the sample trees measured, one can compute an estimate of the total volume of the N trees in the population as

$$\hat{T}_y = T_x \left(\frac{\displaystyle\sum_{i=1}^{n} \frac{Y_i}{X_i}}{n} \right)$$

where $T_x = \displaystyle\sum_{i=1}^{N} X_i$ is now the sum of the observed X_i values.

Note that this estimate of total volume is simply the sum of the predicted volumes for all trees adjusted by the mean ratio of observed over predicted volumes of the sample trees. If any sure-to-be-measured trees are encountered, their volume must be added to $\hat{T}_y$ to obtain an estimate of the overall population total. The symbol n denotes the actual number of sample trees measured. For large samples, the number measured should be close to the number desired (denoted previously as n_e). However, minor differences will occur because of vagaries associated with random numbers in the sample-tree selection process.

If trees with an estimated volume greater than K are encountered, their volume is added to $\hat{T}_y$ when estimating total volume. These sure-to-be-measured trees are not included in the computation of variance, however.

The variance of $\hat{T}_y$ can be estimated by one of several different approaches. One approximating formula that should give satisfactory results is

$$S_{\hat{T}_y}^2 = \frac{\sum_{i=1}^{n}\left(\frac{Y_i T_x}{X_i} - \hat{T}_y\right)^2}{n(n-1)}$$

Additional formulas for approximating the variance of 3P estimates have been presented by Schreuder et al. (1968), Grosenbaugh (1976), and others.

12-8   Numerical Example   In practice, 3P sampling is applied in reasonably large populations. For small populations, a complete enumeration would likely be more economical than a 3P sample. However, a greatly simplified example with a small artificial population can be effectively used to illustrate the application of the 3P concepts and computations just described.

Suppose that we have a forest of 10 trees that we wish to inventory by 3P sampling. In this example, the "3P variable" is board-foot volume. The actual volumes of the trees, although unknown to us, are given in Table 12-1.

One conducts the preliminary steps for the 3P inventory as follows:

1 Determine the number of trees n_e to be precisely measured for volume. For this computation, assume a CV of the ratios of measured to estimated volume of 20 percent, an allowable error A of 10 percent, and a probability level of 67 percent (i.e., $t = 1$), which gives

$$n_e = \frac{(1)^2 (20)^2}{(10)^2} = 4$$

2 Conduct a reconnaissance cruise to estimate the volume of the total population $\hat{T}_x$, the maximum individual-tree volume expected K, and the number N of trees in the population. Assume that the precruise resulted in the following values:

$$\hat{T}_x = 1700 \text{ bd ft}$$
$$K = 350 \text{ bd ft}$$
$$N = 12$$

TABLE 12-1
BOARD-FOOT VOLUMES IN EXAMPLE
FOREST OF 10 TREES

Tree no.	Actual volume, bd ft
1	200
2	80
3	300
4	50
5	400
6	160
7	110
8	40
9	60
10	250
Total	1650

3 Using the precruise information, generate a population-specific set of random numbers from 1 to $K + Z$. The value of Z is computed as

$$Z = \frac{1700}{4} - 350 = 75$$

Since we estimated N to be equal to 12, we will generate 12 random numbers from 1 to $K + Z$ (425), with numbers greater than K (350) being assigned a

TABLE 12-2
ESTIMATED VOLUMES AND RANDOM NUMBERS
GENERATED BY 3P CRUISE

Tree no.	Estimated volume, bd ft	Random number
1	180	112
2	90	327
3	300	311
4	60	−0
5	380	−0
6	150	266
7	100	100
8	50	287
9	80	261
10	300	81
. . . .	. . . .	−0
. . . .	. . . .	74

rejection symbol (–0). Note that this set of random numbers will result in the selection of one 3P sample tree for about every 425 bd ft of volume.

4 Proceed to the field and visit each of the N trees in the population. Suppose that the 3P cruise gave the results shown in Table 12-2. Tree number 5 must be measured because its estimated volume exceeds K. Trees 1, 7, and 10 are 3P sample trees because their estimated volumes are greater than or equal to the corresponding random integer.

A summary of this 3P cruise is given in Table 12-3.

5 Compute an estimate of the total volume and variance of this estimate by using the formulas previously given.

$$\text{Total volume} = \hat{T}_y + \text{(sure-to-be-measured)}$$

$$= T_x \left(\frac{\displaystyle\sum_{i=1}^{n} \frac{Y_i}{X_i}}{n} \right) + \text{(sure-to-be-measured)}$$

$$= 1310(1.015) + 400$$

$$= 1330 + 400$$

$$= 1730 \text{ bd ft}$$

$$\text{Variance} = S_{\hat{T}_y}^2 = \frac{\displaystyle\sum_{i=1}^{n} \left(\frac{Y_i T_x}{X_i} - \hat{T}_y \right)^2}{n(n-1)}$$

$$= \left[\left(\frac{200}{180}(1310) - 1330 \right)^2 + \left(\frac{110}{100}(1310) - 1330 \right)^2 \right.$$

$$\left. + \left(\frac{250}{300}(1310) - 1330 \right)^2 \right] \times \frac{1}{3(2)}$$

$$= 14,148$$

One will recall that only the 3P sample trees enter into the computation of variance; thus n equals 3 in this example.

EXTENSIONS, ATTRIBUTES, AND LIMITATIONS OF BASIC 3P SAMPLING

12-9   Extensions of Basic 3P Sampling   There are many extensions of 3P sampling for situations where it is not feasible to visit every tree in the population. These extensions have generally consisted of multistage sampling designs. For example, work has been done on combining point sampling (first stage) with 3P sampling (second stage) in a two-stage design.

TABLE 12-3
SUMMARY OF 3P CRUISE

Tree no.	Measured volume		Estimated volume	Meas./est.
	3P	Sure-to-measure		
1	200	. . . .	180	1.111
2	. . . .	. . . .	90	. . .
3	. . . .	. . . .	300	. . .
4	. . . .	. . . .	60	. . .
5	. . . .	400	. . .	. . .
6	. . . .	. . . .	150	. . .
7	110	. . . .	100	1.100
8	. . . .	. . . .	50	. . .
9	. . . .	. . . .	80	. . .
10	250	. . . .	300	0.833
Total	. . . .	400	1310	3.044

In such a two-stage design, point-sampling techniques are used to select a subset of trees from the total population. Trees selected by the point sample are then assessed for some 3P variable. One might recall that trees are selected with probability proportional to their basal area (dbh^2) in point sampling. Thus if we make the 3P variable height, measurement trees would be selected with probability proportional to dbh^2 times height, which is highly correlated with volume.

Those interested in extensions of the basic 3P methods described here should consult the literature. References cited at the end of this chapter provide an introduction to this literature.

12-10   Attributes and Limitations of 3P Sampling It should be noted that 3P sampling is not limited to timber inventory alone, that 3P variables other than volume have been used successfully, and that many different techniques and instruments can be applied for obtaining the "measured" value of the 3P variable. The 3P selection rule (probability proportional to prediction) has many potential applications in forest sampling other than just selecting trees for volume measurement, and optical dendrometers are useful in a host of measurement applications besides 3P timber volume estimation. For any inventory, the forester must select the sample design, field measurement methods, and instruments that will provide the required information at an acceptable cost.

In summary, 3P sampling has found its greatest application in timber inventory in situations where:

1 There are relatively few stems per unit area and each individual tree is of relatively high value.

2 Each stem is utilized for several different products.

3 There is no convenient place or time in the harvesting and utilization process where the products can be scaled by conventional methods.

Under the circumstances listed, a 3P sample design similar to the basic example given here may be feasible. For cases where there are numerous stems per unit area, each being of relatively low value, the basic 3P scheme will likely not be efficient.

PROBLEMS

12-1 Prepare a written report on possible applications for 3P sampling procedures, besides the timber sale example discussed in this chapter.

12-2 The following values were determined for a timbered tract that is to be inventoried with 3P sampling:

Number of trees n_e to be measured	140
Estimate of sum of volumes $\hat{T}_x$ of N trees in population	1,800,000 bd ft
Estimate of maximum individual-tree volume expected K	620 bd ft

Compute the value for Z that should be used when generating the random numbers list for this tract.

12-3 Given the following data from a 3P cruise of a population of 15 trees, estimate the total cubic-foot volume and its associated variance for this 3P cruise.

	Measured volume, cu ft		Estimated
Tree no.	3P	Sure-to-measure	volume, cu ft
1	. . . .	. . . .	20
2	14	. . . .	16
3	. . . .	. . . .	31
4	25	. . . .	22
5	. . . .	. . . .	14
6	. . . .	. . . .	7
7	. . . .	41	. . .
8	. . . .	. . . .	16
9	30	. . . .	27
10	. . . .	. . . .	17
11	. . . .	. . . .	38
12	. . . .	46	. . .
13	. . . .	. . . .	5
14	8	. . . .	10
15	. . . .	. . . .	16

12-4 A reconnaissance cruise through an area to be inventoried by 3P sampling yielded the following information:

Estimate of sum of values $\hat{T}_x$ of N trees in population	60
Estimate of maximum individual-tree volume expected K	10
Number N of trees in population	15
Number of trees n_e to be measured (previously calculated)	5

Use the random numbers table shown in the Appendix of this text and the information from the reconnaissance survey to develop a tailor-made list of random numbers for use in this 3P cruise.

12-5 For a given timber tract in your area, conduct a reconnaissance cruise to determine $\hat{T}_x$, K, and N. Submit a written report describing the procedures used and the time required to develop this information.

REFERENCES

Bell, J. F., and Dilworth, J. R. 1990. Log scaling and timber cruising. O.S.U. Bookstores, Inc. Corvallis, Oreg. 394 pp.

Furnival, G. M., Gregoire, T. G., and Grosenbaugh, L. R. 1987. Adjusted inclusion probabilities with 3P sampling. *Forest Sci.* **33**:617–631.

Grosenbaugh, L. R. 1964. Some suggestions for better sample-tree-measurement. *Soc. Amer. Foresters (1963, Boston, Mass.) Proc.*:36–42.

———. 1965. Three-pee sampling theory and program "THRP" for computer generation of selection criteria. *U.S. Forest Serv., Pacific Southwest Forest and Range Expt. Sta., Res. Paper* PSW-21. 53 pp.

———. 1967a. The gains from sample-tree selection with unequal probabilities. *J. Forestry* **65**:203–206.

———. 1967b. STX—Fortran-4 program for estimates of tree populations from 3P sample-tree-measurements. *U.S. Forest Serv., Pacific Southwest Forest and Range Expt. Sta., Res. Paper* PSW-13. 2d ed., rev. 76 pp.

———. 1974. STX-3-3-73: Tree content and value estimation using various sample designs, dendrometry methods, and V-S-L conversion coefficients. *U.S. Forest Serv., Southeast. Forest Expt. Sta., Res. Paper* SE-117. 112 pp.

———. 1976. Approximate sampling variance of adjusted 3P sampling estimates. *Forest Sci.* **22**:173–176.

Hartman, G. B. 1967. Some practical experience with 3-P sampling and the Barr and Stroud dendrometer in timber sales. *Soc. Amer. Foresters (1966, Seattle, Wash.) Proc.*:126–130.

Johnson, F. A., and Hartman, G. B., Jr. 1972. Fall, buck, and scale cruising. *J. Forestry* **70**:566–568.

———, Dahms, W. G., and Hightree, P. E. 1967. A field test of 3P cruising. *J. Forestry* **65**:722–726.

Mesavage, C. 1971. STX timber estimating with 3P sampling and dendrometry. *U.S. Dept. Agr. Handbook* 415, Government Printing Office, Washington, D.C. 135 pp.

Schreuder, H. T., Brink, G. E., and Wilson, R. L. 1984. Alternative estimators for point-Poisson sampling. *Forest Sci.* **30**:803–812.

———, Sedransk, J., and Ware, K. D. 1968. 3-P sampling and some alternatives, I. *Forest Sci.* **14**:429–453.

Wiant, H. V., Jr. 1976. Elementary 3P sampling. *West Va. Univ. Agr. and Forestry Expt. Sta. Bull.* 650T. 31 pp.

Yocom, H. A., and Bower, D. R. 1975. Estimating individual tree volumes with Spiegel Relaskop and Barr and Stroud dendrometers. *J. Forestry* **73**:581–582, 605.

CHAPTER 13

USING AERIAL PHOTOGRAPHS

13-1 Purpose of Chapter   Aerial photographs are useful tools of the forest manager. A basic knowledge of the location and extent of the forest is critical to the management of forest resources. Aerial photographs are used to develop maps, but these maps have somewhat different properties than those described in Chap. 3. Aerial photographs are also useful in designing and conducting field inventories (Chaps. 9 through 12), and can actually be used to estimate tree and stand characteristics directly in ways analogous to field methods.

This chapter includes information on how aerial photographs can be used to identify basic forest cover types and how various stand parameters can be estimated from aerial photographs. Only the most elementary techniques are described here. Readers interested in nonphotographic imagery (e.g., satellites, thermal scanners, radar, and aerial videography) and detailed photogrammetric procedures should consult the references cited at the end of the chapter.

13-2 Types of Aerial Photographs   As a general rule, foresters are primarily concerned with *vertical photographs,* i.e., those taken with an aerial camera pointed straight down toward the earth's surface. Consecutive exposures in each flight line are overlapped to allow three-dimensional study with a stereoscope. Although few (if any) aerial photographs are truly vertical views, they are usually presumed to be vertical when exposures are tilted no more than 3°. Unless otherwise specified, the terms *photo* and *photograph* as used in this chapter will denote vertical aerial photographs.

Oblique photographs are exposures made with the camera axis pointed at an angle between the vertical and the horizon. Although obliques are useful for panoramic views, they are not easily adapted to stereoscopic study or to photo measurement; hence they are seldom used for forest inventory purposes.

Mosaics are assembled by cutting, matching, and pasting together portions of individual vertical exposures; the result is a large photograph that appears to be a single print. Controlled mosaics, i.e., those compiled at a uniform scale from ground reference points, provide good map approximations. However, controlled mosaics are quite expensive and cannot be viewed three dimensionally. Except for pictorial displays, their use by foresters is limited.

13-3 Black-and-White Aerial Films Selection of the proper film is an important factor in distinguishing vegetation classes on aerial photos. Foresters usually rely on two types of black-and-white films, panchromatic and infrared, and two kinds of color films, normal color and infrared color.

Images on *panchromatic* film are rendered in varying shades of gray, with each tone comparable to the density of an object's color as seen by the human eye. Panchromatic is a superior black-and-white film for distinguishing objects of truly different colors; it is recommended for such projects as highway route surveys, urban planning, mapping, and the locating of property ownership boundaries (Fig. 13-1). Since old roads and trails are easily seen, panchromatic prints are also useful as field maps for the land manager who must find routes over unfamiliar terrain.

Because panchromatic film is only moderately sensitive to the green portion of the spectrum, most healthy vegetation appears in similar gray tones on the prints. Where only coniferous (needleleaf) trees are present, panchromatic film may be preferred for the classification of forest vegetation. On the other hand, if deciduous (broadleaf) vegetation is interspersed with coniferous trees, separation of the different types can be more easily accomplished by using black-and-white infrared film.

Gray tones on *infrared* film apparently result from the degree of infrared reflection of objects rather than from their true colors. For example, broadleaf vegetation is highly reflective and, therefore, photographs in light tones of gray; coniferous or needleleaf vegetation tends to be less reflective in the near-infrared portion of the spectrum and, consequently, registers in much darker tones. This characteristic makes infrared film the preferred black-and-white emulsion for delineating timber types in mixed forests.

Bodies of water absorb infrared light to a high degree and, thus, register quite dark on the film unless heavily silt-laden. This rendition is useful for determining the extent of river tributaries, canals, tidal marshes, swamps, and shorelines. Infrared films are superior to panchromatic materials for penetration of haze.

FIGURE 13-1
Panchromatic stereogram of a forested area in eastern Texas being developed for housing
(above). Scale is about 1:12,000. Below is a large-scale photograph covering the block outlined on the stereogram. *(Courtesy United Aerial Mapping.)*

13-4   Color Aerial Films   *Normal color film* is sensitized to all visible colors; it provides imagery with natural color rendition when properly exposed and processed. The emulsion has proved especially valuable for identifying soil types, rock outcrops, and forest vegetation. Normal color can also penetrate water, and it is therefore useful for subsurface exploration, hydrographic control, and the delineation of shoreline features. As is the case with most color films, correct exposures require conditions of bright sunlight.

Color films are usually processed to produce positive color transparencies rather than ordinary paper prints. In most instances, conventional color photographs are superior to black-and-white pictures for studies of natural vegetation, because the human eye can distinguish many more color variations than gray tones.

Infrared color film has proved especially valuable for carrying out a variety of forest-survey projects, such as the early detection of disease and insect outbreaks in timber stands, and for detecting soil and vegetation patterns disturbed by human activity. Basic to such application is the identification of tree species or forest cover types, a task that requires information on the infrared reflectivity of various types of foilage. Since healthy broadleaf trees have a higher infrared reflectivity than healthy needleleaf trees, their photographic images can usually be separated on these kinds of exposures.

Obviously, no single film emulsion serves all purposes. Instead, the varied tones and patterns produced by differing ranges of film sensitivity complement each other, and the maximum amount of information can be extracted only when several types of imagery covering the same area are studied simultaneously.

13-5   Seasons for Aerial Photography   The best season for obtaining aerial photographs depends on the nature of the features to be identified or mapped, the film to be used, and the number of days suitable for photographic flights during a given time of the year. Other factors being equal, aerial surveys are likely to be less expensive in areas where sunny, clear days predominate during the desired photographic season.

Foresters interested in vegetation studies will usually specify that aerial photographs be taken during the growing season, particularly when deciduous plants constitute an important component of the vegetative cover. When it is essential that deciduous trees, evergreens, and mixtures of the two groups be delineated, either black-and-white infrared or color infrared film is frequently specified. In regions where evergreen plants predominate, however, panchromatic film or normal color film may be used with equal success.

Several research projects involving black-and-white photography of forest areas have indicated that the best timing for infrared coverage is from mid-spring to early summer—after all trees have produced some foliage but prior to maximum leaf pigmentation. Successful panchromatic photographs of timberlands can be made throughout the year, but in northern United States, and in

Canada, best results have been obtained in late fall—just before deciduous tree species, such as aspen or tamarack, shed their leaves. For a brief period of perhaps 2 weeks' duration, foliage color differences will provide good photographic contrasts between most of the important timber species in this region.

13-6 Determining Photographic Scales The vertical aerial photograph presents a true record of angles, but measures of horizontal distances vary widely with changes in ground elevations and flight altitudes. The nominal scale (as 1:20,000) is representative only of the datum, an imaginary plane passing through a specified ground elevation above sea level. Calculation of the average photo scale will increase the accuracy of subsequent photo measurements.

Aerial cameras in common use have focal lengths of 6, 8.25, or 12 in. (0.5, 0.6875, or 1 ft). This information, coupled with the altitude of the aircraft above ground datum, makes it possible to determine the representative fraction (RF) or natural scale:

$$RF = \frac{\text{focal length, ft}}{\text{flying height above ground, ft}}$$

The exact altitude of the aircraft is rarely known to the interpreter, however, and photo scale is more often calculated by

$$RF = \frac{\text{photographic distance between two points, ft}}{\text{ground or map distance between same points, ft}}$$

As an example, the distance between two road intersections might be measured on a vertical photograph as 3.6 in. (0.3 ft). If the corresponding ground distance is measured as 3960 ft, the representative fraction would be computed as

$$RF = \frac{0.3}{3960} = \frac{1}{13,200} \qquad \text{or } 1:13,200$$

It is not essential to calculate the scale of every photograph in a flight strip. In hilly terrain, every third or fifth print may be used; in flat topography, every tenth or twentieth may be used. Scales of intervening photos can be obtained by interpolation.

13-7 Photogeometry It is important to note here that although aerial photographs are used for mapping purposes, they are not maps in the true sense of the word. Maps have a single known scale, and if no drafting error is made, objects on maps are shown in their correct locations relative to one another. Aerial photographs, however, have image displacements, and distortions created by tilt of the aircraft, varying elevations, and imperfect camera lenses, in addition to the scale changes discussed in Sec. 13-6. While the effects of these displacements and other factors on maps are usually small, they can create significant problems if the terrain is rugged, if flying heights are very low, or if the aerial camera is excessively tilted.

Orthophotographs are, however, very similar to maps. An *orthophotograph* is a reproduction, prepared from ordinary perspective photographs, in which image displacements caused by tilt and relief have been entirely removed. When these unique photographs are assembled into an *orthophotomosaic,* the result is a picture map with both scale and planimetric detail of high reliability. When such mosaics are overprinted onto standard mapping quadrangles, they are referred to as *orthophotoquads.*

13-8   Aligning Prints for Stereoscopic Study   Photographic flights are planned so that prints will overlap about 60 percent of their width in the line of flight and about 30 percent between flight strips. For effective stereoviewing, prints must be trimmed to a nominal 9-in. × 9-in. size, preserving the four fiducial marks in the photo corners or at the midpoint of each of the edges. The principal point (PP) is at the center of the photo and is located by the intersection of lines drawn from opposite sets of fiducial marks. The conjugate principal points (CPPs, or points that correspond to PPs of adjacent photos) are located by stereoscopic transfer from overlapping prints. Each photo, thus, has one PP and two CPPs, except that prints at the ends of flight lines have only one CPP.

To align the photographs for stereoscopic study, a print is selected and fastened down with shadows toward the viewer. The adjacent photo is placed with its CPP about 2.2 in. (average interpupillary distance) from the corresponding PP on the first photo. With flight lines superimposed, the second photo is positioned. A lens stereoscope is placed with its long axis parallel to the flight line and with the lenses over corresponding photo images. In this way an overlapping strip 2.2 in. wide and nearly 9 in. long can be viewed by moving the stereoscope up and down the overlap area (Fig. 13-2).

COVER-TYPE IDENTIFICATION AND MAPPING

13-9   Forest Type Recognition   Interpreters of forest vegetation should be well versed in plant ecology and the various factors that influence the distribution of native trees and shrubs. Field experience in the region of interest is also a prime requisite, because many cover types must be deduced or inferred from associated factors instead of being recognized directly from their photographic images.

An inferential approach to cover-type identification becomes more and more important as image scales and resolution qualities are reduced. Range managers may rely exclusively on this technique where they must evaluate the grazing potential for lands obscured by dense forest canopies.

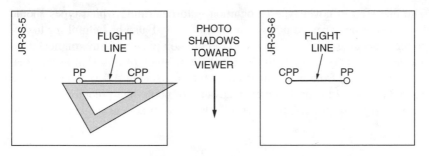

A. PRELIMINARY PHOTO ORIENTATION

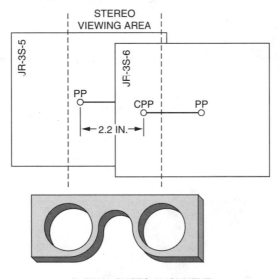

B. FINAL PHOTO ALIGNMENT

FIGURE 13-2
Alignment of 9-in. by 9-in. vertical photographs for viewing with a lens stereoscope.

The degree to which cover types and plant species can be recognized depends on the quality, scale, and season of photography; the type of film used; and the interpreter's background and ability. The shape, texture, and tone (color) of plant foliage as seen on vertical photographs can also be influenced by stand age or topographic site. Furthermore, such images may be affected by time of day, sun angle, atmospheric haze, clouds, or inconsistent processing of negatives and prints. In spite of insistence on rigid specifications, it is often impossible to obtain uniform imagery of extensive land holdings. Nevertheless, experienced interpreters can reliably distinguish cover types in diverse vegetation regions when photographic flights are carefully planned to minimize the foregoing limitations.

The first step in cover-type recognition is to determine which types should and should not be expected in a given locality. It will also be helpful for the interpreter to become familiar with the most common plant and environmental associations of those types most likely to be found. Much of this kind of information can be derived from generalized cover-type maps (Fig. 13-3) and by ground or aircraft checks of the project area in advance of photo interpretation.

In some regions, photo-interpretation keys are available as aids in the recog-

FIGURE 13-3
Generalized forest type map of California. *(U.S. Forest Service drawing.)*

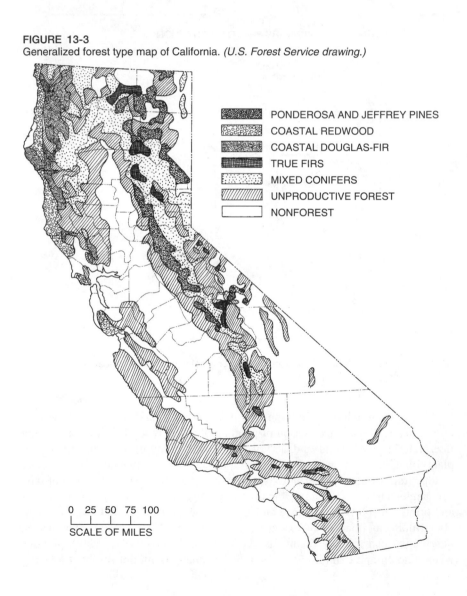

PONDEROSA AND JEFFREY PINES
COASTAL REDWOOD
COASTAL DOUGLAS-FIR
TRUE FIRS
MIXED CONIFERS
UNPRODUCTIVE FOREST
NONFOREST

0 25 50 75 100
SCALE OF MILES

nition of forest cover types; such keys can be especially useful when they are il-lustrated with high-quality stereograms. Vegetation keys for Anglo-American types are most easily constructed for northern and western forests where conifers predominate. In these regions, there are relatively few species to be considered and crown patterns are fairly distinctive for each important group. It should also be noted that forest type recognition can often be improved by using color pho-tographs instead of black-and-white imagery.

13-10   Identifying Individual Species   The identification of an individual species through aerial photography is most easily accomplished when that species occurs naturally in pure, even-aged stands. Under such circumstances, the cover type and the plant species are synonymous. Therefore, reliable delin-eations *may* be made from medium-scale aerial photography. As a rule, howev-er, individual plants can be identified only from large-scale photography (Fig. 13-4). Table 13-1, based on a synopsis of several research reports, illustrates the relationship between photo scale and expected levels of plant recognition.

Photographic identification of individual plants requires that interpreters be-come familiar with a large number of species *on the ground.* For example, there are more than a thousand species of woody plants that occur naturally in the United States; foresters and range managers can rarely identify more than one-fourth of this number.

13-11   Timber Type Maps   Type maps are no longer considered essential by all foresters, but at times their cost may be justified. A general ownership

FIGURE 13-4
Stereogram showing live white pines (large, dark crowns) and dead balsam firs killed by the spruce budworm (smaller, whitish crowns). Scale is about 132 ft per in. *(U.S. Forest Service photograph.)*

TABLE 13-1
RELATIONSHIP BETWEEN PHOTOGRAPHIC SCALE AND EXPECTED LEVELS OF
PLANT RECOGNITION

Photographic scale	General level of plant discrimination
1:30,000–1:100,000	Recognition of broad vegetative types, largely by inferential processes
1:10,000–1:30,000	Direct identification of major cover types and species occurring in pure stands
1:2500–1:10,000	Identification of individual trees and large shrubs
1:500–1:2500	Identification of individual range plants and grassland types

map showing principal roads, streams, and forest types may be desired for management planning and illustrative purposes. When making a photo-controlled ground cruise where precise forest-area estimates are required, it may be necessary to measure stand areas on controlled maps of known scale rather than directly on contact prints; this is particularly important when topography causes wide variation in photo scales.

The wise interpreter will delineate only those forest types that can be consistently recognized. For maximum accuracy, type lines should be drawn under the stereoscope. Wherever feasible, it is recommended that forest cover types be coded according to the system devised by the Society of American Foresters (Eyre, 1980). In the past, timber type maps have assumed a wide variety of forms from one locale to another, but it is advantageous to employ uniform symbols for designating tree species and stand-size classes.

13-12   Using Photos for Field Travel   Although photographic flights are planned to run either north-south or east-west, few prints will be oriented precisely with the cardinal directions. If one wishes to travel cross-country with the aid of a vertical photograph, it is usually necessary to first establish a line of known compass direction on the print. This reference line may be transferred to the photo from existing maps or located directly on the ground by taking the compass bearing of any straight-line feature, e.g., a road or field edge.

Once the reference line is drawn on the photograph, it should be extended so that it intersects the proposed line of cross-country travel. The angle between the two lines is then measured with a protractor to establish the bearing of the travel route. If one wishes to travel to a specific point or field plot along the proposed line, the photo scale should be determined as precisely as possible; then the travel distance can be determined directly on the print with an engineer's scale.

BASIC FOREST MEASUREMENTS

13-13 Measuring Area and Distance Areas, distances, and direction are measured on photographs just as they are on maps (Chap. 3). Area is estimated with a dot grid or planimeter. Ground distances are calculated by measuring a photo distance and converting it to ground units using photo scale. Compass bearings and azimuths can be determined on photographs if a base line of known direction can be established on the photo. It should be remembered, however, that photographs are not maps, and the displacements and variation present on photos occasionally degrade these measurements. Area measurements are typically affected most and thus warrant extra care.

13-14 Measuring Heights by Parallax To determine heights of objects on stereopairs of photographs, it is necessary to measure or estimate (1) absolute stereoscopic parallax and (2) differential parallax. Absolute stereoscopic parallax, measured parallel to the line of flight, is the algebraic difference of the distances of the two images from their respective PPs. Except in mountainous terrain, the average photo base length is ordinarily used as an approximation of absolute stereoscopic parallax. It is measured as the mean distance between the PP and CPP for an overlapping pair of photographs.

Differential parallax is the difference in the absolute stereoscopic parallax at the top and the base of the object, measured parallel to the flight line. The basic formula for conversion of parallax measurements on aerial photographs is

$$h = (H)\frac{dP}{P + dP}$$

where h = height of measured object
H = height of aircraft above ground datum
dP = differential parallax
P = absolute stereoscopic parallax at base of object being measured

If object heights are to be determined in feet, the altitude of the aircraft must also be in feet. Absolute stereoscopic parallax and differential parallax must be expressed in the same units; ordinarily, these units will be in thousandths of inches or hundredths of millimeters.

13-15 Parallax-Measuring Devices Differential parallax dP is usually measured stereoscopically with a parallax wedge or with a stereometer employing the "floating-mark" principle; use of the stereometer is detailed here.

The typical stereometer (or "parallax bar") has two lenses attached to a metal frame that houses a vernier and a graduated metric scale. The left lens contains a fixed reference dot; the dot on the right lens can be moved laterally by means of the vernier. The stereometer is placed over the stereoscopic image parallel to the line of flight (Fig. 13-5). The right-hand dot is moved until it fuses with the reference dot and appears as a single dot resting on the ground, and the vernier reading is recorded to the nearest 0.01 mm. Then the vernier is turned until the fused dot appears to "float" at the elevation of the top of the object. A second vernier reading is taken, and the difference between the two readings is the differential parallax dP. This value can be substituted in the parallax formula without conversion if the absolute parallax P is also expressed in millimeters.

As an example, assume that the two stereometer readings for a building were 10.75 mm (ground) and 9.63 mm (top). The differential parallax is therefore 1.12 mm. If we have an average photo base P of 91.44 mm and an aircraft flying height of 3600 ft, the height of the building would be computed as

$$h = (3600)\frac{1.12}{91.44 + 1.12} = 43.56 \text{ ft}$$

FIGURE 13-5
Lens stereoscope with attached stereometer. This "floating-dot" instrument can be used to measure the heights of objects on vertical stereopairs of photographs. *(Courtesy of Carl Zeiss, Inc.)*

Once the photographic specifications are fixed, the expected precision of height measurement is largely dependent on the stereoscopic perception of the individual interpreter. At photo scales of 1:10,000 to 1:15,000, skilled interpreters can determine the heights of clearly defined objects within ±5 to 10 ft. Measurement precision tends to improve as photo scales become larger, i.e., as aircraft heights above the ground datum decrease.

13-16 Tree-Crown Diameters Measurements of tree-crown diameters are of interest because, for many species, these measurements may be closely correlated with stem diameters. Such relationships have been verified for a number of conifers, notably those occurring in open-grown, even-aged stands. As a result, tree volume tables based on crown diameter (in lieu of dbh) can be constructed for use with aerial photographs (Fig. 13-6).

Crown diameters may be measured either on the ground or on aerial photographs. The difference in perspective afforded by the two measurement techniques can lead to varying results, even for the same trees. If the crowns are im-

FIGURE 13-6
Comparison of ground and photographic measurements in the determination of individual-tree volumes.

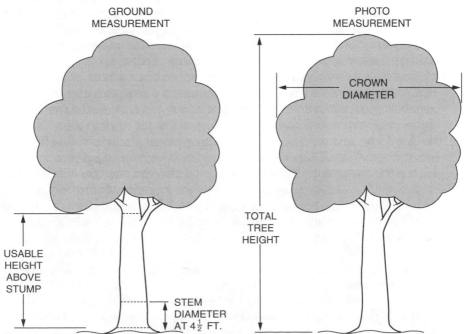

aged on small-scale photographs, for example, only that part of the diameter visible from above is measured; narrow, single branches and irregular crown perimeters may not be resolved by the photographic system. Therefore, photo-measured crown diameters are sometimes smaller than crown measurements made on the ground.

13-17   Tree Counts   One of the easiest and most common forest measurements made on aerial photographs is number of trees per unit area, or tree count. Number of trees per unit area is a commonly used measure of stand density (Chap. 14). Tree counts are used to expand individual-tree volume estimates to a unit-area basis, and, in some instances, as an input variable for growth and yield models. If tree count can be reliably estimated from aerial photographs, the time savings may be significant since it may take 25 to 50 times longer to measure a plot in the field than on an aerial photograph.

The procedure for estimating the number of trees per unit area on aerial photographs is simple. Since it is generally not feasible to count every tree in the area of interest, a sampling process analogous to field procedures is used. A plot of known dimension is constructed on clear material and placed on the photo in a random (or systematic) fashion within the boundaries of the tract of interest. Next, the number of crowns within the plot is tallied and then expanded to a per-unit-area basis using the scale of the photographs. The process is repeated until an adequate sample size is attained, and averages are computed. Some confusion often occurs in the counting process because individual crowns may be difficult to identify and because the variable of interest is stems per unit area, not crowns per unit area. In many forest stands, there is not a one-to-one correspondence between the total number of stems per unit area and the number of visible crowns per unit area.

The accuracy of aerial-photo-based tree counts varies considerably and is highly dependent upon stand age, stand structure, and the characteristics of the aerial photographs used. Photo-based tree counts typically decrease in number and accuracy on smaller-scale photos because smaller crowns merge and become difficult to discern. Thus, scale of the photographs must be carefully selected. Season of acquisition and film type can also affect the accuracy of tree counts. The most successful applications of photo-based tree counts have been in relatively young, even-aged stands of intolerant species, such as southern pines or Douglas-fir. For instance, the number of surviving stems in 3-, 4-, and 5-year-old loblolly pine plantations in Virginia was estimated within about 10 percent using large-scale, 35-mm aerial photographs. The least successful situations involve tolerant species where significant numbers of smaller stems are obscured by taller, upper-canopy-level trees. Even in the best of situations, it is virtually impossible to locate and tally every tree in photo plots. Adjustment factors or equations have been successfully used to correct for these inherent omissions.

If an aerial volume table is based on *photo measurements* of tree crowns, the biases and/or errors of the interpreter are incorporated into the table. This will be acceptable, provided other interpreters with different biases do not have to use the same table. If *ground measurements* of crown diameters are used instead, each interpreter first measures a group of "test trees" to establish an individual photo/ground adjustment ratio. Such tables can then be utilized by large numbers of interpreters.

The assessment of crown diameter is a simple linear measure. On the ground, two persons with steel tape and plumb bobs align themselves at opposite edges of the tree crown by using a vertical sighting device such as a periscope. Two or more diameter measurements are taken for each tree. Very small individual branches and minor crown irregularities are usually ignored. The nearer the widest part of the crown is to ground level, the more accurate the resulting measurement.

Various scales, tube magnifiers, and "wedges" are available for measuring crown diameters on aerial photographs (Fig. 13-7). Careful observers can measure within ±0.1 mm, and so accuracy is dependent on image scale, film resolution, and individual ability. Care must be exercised to avoid the inclusion of crown shadows as part of the measurement. It has been observed that most aerial-photo measurements tend to improve as one changes from paper prints to black-and-white film diapositives to color transparencies. Improvements are partially due to higher resolution and, with color transparencies, better contrast between tree images and backgrounds.

13-18 Individual-Tree Volumes Multiple-entry tree volume tables based on dbh and total height can be converted to aerial volume tables when correlations can be established between crown diameters and stem diameters. Photo determinations of crown diameter are substituted for the usual ground measures of dbh, and total heights are measured on stereoscopic pairs of photographs by the parallax method (Table 13-2).

Volume estimates based on this approach or on prediction equations will generally have a lower precision (i.e., greater standard error) than those estimates based on dbh and height because tree volumes are more closely correlated with

FIGURE 13-7
A simple dot-type scale for measuring crown diameters on vertical photographs. Such scales are usually printed on transparent film. *(U.S. Forest Service drawing.)*

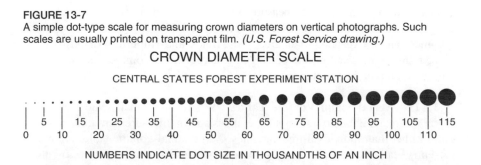

CROWN DIAMETER SCALE

CENTRAL STATES FOREST EXPERIMENT STATION

5	15	25	35	45	55	65	75	85	95	105	115
0	10	20	30	40	50	60	70	80	90	100	110

NUMBERS INDICATE DOT SIZE IN THOUSANDTHS OF AN INCH

TABLE 13-2
INDIVIDUAL-TREE VOLUME TABLE FOR SECOND-GROWTH SOUTHERN PINES,
CU FT*

Crown diameter class, ft	Total tree height, ft						
	50	60	70	80	90	100	110
10	9.5	11.5	12.5	15.0	17.5	19.5	. . . .
12	12.5	14.5	16.5	18.0	20.5	22.5	. . . .
14	15.0	17.0	19.0	23.5	25.0	27.5	30.5
16	17.5	20.5	24.0	27.5	30.5	33.0	36.0
18	. . . .	23.5	27.0	30.5	34.5	38.0	42.5
20	. . . .	28.0	33.5	36.0	40.0	45.5	49.0
22	. . . .	32.5	37.0	42.5	46.5	52.0	57.5
24	. . . .	37.0	42.5	48.5	54.5	60.0	66.0
26	. . . .	42.5	47.5	54.0	61.0	67.5	75.5
28	. . . .	. . . .	53.0	60.5	70.5	76.0	83.0
30	. . . .	. . . .	60.5	68.0	78.0	85.5	94.5

*Based on 324 trees in Arkansas, Louisiana, and Mississippi. Gross volumes are inside bark and include the merchantable stem to a variable top averaging 6 in. ib.

dbh than with crown diameter. And, of course, dbh can be measured with greater precision. Nevertheless, aerial tree volume tables have proved to be cost-efficient for many inventories, particularly for coniferous species growing in relatively inaccessible regions.

As an example, the Forest Management Institute of Ottawa, Canada, has successfully used large-scale aerial photography to inventory a forest of approximately 2700 km² in the Mackenzie River Delta, Northwest Territories. An equation was developed to estimate tree volumes (largely white spruce) on the delta; it was applied to all trees on selected photo sample plots to provide plot-volume estimates. An inventory objective was to estimate total timber volume within ±20 percent at a probability level of 0.95. The aerial-photo approach proved to be the most cost-efficient means of achieving this objective.

Large-scale, high-quality aerial photography is essential for obtaining reliable crown-diameter and height measurements of individual trees. Furthermore, image resolution must be sufficient to permit reliable *stem counts* so that tree volumes can be expanded to an area basis. Black-and-white film diapositives or color transparencies at a scale of 1:5000 or larger are recommended for consistent photographic interpretation results.

13-19   Aerial Stand-Volume Tables   Where only small-scale aerial photographs are available to interpreters, emphasis is on measurement of *stand variables* rather than individual-tree variables. Aerial stand-volume tables are multiple-entry tables that are usually based on assessments of two or three photographic characteristics of the dominant-codominant crown canopy: average stand height, average crown diameter, and percent of crown closure. These tables are usually derived by multiple regression analysis; photographic measurements of the independent variables are made by several skilled interpreters when developing the volume-prediction equation.

13-20 Crown Closure Photographic and ground measurements of tree heights and crown diameters have been previously described. Crown closure, also referred to as *crown cover* and *canopy closure,* is defined by photo interpreters as the percent of a forest area occupied by the vertical projections of tree crowns. The concept is primarily applied to even-aged stands or to the dominant-codominant canopy level of uneven-aged stands. When used in this context, the maximum value possible is 100 percent.

In theory, crown closure is associated with stand volume because such estimates are approximate indicators of stand density, e.g., number of stems per acre. Since basal areas and numbers of trees cannot be determined directly from small-scale photography, crown closure is sometimes substituted for these variables in stand-volume–prediction equations. Photographic estimates of crown closure are normally used because reliable ground evaluations are much more difficult to obtain (Fig. 13-8).

FIGURE 13-8
An indication of tree-crown closure as seen from a ground-level camera. *(Courtesy Ben Jackson, Louisiana State University.)*

At photo scales of 1:15,000 and smaller, crown-closure estimates are usually made by ocular judgment, and stands are grouped into 10 percent classes. Ocular estimates are easiest in stands of low density, but they become progressively more difficult as closure percentages increase. Minor stand openings are difficult to see on small-scale photographs, and they are often shrouded by tree shadows. These factors can lead to overestimates of crown closure, particularly in dense stands. And if ocular estimates are erratic, the variable of crown closure may contribute very little to the prediction of stand volume.

With high-resolution photographs at scales of 1:5000 to 1:15,000, it may be feasible to derive crown-closure estimates with the aid of finely subdivided dot grids. Here, the proportion of the total number of dots that falls on tree crowns provides the estimate of crown closure. This estimation technique has the virtue of producing a reasonable degree of consistency among various photo interpreters; it is therefore recommended wherever applicable.

Table 13-3, compiled for Douglas-fir stands, is based on the variables of total stand height and crown-closure percent.

13-21 Stand-Volume Estimates Once an appropriate aerial stand-volume table has been selected (or constructed), there are several procedures that can be employed to derive stand volumes. One approach is as follows:

1 Outline tract boundaries on the photographs, utilizing the effective area of every other print in each flight line. This assures stereoscopic coverage of the area on a minimum number of photographs and avoids duplication of measurements by the interpreter.

2 Delineate important forest types. Except where type lines define stands of relatively uniform density and total height, they should be further broken down into homogeneous units so that measures of height, crown closure, and crown diameter will apply to the entire unit. Generally, it is unnecessary to recognize stands smaller than 5 to 10 acres.

3 Determine the area of each condition class with dot grids or a planimeter. This determination can sometimes be made on contact prints.

4 By stereoscopic examination, measure the variables for entering the aerial stand-volume table. From the table, obtain the average volume per acre for each condition class.

5 Multiply volumes per acre from the table by condition-class areas to determine gross volume for each class.

6 Add class volumes for the total gross volume on the tract.

13-22 Adjusting Photo Volumes by Field Checks Aerial volume tables are not generally reliable enough for purely photographic estimates, and some allowance must be made for differences between gross-volume estimates and actual net volumes on the ground. Therefore, a portion of the stands (or condition

TABLE 13-3
AERIAL-PHOTO STAND-VOLUME TABLE FOR EVEN-AGED DOUGLAS-FIR IN THE
PACIFIC NORTHWEST, IN 100 CU FT PER ACRE*

Stand height, ft†	Crown closure, %‡								
	15	25	35	45	55	65	75	85	95
40	5	8	11	13	14	15	15	14	13
50	7	11	15	18	20	21	22	21	20
60	9	15	20	24	27	29	30	29	28
70	12	19	25	31	34	37	39	39	38
80	14	24	31	38	43	46	48	49	49
90	17	28	38	45	52	56	59	61	61
100	21	33	45	54	61	67	72	74	75
110	24	39	52	63	72	79	85	88	90
120	28	45	60	73	83	92	99	103	106
130	31	51	68	83	95	106	114	120	124
140	35	57	77	94	108	121	130	138	143
150	40	64	86	105	122	136	148	157	163
160	44	71	96	118	136	153	166	177	185
170	49	79	106	130	152	170	185	198	208
180	54	87	117	144	168	188	206	220	232
190	59	95	128	158	184	208	227	244	257
200	64	104	140	173	202	228	250	269	284
210	70	113	152	188	220	249	274	295	313
220	75	122	165	204	239	271	298	322	342
230	81	132	178	220	259	293	324	351	373
240	87	142	192	238	280	317	351	380	406
250	94	152	206	256	301	342	379	411	439
260	100	163	221	274	323	367	408	443	474

*Gross volume, in trees 5 in. and larger, from stump to top limit of 4 in. dib. (Reprinted from Pope, 1962.)
†Average height of dominants and codominants, as measured in the field.
‡Includes all trees in the major canopy; average photo estimate of several experienced interpreters.
Table based on 282 one-fifth-acre plots, largely in western Oregon.

classes) that are interpreted should be checked in the field. If field volumes average 600 cu ft per acre as compared with 800 cu ft per acre for the photo estimates, the adjustment ratio would be 600/800, or 0.75. When the field checks are representative of the total area interpreted, the ratio can be applied to photo volume estimates to determine adjusted net volume. It is desirable to compute such ratios by forest types, because deciduous, broad-leaved trees are likely to require larger adjustments than conifers.

The accuracy of aerial volume estimates depends not only upon the volume tables used but also on the ability of interpreters who make the essential photographic assessments. Since subjective photo estimates often vary widely among individuals, it is advisable to have two or more interpreters assess each of the essential variables.

OBTAINING AERIAL PHOTOGRAPHS

13-23   The Options   Aerial photographs can be obtained by purchasing existing imagery, by taking the necessary pictures yourself, or by contracting for new coverage through a private aerial-survey company. Each solution has its own advantages and limitations. For example, existing photography has the advantage of low cost, but it may be outdated or available only as black-and-white prints. Do-it-yourself photography is often sufficient for small areas of spot coverage, but the amateur rarely has the equipment and professional expertise to photograph large land areas with required standards of precision. And contracting for a special aerial survey, the choice most likely to result in superior pictures, may be rather costly for small or irregularly shaped land areas.

13-24   Photography from Commercial Firms   A wide selection of photographic negatives are held by private aerial-survey companies in the United States and Canada. In many instances, prints can be ordered directly from these companies after permission is obtained from the original purchaser. A large share of the available coverage has been obtained on panchromatic film with aerial cameras having distortion-free lenses. As a result, photographs are ideally suited for stereoscopic study because of fine image resolution and a high degree of three-dimensional exaggeration. Scales are usually 1:25,000 or larger for recent photography. In addition to contact prints and photo index sheets, most aerial-mapping organizations will also sell reproductions of special atlas sheets or controlled mosaics. These items can be useful for pictorial displays and administrative planning.

Prints purchased from private companies may cost more than those from public agencies, but they are often of higher quality and at larger scale—factors that may offset any price differential. Quotations and photo indexes can be obtained by direct inquiry to the appropriate company. Names and addresses of leading photogrammetric concerns are available in current issues of *Photogrammetric Engineering and Remote Sensing*.

13-25   Photography from the U.S. Government   Most of the United States has been photographed in recent years for various federal agencies. Although there is no central repository that can supply prints of all government imagery, a large portion can be purchased at reasonable prices through the U.S. Departments of Agriculture or Interior. The U.S. Geological Survey maintains a data base of existing aerial photography called the Aerial Photography Summary Record System. Search requests are free and available to the public through the U.S.G.S. As a rule, most photographs purchased through federal agencies range in scale from about 1:12,000 to 1:40,000. Panchromatic film is commonly used, although infrared and color films are increasing in popularity. Also, there is an increasing amount of high-altitude color photography (scale 1:60,000 to 1:125,000) available from some agencies.

The age of existing photography usually varies from about 2 to 8 years, with agricultural regions, urban areas, and large reservoirs being rephotographed at the most frequent intervals. Photo index sheets of existing photography (Fig. 13-9)

FIGURE 13-9
Portion of an aerial-photo index sheet for Montgomery County, Texas. *(Courtesy U.S. Department of Agriculture.)*

can be viewed at local or regional offices of the Agricultural Stabilization and Conservation Service, the Forest Service, or the Geological Survey.

13-26 Photography from the Canadian Government The National Air Photo Library is the central storehouse for the Canadian government's aerial photography, including the Yukon, the Northwest Territories, and the Provinces of Newfoundland, Labrador, Nova Scotia, Prince Edward Island, Manitoba, and Saskatchewan. Photography obtained specifically for other provinces may be available through various agencies within those provinces. Written inquiries should include a map of the area involved, a statement regarding the proposed use of the photography, and specifications as to whether stereoscopic coverage is desired.

13-27 Taking Your Own Pictures If oblique or near-vertical photographs taken with small-format cameras are sufficient for supplementary coverage of project areas, the do-it-yourself approach provides an alternative for limited types of aerial surveys. High-wing monoplanes offer good side visibility and low stalling speeds; such aircraft can be rented (with pilot) at reasonable hourly rates. Under certain circumstances, it may be permissible to remove the aircraft door on the passenger's side to obtain even better visibility during flights.

For oblique exposures taken through aircraft windows, standard press cameras work quite well. When cameras must be exposed to the aircraft slipstream, however, rigidly designed lens systems should be used instead of those with folding bellows. Surplus, military-reconnaissance cameras can sometimes be rented or purchased, but most scientists seem to prefer conventional 35-mm or 70-mm formats because films are readily available and inexpensive and cameras can be equipped with interchangeable lenses and motorized film drives. Furthermore, imagery in these two formats can be optically enlarged (e.g., for map revision and updating) by use of ordinary slide projectors.

The negative scales of do-it-yourself aerial photography will generally range from about 1:2500 to 1:25,000. Additional suggestions on techniques are provided by references listed at the end of the chapter.

13-28 Contracting for New Photography This is the preferred method of obtaining photographic coverage, since specifications such as film, filter, scale, and season can be placed under the control of the buyer. As a rule, the cost per square mile for special coverage decreases as photo scales become smaller and as land areas become larger. For coverage at a scale of 1:15,840 (4 in. per mile), a good camera crew can photograph 750 to 900 square miles in 5 to 8 hours of flying time. Thus photography of small land areas is dictated more by the cost of relocating the photographic aircraft and crew than by the actual flying time required.

When there is a choice to be made between two or more aerial-survey companies, the purchaser is advised to request photographic samples from each company. Such print samples provide useful guidelines to the quality of work that may be expected. Foresters who feel unqualified to evaluate sample photography should retain the services of a consultant to assist in drawing up photographic specifications, defining areas to be covered, and inspecting the finished product. When special flights are justified, contract specifications should be thoroughly discussed by buyer and contractor prior to actual flights. A few extra days of advance planning will sometimes alleviate the need for reflights and will help prevent possible disputes arising from definitions of stereoscopic coverage, exposure quality, or film-processing deficiencies.

13-29 Other Remote-Sensing Tools   Aerial photographs have been used successfully by natural resource managers for decades. Aerial photography is only one of the many products of the field called *remote sensing*. In the past 20 years, the remote-sensing discipline has produced other tools that may be helpful to foresters in certain situations. Satellite imagery has been applied in British Columbia to update cutting records and maps. Aerial infrared video has been used to monitor the spread of insects and disease. Thermal scanners have been used to locate hot spots and fire position in the western United States. While each of these tools is specialized in its application and has limitations, resource managers should consider all available tools when gathering data and gaining information needed to make informed decisions.

PROBLEMS

13-1 Using a set of aerial photographs from your own locality, determine the average scale (a) as a representative fraction, (b) in feet or chains per inch, and (c) in acres per square inch. Then establish a line of known compass bearing on the photographs.

13-2 Determine the heights of 10 trees, buildings, or other objects from parallax measurements. After completion of photographic estimates, obtain ground measurements of the same objects with an Abney level or other hypsometer. Compare results, and explain reasons for differences noted.

13-3 Delineate the principal forest types found on photographs in your locality; code types by the Society of American Foresters' designations, and verify by ground reconnaissance. Which types are easily recognized? Which types are particularly difficult to identify?

13-4 Obtain 50 to 100 paired measurements of dbh and crown diameter for a coniferous species. If the variables appear to be linearly correlated, fit a simple linear regression to the data by the method of least squares. Then use the relationship to convert a single-entry tree volume table into a "local" aerial volume table.

13-5 Using local photographs, stratify a forest area by three to five volume or type categories. Design ground cruises based on (a) stratified random sampling and (b) simple random sampling. Use the same number of field plots for both cruises. Compare results and relative efficiencies of the two systems.

13-6 Using the measurement data that follow, determine the height of each tree by applying the parallax formula from Sec. 13-14. In the following listings,

$$H = \text{height of airplane above ground datum}$$
$$P = \text{absolute stereoscopic parallax at base of tree}$$
$$dP = \text{differential parallax}$$

Determine the height of each tree in the units used for H.

a $H = 5000$ ft; $P = 90.00$ mm; $dP = 1.05$ mm.
b $H = 4200$ ft; $P = 84.62$ mm; $dP = 1.10$ mm.
c $H = 1000$ m; $P = 92.01$ mm; $dP = 1.17$ mm.
d $H = 7000$ ft; $P = 3.602$ in.; $dP = 0.041$ in.
e $H = 1800$ m; $P = 4.116$ in.; $dP = 0.052$ in.

REFERENCES

Aldred, A. H., and Hall, J. K. 1975. Application of large-scale photography to a forest inventory. *Forestry Chron.* **51:**1–7.

———, and Sayn-Wittgenstein, L. 1972. Tree diameters and volume from large-scale aerial photographs. *Can. Forest Serv., Ottawa, Information Report* FMR-X-40. 39 pp.

Aldrich, R. C. 1979. Remote sensing of wildland resources: A state of the art review. *U.S. Forest Serv., Rocky Mt. Forest and Range Expt. Sta., Gen. Tech. Report* RM-71. 80 pp.

American Society of Photogrammetry. 1968. *Manual of color aerial photography.* Banta Publishing Co., Menasha, Wis. 550 pp.

———. 1975. *Manual of remote sensing.* 2 vols. A.S.P., Falls Church, Va. R. G. Reeves, ed. 2144 pp.

Avery, T. E. 1978. Forester's guide to aerial photo interpretation (revised). *U.S. Dept. Agr. Handbook* 308, Government Printing Office, Washington, D.C. 41 pp., .

———, and Berlin, G. L. 1992. *Fundamentals of remote sensing and airphoto interpretation.* Macmillan Publishing Co., Riverside, N.J. 472 pp.

———, and Canning, J. 1973. Tree measurements on large-scale aerial photographs. *New Zealand J. Forestry* **18:**252–264.

Bonner, G. M. 1968. A comparison of photo and ground measurements of canopy density. *Forestry Chron.* **44:**12–16.

Ciesla, W. M. 1974. Forest insect damage from high-altitude color-IR photos. *Photogramm. Eng.* **40:**683–689.

Driscoll, R. S., Better, D. R., and Parker, H. D. 1978. Land classification through remote sensing—Techniques and tools. *J. Forestry* **76:**656–661.

Eyre, F. H. (ed.). 1980. *Forest cover types of the United States and Canada.* Society of American Foresters, Washington, D.C. 148 pp.

Hitchcock, H. C., III. 1974. Constructing an aerial volume table from existing tarif tables. *J. Forestry* **72:**148–149.

Johnson, E. W., and Sellman, L. R. 1974. Forest cover photo-interpretation key for the Piedmont habitat region in Alabama. *Auburn University, For. Dep. Series* 6. 51 pp.

Kippen, F. W., and Sayn-Wittgenstein, L. 1964. Tree measurements on large-scale, vertical 70-mm air photographs. *Forest Res. Br., Can. Dep. For. Pub.* 1053. 16 pp.

Kirby, C. L. 1980. A camera and interpretation system for assessment of forest regeneration. *Environ. Can., Can. Forest Serv., North. Forest Res. Cent., Edmonton, Alberta, Information Report* NOR-X-221. 8 pp.

Lillesand, T. M., and Kiefer, R. W. 1987. *Remote sensing and image interpretation.* 2d ed. John Wiley & Sons, New York. 721 pp.

Madill, R. J., and Aldred, A. H. 1977. Forest resource mapping in Canada. *The Can. Surv.* (March), pp. 9–20.

Paine, D. P. 1981. *Aerial photography and image interpretation for resource management.* John Wiley & Sons, New York. 571 pp.

Pope, R. B. 1962. Constructing aerial photo volume tables. *U.S. Forest Serv., Pacific Northwest Forest and Range Expt. Sta., Res. Paper* 49. 25 pp.

Smith, J. L., and Mead, R. A. 1981. A comparison of two aerial photo volume tables for pine stands in Mississippi. *So. J. Appl. For.* **5:**92–96.

————, Zedaker, S. M., and Heer, R. C. 1989. Estimating pine density and competition conditions in young pine plantations using 35mm aerial photography. *So. J. Appl. For.* **13:**107–112.

Woodcock, W. E. 1976. Aerial reconnaissance and photogrammetry with small cameras. *Photogramm. Eng. and Remote Sensing* **42:**503–511.

CHAPTER 14

SITE, STOCKING, AND STAND DENSITY

14-1 The Concept of Site As defined by the Society of American Foresters (1971), *site* refers to "an area considered in terms of its environment, particularly as this determines the type and quality of the vegetation the area can carry." If required, site may be classified qualitatively into site *types,* by their climate, soil, and vegetation, or quantitatively into site *classes,* by their potential to produce primary wood products.

Insofar as foresters are concerned, the primary purposes of site measurement are (1) to identify the potential productivity of forest stands, both present and future, and (2) to provide a frame of reference for land management diagnosis and prescription. In the United States, most attention has been given to the first purpose, while little attention has specifically been directed toward the second (Jones, 1969).

Theoretically, it should be possible to measure site directly by analyzing the many factors affecting the productivity of forests, such as soil nutrients and moisture, temperature regimes, available light, topography, and so on. Although attempts at direct measurement of site have been made, such an approach may not be of immediate value to the practicing forester; consequently, indirect estimates of site are frequently employed.

14-2 Direct Measurement of Forest Productivity When available, historical yield records of forest productivity provide a direct measure of site quality. Averaging productivity from multiple rotations of fully stocked stands of the desired species provides excellent information on given sites. Unfortunately, productivity data like these do not exist for most forest sites, and the actual yield can be affected by such factors as genetic composition, stand density, competing vegetation, pests, and the climate experienced during the period over which the growth was measured. Consequently, indirect methods of evaluating site quality are generally used. The most common indirect method for assessing site quality for wood-producing purposes involves measurement of tree heights on the site.

14-3   Tree Height as a Measure of Site Quality   Of all the commonly applied indirect measures of site, tree height in relation to tree age has been found the most practical, consistent, and useful indicator. Theoretically, height growth is sensitive to differences in site quality, little affected by varying density levels and species compositions, relatively stable under varying thinning intensities, and strongly correlated with volume. This measure of site is termed *site index*. Site index is the most widely accepted quantitative measure of site in the United States.

As generally applied, site index is estimated by determining the average total height and age of dominant and codominant trees in well-stocked, even-aged stands. When these two variables (total height and age) have been ascertained for a given species, they are used as coordinates for interpolating site index (height at a specified index age, such as 25, 50, or 100 years) from a specially prepared set of curves (Fig. 14-1), or for substitution into a site-index equation.

In preparing site-index curves for various species, either age at breast height or total age may be used as the independent variable. Site-index curves for plantations are generally based on number of years since planting as the age variable. Thus age can be obtained from planting records, eliminating the need for increment borings. For trees growing in natural stands, age at breast height is often preferable, because this is a standard point of tree-diameter measurement and a convenient height for making increment borings. When total age is used, it is necessary to estimate the number of years required for the tree to grow from seed to the height where an increment boring is made; this number, which is especially variable for more tolerant species, is then added to the annual ring count to obtain total age. Use of age at breast height in lieu of total age eliminates the need for such arbitrary correction factors.

14-4   Field Measurement of Site Index   To determine the site index of a forest stand, average total height and age are determined from measurements obtained from *site trees*. Site trees should meet certain specifications, such as being dominant or codominant and even-aged, showing no evidence of crown damage, disease, sweep, crook, forking, or prolonged suppression.

Measurements of total height are commonly made with a hypsometer, while age may be determined by extracting an increment core. The number of trees measured depends upon the variability of total heights and ages in the stand being evaluated. The number of sample units required to estimate total height

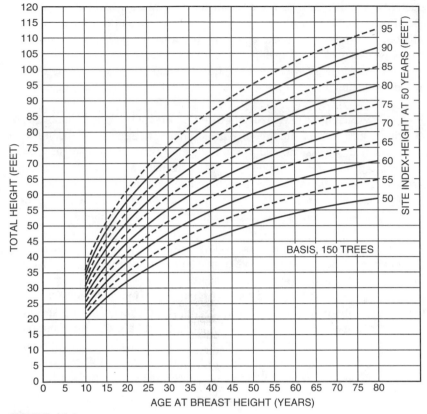

FIGURE 14-1
Site-index relationships for East Texas longleaf pine. The index age is 50 years. *(From Larson and Moehring, 1972.)*

and age for a given confidence interval and probability level can be determined by application of the formula for computing sampling intensity (Chap. 8). As an example, if the average total height is to be measured within ±5 ft at a confidence probability of 95 percent (*t* value of 2) and a preliminary sample of five total-height measurements indicates a standard deviation of ±7.5 ft, the required number of height measurements may be calculated as

$$n = \left[\frac{(2)(7.5)}{5}\right]^{2} = 9 \text{ measurements}$$

For the site index to be expressed on a standard basis, an index age must be presumed. For most regions, the period in the life of the stand that approximates the culmination of mean annual growth (Chap. 15) in well-stocked stands has been selected as the index age Accordingly, 100 years has been used for most western species and 50 years for eastern species. Special site-index curves based on an index age of 25 years are available for plantations that are managed on rotations shorter than 50 years.

14-5   Construction of Site-Index Curves   If one measures a stand that is at an index age, the average height of dominants and codominants is the site index. In most instances, however, stands measured are less than or greater than the index age. Consequently, a set of curves (Fig. 14-1) or an equation is needed to project the dominant stand height to the standard reference age.

Early site-index curves were constructed by graphical techniques. Data on heights and ages were collected from a variety of stands on different site-quality land and of varying ages. These paired height-age values were then plotted on graph paper and a "guide" curve was drawn to depict the general trend in the data. All other site-index curves were then proportional to the guide curve. For example, if the guide curve passed through height 100 ft at index age, then the site-index-90 curve was drawn by multiplying values on the guide curve by 0.9, the site-index-110 curve by multiplying the guide curve by 1.1, etc. When such curves for different site-index classes are proportional, they are termed *anamorphic* (i.e., all curves have the same shape).

Anamorphic site-index curves are now constructed by regression techniques (Chap. 2). Plotting height over age for pure, even-aged stands results in a generally sigmoid shape; thus some transformation of the variables is needed if linear regression methods are applied. The most common transformation is

$$\log H_d = b_0 + b_1 A^{-1}$$

This is a special case of the simple linear regression model

$$Y = b_0 + b_1 X$$

where Y is the logarithm of the height of dominants and codominants H_d and X is the reciprocal of stand age A.

The guide curve for a set of anamorphic site-index curves can be established by fitting the model of the logarithm of the height and the reciprocal of age to data from stands of varying site qualities and ages. To avoid bias in the guide curve, it is important that, insofar as possible, all site-index classes of interest be represented approximately equally at all ages. If only poor-quality sites are sampled for the older ages, for example, the guide curve will tend to "flatten" too quickly and bias the entire family of site-index curves.

After the guide curve is estimated, an equation for site index as a function of measured age and height can be constructed by noting that when age is equal to index age A_i, height is equal to site index S; that is,

$$\log S = b_0 + b_1 A_i^{-1}$$

This implies that

$$b_0 = \log S - b_1 A_i^{-1}$$

Substituting the implied definition for b_0 into the original guide-curve equation results in

$$\log H_d = \log S - b_1 A_i^{-1} + b_1 A^{-1}$$
$$= \log S + b_1 (A^{-1} - A_i^{-1})$$

This can be used to generate site-index curves, and the equation can be algebraically rearranged as

$$\log S = \log H_d - b_1 (A^{-1} - A_i^{-1})$$

This is the form used to estimate site index (height at index age) when age and height measurements are given.

Site-index curves constructed by the anamorphic method just described may not provide fully satisfactory results. The guide-curve technique is sound only if the average site quality in the sample data is approximately the same for all age classes. If the average site quality varies systematically with age, the guide curve will be biased. In many timber types, younger stands are associated with generally better sites, whereas older stands are concentrated on the poorer sites left to harvest last. When a negative age–site quality correlation exists in the data, the height-age guide curve will be biased—showing excessive growth at younger ages and insufficient growth at older ages. Because all site-index curves are proportional to the guide curve, if it is biased the entire family of anamorphic curves will be biased. Consequently the estimation of an unbiased guide curve is of primary importance when the anamorphic technique is applied to develop site-index curves.

Another weakness of anamorphic site-index curves is the assumption of a common shape for all site classes. For many species, the height-curve shape varies with site quality—higher-quality lands generally exhibit more pronounced sigmoid shapes and lower-quality lands produce height-growth patterns that are "flatter." Families of site-index curves that display differing shapes for different site-index classes are termed *polymorphic*. In recent years, polymorphic site-index curves have been constructed for many different species. When available, polymorphic curves generally reflect height-growth trends across a wide range of site qualities more accurately than anamorphic curves and, thus, are generally preferred.

When constructing polymorphic site-index equations, height is expressed as a function of age and some other variable(s) such as site index. For example, the following function, which describes height H_d as a function of stand age A and site index S, base age 50 years, was used to develop polymorphic site-index curves for eastern white pine (Beck, 1971b):

$$H_d = [63.36 + 0.68208S][1 - e^{(-0.01007 + 0.00030S)A}]^{1.9094}$$

It should be noted that, in application, most polymorphic curves require that one measure only the stand height and age to estimate site index.

14-6 Interspecies Site-Index Relationships Direct application of the site-index method requires that suitable site trees of the species of interest be present. For situations where this requirement is not met but there are suitable site trees of another species available for measurement, interspecies site-index relationships have been developed. An equation relating the site index of two species can be developed by measuring the site index on areas where site trees for both species occur together and then computing a regression to predict the site index of one species from that of the other.

Olson and Della-Bianca (1959) investigated site-index relationships among several species and species groups occurring in the Piedmont region of Virginia, North Carolina, and South Carolina. They related the site index of several species to that of yellow-poplar, which is a site-sensitive species. For example, Olson and Della-Bianca's regression equation to relate site index of black oak to that of yellow-poplar is

$$Y = 39.7 + 0.45X$$

where Y = site index of black oak (feet at base age 50 years)
$\quad\quad X$ = site index of yellow-poplar (feet at base age 50 years)

While equations relating site index between species have been useful in forest management, these predictions must be used with care. In order to develop equations relating the site index of two species, the species must be found together. Some species will occur together on certain site types but not on others. Data sets used to derive species site-index conversion equations, by necessity, only contain the types of sites where the two species naturally occur in mixture. Thus the regression equation computed from the data may not be applicable to sites where the species of interest is not present but the predictor species is.

14-7   Periodic Height Growth   An alternative to basing site-quality evaluation on current stand height is to use information on periodic height growth. The use of height growth for some relatively short period during the life of the stand to assess site quality is commonly referred to as the *growth-intercept method.* Growth-intercept values can be used directly as measures of site quality, but they are usually used to calculate site-index estimates. This approach was developed primarily for use in young stands because site-index curves have been found unreliable for juvenile stands. The growth-intercept method is most feasible for species that put on a single well-defined whorl of lateral branches each growing season.

The best indication of current growing conditions is current height increment, which may be discerned from measurement of the terminal leader (Spurr and Barnes, 1980). For taller trees, however, this can be a difficult and time-consuming task. Very early height growth can be greatly affected by herbaceous competition, and thus it would not be a useful indicator of site potential. For the middle part of the height-age curve, height growth is relatively constant; thus it is this middle portion that is used when applying the growth-intercept approach. Although the details vary in the past applications, all growth-intercept methods involve the measurement of length of a specified number of successive annual internodes, beginning at some defined point on the stem. A growth-intercept definition applied in several instances "is the height growth for the 5-year period that begins when breast height is reached." However, Alban (1972) found that the precision of estimating site index for red pine was much improved by measuring the 5-year growth intercept beginning at 8 feet above ground rather than at breast height. Alban's growth-intercept equation, developed from natural stands of red pine in Minnesota, follows:

$$SI = 32.54 + 3.434X$$

where SI = site index (feet at base age 50 years)

X = 5-year growth intercept in feet (height growth for the 5-year period that begins from the first whorl above 8 ft from ground level)

In application, a sample of trees is selected and growth intercept is measured on the sample trees; then the average growth intercept is computed and substituted into the site-index prediction equation.

In addition to having been found more useful than total height for estimating site index in young stands of certain species, growth intercept has the advantage of eliminating the need to measure age. Age measurements can be costly to obtain and are subject to error. The growth-intercept method also eliminates the need to measure total tree heights, which in dense stands can be a difficult and time-consuming measurement. The method suffers from the disadvantage that short-term climatic fluctuations may render the results inaccurate and that sometimes the early growth of a stand does not accurately reflect later growth.

14-8 Physical-Factors Approach Foresters often need an estimate of site quality on areas where forest stands do not presently exist or where the extant stands do not contain suitable site trees. In these situations an alternative to measuring tree heights and ages is necessary. One alternative that has received widespread attention involves using environmental factors (e.g., topographic and/or soils variables) to predict site quality (generally expressed as the site index for a particular species of interest). A productivity-estimation method based on the relatively permanent features of soil and topography can be used on any site, regardless of the presence, absence, or condition of the vegetation.

The environmental-factors approach typically involves measuring site index and topographic and soils variables in a sample of stands throughout some defined physiographic region and climatic zone. (Climatic variables are sometimes included as well.) The measured site index is then related to the topographic and soils variables by means of linear regression analysis. The prediction equation developed from data from sites where suitable stands for site-index determination are found is then applied to sites where site index cannot be measured because of a lack of suitable site trees. While this approach has a strong intuitive appeal, validation of the resultant equations has often shown that the site-index predictions are not reliable—especially if the equation is applied outside the geographic region in which it was developed.

An example of the physical-factors approach is the equation developed by Sprackling (1973) to estimate the site index of Engelmann spruce growing on granitic soils in southern Wyoming and northern Colorado. The following soil and topographic factors were selected for possible inclusion in the regression equation to estimate site index: (1) aspect, (2) slope percent, (3) slope position, (4) elevation, (5) soil depth to the C horizon, and (6) texture of the B horizon. Analysis indicated that two variables can be used to estimate the site index of Engelmann spruce (Fig. 14-2):

$$Y = -106.63509 + 62.46021 \ (X1) + 809{,}396.2 \ (X2)$$
$$R^2 = 0.646 \qquad S_{y.x} = 9.00 \text{ ft}$$

where Y = site index (feet at base age 100 years)
  $X1$ = logarithm base 10 of soil depth in inches to the top of the C horizon
  $X2$ = 1/elevation in ft

14-9 Indicator-Plant Approach The presence, abundance, and size of understory plants can serve as useful indicators of forest site quality. While understory plants are often more affected by overstory density and composition,

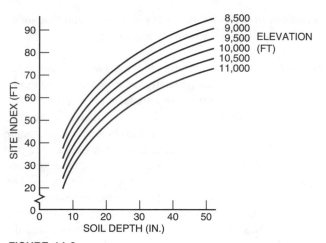

FIGURE 14-2
Site index for Engelmann spruce on granitic soils in northern Colorado and southern Wyoming in relation to soil depth and elevation. *(From Sprackling, 1973.)*

site history, and localized disturbance than overstory trees, they also tend to recover from disturbance more quickly. In cases where the understory plants have a relatively narrow ecological tolerance, they can serve as useful indicators of growing conditions for trees.

Understory plant communities of northern coniferous forests are relatively simple; only a few plant species occur in the understory, and communities are distinct and easily recognized (Carmean, 1975). Consequently, the indicator-plant approach has been found more applicable in the forests of northern Europe and in Canada than in other areas. Although attempts have been made to develop indicator-plant methods in more southerly latitudes of North America, the results have not been very successful because the sites have often been repeatedly disturbed and there is a complex variety of understory plant communities.

In some instances the indicator-plant approach has been combined with the height-age concept, and site-index curves have been developed for distinct understory vegetal types. Thus, despite the limitations and complexities of applying the technique, indicator plants can supplement other site-quality evaluation methods.

Practicing foresters should be aware of indicator-plant–site-quality relationships, since this information can provide a quick appraisal or useful supplementary information for assessing site quality. However, the following limitations must be kept in mind: (1) the method permits site evaluation only in relative or qualitative terms, (2) the understory characteristics are generally quite sensitive to disturbances such as fire or grazing, (3) understory vegetation generally reflects only the fertility of the topmost horizons of the soil profile—deeper horizons may have little impact on understory vegetation but greatly influence site quality for tree growth—and (4) a sound background in plant ecology is a prerequisite for reliable classifications.

14-10 Limitations of Site Index The main drawbacks that have been cited regarding the use of site index as a measure of forest productivity are as follows:

1 Exact stand age is often difficult to determine, and small errors can cause relatively large changes in the site-index estimate.

2 The concept of site index is not well suited for uneven-aged stands, areas of mixed species composition, or open lands.

3 Effects of stand density are not considered except by arbitrary selection of site trees in well-stocked stands that have been unaffected by past suppression. Other variables associated with stand volume (i.e., dbh and stem form) are not directly taken into account. As a result, an index based on total height and age alone may not provide a valid estimate of the growing capacity for a particular site.

4 Site index is not a constant; instead, it may change periodically due to environmental and climatic variations or management activities.

5 Except in limited instances, the site-index value for one species cannot be translated into a usable index for a different species on the same site (Doolittle, 1958).

In spite of the foregoing limitations, site index is a useful tool because it provides a simple numerical value that is easily measured and understood by the practicing forester. Its use will apparently be continued until the day when the varied factors affecting the productivity of forests can be reduced to an equally simple and quantitative measurement.

STOCKING AND STAND DENSITY

14-11 Definitions Although stocking and stand density are terms that are often applied interchangeably, the two terms are not synonymous. *Stand density* denotes a quantitative measurement of the stand, whereas *stocking* refers to the adequacy of a given stand density to meet some management objective. Accordingly, stands may be referred to as understocked, fully stocked, or overstocked. A stand that is "overstocked" for one management objective could be "understocked" for another.

Stand density is a quantitative term describing the degree of stem crowding within a stocked area; it can be expressed in absolute or relative terms. Absolute measures of density are determined directly from a given stand without reference to any other stand. For example, number of trees per acre is an absolute measure that expresses the density of trees on an area basis. Relative density is based on a selected standard density. If, for instance, "fully stocked" is defined on a basal-area basis, the ratio of the measured basal area in a stand to that of the fully stocked ideal is a relative measure of stand density. The problem of what constitutes full stocking makes application of relative density measures difficult.

14-12 Measures of Stocking The main difficulty arising from the application of stocking concepts is that of deciding just what should constitute full stocking for a particular species on a given site. As outlined by Bickford et al. (1957), "The stocking that results in maximum yield is the ideal that every forest manager would like to have if he only knew what it was and could recognize it if he saw it." Although stocking can also be specified in terms of the capacity of an area to support trees, most foresters think of stocking in terms of "best growth" rather than as a measure of site occupation.

As a holdover from European forestry practices, stands that are fully stocked have also been referred to as *normal* stands. The theory was developed that maximum volume increment would be obtained with full or normal stocking. Thus an ideal and regulated normal forest would be composed of a normal distribution of age classes, normal growing stock, and, consequently, a normal increment. In such a hypothetical forest, tree crowns are fitted together so that no sunlight is wasted, and each crown is matched with a root system that fully utilizes the soil (Bickford et al., 1957).

As a follow-up to the foregoing concept, normal yield tables were compiled to describe the expected production of normal forests (Chap. 16). Such tables were based on the "average best," pure, even-aged stands that could be located for various species and site conditions. In brief, the normal forest represented a paradox—a goal to be sought by forest managers, but one that was both unrealistic and unattainable, if not undesirable.

Fortunately, the elusive concept of normality has largely been erased from American forest management. Even under the hazardous assumption that a normal forest can be attained (and recognized when existent), it has become increasingly apparent that so-called full stocking does not necessarily imply maximum volume growth. Furthermore, the utility of normal yield tables is severely handicapped by the fact that no reliable methods are available for predicting yields of nonnormal or understocked stands.

Stocking levels are of prime concern to the forest manager because controlled changes in these levels may allow the forester to shorten or lengthen the rotation, favor desired species, and maximize the yield of selected timber products. Although the extremes of stocking can be easily recognized, full stocking can only be defined as a closed canopy stand that represents the "average best" to be found. Understocked stands are characterized by trees of rough form, excessive taper, and a high live-crown ratio. Overstocked stands may represent a stagnated condition, with trees exhibiting a low live-crown ratio and numerous dead stems. In both instances, the result is a reduction in net volume increment of wood products of interest from the "fully stocked" ideal.

Because of difficulties with defining stocking in a forest management context, quantitative measures of stand density are generally used to derive silvicultural prescriptions and to predict growth and yield.

14-13 Basal Area per Acre Because it is simple to measure by point-sampling techniques, objective, and easily understood, basal area (BA) per acre provides a logical expression of stand density. Stand BA is the cross-sectional area (in square feet at dbh) of all stems, or of some specified portion of the stand, expressed on a per-acre basis. In countries that employ the metric system, BA is stated in square meters per hectare.

Stand BA is highly correlated with the volume and growth of forest stands. Numerous variable-density yield tables have been derived by regression analysis by using the variables of BA per acre, site index, and stand age for a given species. In addition, many silvicultural considerations, such as thinning intensity, are commonly based on BA measurements.

14-14 Trees per Acre As an expression of stand density, number of trees per acre has limited value in natural stands, but it has been extensively used in planted stands. Many of the variable-density yield tables for planted stands are based on number of trees per unit area, and silvicultural prescriptions for plantations are often made in terms of tree numbers rather than BA per acre.

14-15 Stand-Density Index Measures of density based on the two component parts of BA—number of trees per unit area and diameter of the tree of average BA—have been called *stand-density indices* (Spurr, 1952). In 1933 Reineke pointed out that plotting the logarithm of number of trees per acre against the logarithm of average diameter of fully stocked stands generally resulted in a straight-line relationship. He also found that the same slope could, in most cases, be used to define the limits of maximum stocking. This negatively sloping line, termed the *reference curve,* was expressed by

$$\log N = -1.605 \log D + k$$

where N = number of trees per acre
D = diameter of tree of average BA
k = a constant varying with species

In this section, all logarithms are base 10.

By definition, when D equals 10 in., log N equals log SDI; that is,

$$\log \text{SDI} = -1.605(1) + k$$

which implies that

$$k = \log \text{SDI} + 1.605$$

Substituting the implied definition for k into the reference-curve formula gives

$$\log N = \log \text{SDI} + 1.605 - 1.605 \log D$$

which can be rearranged as

$$\log \text{SDI} = \log N + 1.605 \log D - 1.605$$

to provide an expression for computing stand-density index from number of trees per acre and diameter of the tree of average BA.

One might assume that a stand with BA 120 sq ft and 480 trees per acre is measured. The diameter of the tree of average basal area D is

$$\sqrt{\frac{120 / 480}{0.005454}} = 6.77 \text{ in.}$$

Substituting into the stand-density index formula gives

$$
\begin{aligned}
\log \text{SDI} &= \log (480) + 1.605 \log (6.77) - 1.605 \\
&= 2.4093 \\
\text{SDI} &= 257
\end{aligned}
$$

Stand-density index is reasonably well correlated with stand volume and growth, and several variable-density yield tables have been constructed using it. However, in most cases BA is fully as satisfactory as stand-density index, and because it is more simply obtained, it is preferred as a measure of density.

14-16   3/2 Law of Self-Thinning   The so-called 3/2 law of self-thinning, like Reineke's stand-density index, is based on the concept of a maximum size-density relationship. In the case of the 3/2 law of self-thinning, the logarithm of mean tree volume or weight is plotted against the logarithm of the number of trees per unit area. For pure, even-aged stands that are sufficiently crowded such that competition-induced mortality ("self-thinning") is occurring, the slope of the line of logarithm of mean volume versus logarithm of trees per unit area has been found to be approximately −3/2, but the intercept varies by species. That is,

$$\log \bar{V} = -3/2 \log N + a$$

where $\overline{V}$ = mean tree volume

$\qquad N$ = number of trees per unit area, acre or ha

$\qquad a$ = a constant varying with species

Obviously, the 3/2 law of self-thinning is closely related to the stand-density index—in fact, the two can be shown to be mathematically equivalent. Reineke's stand-density index was developed with log N on the left-hand side of the equation. Rearranging the 3/2 relationship with log N on the left-hand side gives

$$- 3/2 \log N = \log \overline{V} - a$$

Assuming that mean tree volume is proportional to the diameter of the tree of average BA raised to the power of 2.4 (Bredenkamp and Burkhart, 1990), that is,

$$\overline{V} = c\overline{D}^{2.4}$$

where c is a constant, and substituting the definition for $\overline{V}$ into the rearranged equation for the 3/2 relationship, one obtains

$$-3/2 \log N = \log c + 2.4 \log \overline{D} - a$$

Multiplying the preceding equation by $-2/3$ and combining the constants into a single term designated k results in

$$\log N = - 1.6 \log \overline{D} + k$$

which is the stand-density index reference line.

Although the two concepts are mathematically equivalent, due to measurement considerations and historical precedent the stand-density index has been applied widely in forestry, whereas the 3/2 law of self-thinning has been utilized prevalently by plant ecologists. There are examples of applications of the 3/2 relationship in forestry, such as the stand-density management diagrams for plantations of coastal Douglas-fir developed by Drew and Flewelling (1979).

14-17 Relative Spacing The average distance between trees divided by the average height of the dominant canopy has been termed *relative spacing*. Assuming square spacing, the average distance between trees can be computed as the square root of the number of square feet per acre (43,560) divided by the number of trees per acre. This average distance between trees in feet is then divided by the average height of the dominant canopy in feet to compute relative spacing. In formula form, relative spacing (RS) is computed as

$$RS = \frac{\sqrt{43,560/N}}{H_d}$$

where N = number of trees per acre
$\quad H_d$ = average height of the dominant canopy, ft

The comparable formula in metric units is

$$RS = \frac{\sqrt{10,000/N}}{H_d}$$

where N = number of trees per hectare
$\quad H_d$ = average height of the dominant canopy, m

The constant 10,000 is the number of square meters per hectare. Because RS is a pure number (no units), its value will be the same for any given stand regardless of whether English or metric units are used.

For even-aged stands, relative spacing initially drops rapidly; then it levels off at a lower limit (Fig. 14-3). After reaching the lower limit, RS will increase somewhat if the stand is carried to an advanced age. The lower limit of relative spacing is fairly constant for a given species regardless of the site quality and the initial density. Loblolly pine, for example, achieves a minimum relative spacing of approximately 0.15 (Lemin and Burkhart, 1983), whereas *Eucalyptus grandis* reaches a much lower value of around 0.05 (Bredenkamp and Burkhart, 1990).

FIGURE 14-3
Time trends of relative spacing in stands of different density. *(After Clutter et al., 1983.)*

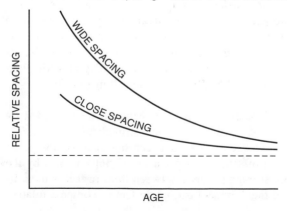

Although it may not be immediately obvious, relative spacing is closely related to stand-density index. Height has been found to be proportional to diameter raised to the power of 0.8 (Curtis, 1970); that is,

$$H = aD^{0.8}$$

Assuming that the height of the dominant canopy (H_d) can be related to the quadratic mean diameter (D) by the preceding relationship, then H_d in the relative spacing formula can be replaced by $aD^{0.8}$, giving

$$RS = \frac{\sqrt{b/N}}{aD^{0.8}}$$

where b is an appropriate constant (43,560 or 10,000). After combining the constants b and a into a single constant denoted c, the RS formula can be rewritten as

$$RS = \sqrt{cN^{-1}D^{-0.8}}$$

Squaring both sides gives

$$RS^2 = cN^{-1}D^{-1.6}$$

Taking the logarithm of both sides yields

$$2 \log RS = \log c - \log N - 1.6 \log D$$

Rearranging the above equation gives

$$\log N = \log c - 2 \log RS - 1.6 \log D$$

If RS is at or near its lower limit, it can be assumed constant, and the terms $\log c - 2 \log RS$ can be set equal to a constant called k, giving

$$\log N = -1.6 \log D + k$$

which is, of course, the stand-density index reference line.

Relative spacing has been found useful for predicting mortality and for deriving thinning schedules (Wilson, 1979).

14-18   Crown Competition Factor   Developed by Krajicek et al. (1961), crown competition factor (CCF) is a measure of stand density rather than of crown cover. CCF reflects the area available to the average tree in a stand in relation to the maximum area it could use if it were open-grown.

To compute CCF values, the crown-width/dbh relationship for open-grown trees of the species of interest must be established. Generally, a simple linear regression of the form

$$CW = b_0 + b_1 \text{ dbh}$$

suffices to establish this relationship. Alexander (1971) computed the following equation for open-grown Engelmann spruce trees in Colorado and southern Wyoming:

$$CW = 4.344 + 1.029 \text{ dbh}$$

where CW is in feet and dbh is in inches.

Assuming that the crowns of open-grown trees are circular in shape, the maximum crown area (MCA), expressed as percent of an acre, that can be occupied by the crown of a tree with a specified bole diameter is computed as

$$MCA = \frac{\pi(CW)^2(100)}{(4)(43, 560)} = 0.0018(CW)^2$$

Inserting the equation for Engelmann spruce results in

$$MCA = 0.0018(4.344 + 1.029 \text{ dbh})^2$$
$$= 0.0340 + 0.0161 \text{ dbh} + 0.0019 \text{ dbh}^2$$

CCF for a stand is computed from a stand table by summing the MCA values for each diameter class and dividing by the area in acres. In formula form, the expression for CCF for this example is

$$CCF = \frac{1}{a}(0.0340 \sum n_i + 0.0161 \sum \text{dbh}_i n_i + 0.0019 \sum \text{dbh}_i^2 n_i)$$

where a = plot or stand size, acres
n_i = number of trees in ith dbh class
dbh_i = midpoint of ith dbh class, in.

As an example, the following stand table resulted from the measurement of three $^1/_{10}$-acre plots (i.e., $a = 0.3$ acre):

dbh	n_i	$dbh_i n_i$	$dbh_i^2 n_i$
4	50	200	800
5	45	225	1125
6	43	258	1548
7	20	140	980
8	17	136	1088
9	11	99	891
10	5	50	500
Total	191	1108	6932

CCF would be computed as

$$CCF = \frac{1}{0.3}[0.0340(191) + 0.0161(1108) + 0.0019(6932)]$$
$$= 125$$

Although not as widely used as BA, CCF has proved useful in comparing different measures of stand density, and it has been found to be highly correlated with growth and yield for various species.

14-19 Stocking Guides Measures of stand density have been used to derive stocking charts. These charts, or guides, are based on the precept that gross increment varies little over a fairly wide range of stand density. The U.S. Forest Service, and many other organizations, has adopted the basic approach of Gingrich (1967) when developing stocking guides for timber management purposes.

As an example, a stocking guide for eastern white pine is presented in Fig. 14-4. This stocking guide, like many others, is a nomogram depicting the relationship between basal area, trees per acre, and the quadratic mean diameter at breast height. On this chart, the A curve is considered the upper limit in stocking for practical management. The B curve represents minimum stocking for full utilization of the site. Stands above the A curve are considered overstocked, stands between the A and B curves are deemed adequately stocked, and stands below the B curve are regarded understocked (Philbrook et al., 1973).

Stocking charts can be used to guide thinning schedules. For instance, stands might be considered for thinning when stocking approaches the A curve. Stocking after thinning should be near the B curve. For stands at or above the A curve, it may be best to reduce the growing stock down near the B curve in several successive thinnings. While stocking charts do not define thinning schedules, they have been found useful for focusing attention on relationships between stand density and tree size. Also, there are no time dimensions on stocking charts of the type described here. However, if growth relationships are available (such as

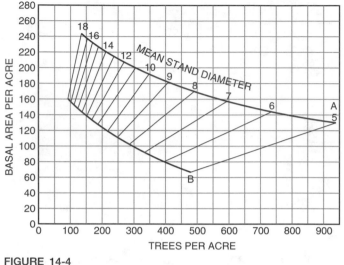

FIGURE 14-4
Stocking guide for nearly pure even-aged white pine stands in the northeastern United
States. *(After Philbrook et al., 1973.)*

by application of growth models, Chap. 16), one can inscribe lines on stocking
charts that describe the development and thinning of a stand over time.

14-20   Measures of Point Density   The stand-density measures discussed
thus far are aimed at providing an estimate of the "average" competition level in
stands. Point-density measures attempt to quantify the competition level at a
given point or tree in the stand. These competition indices provide an estimate of
the degree to which growth resources (e.g., light, water, nutrients, and physical
growing space) may be limited by the number, size, and proximity of neighbors.
The actual competition processes among trees are much more complex than can
be described by a reasonably simple mathematical index. However, these indices
have been found useful for predicting tree mortality and growth.

A large number of competition indices have been developed. Three classes or
types of competition indices described here are (1) area-overlap measurement, (2)
distance-weighted size ratio indices, and (3) area-available (or polygon) indices.

Area-overlap measures are based on the concept that there is a competition-
influence zone around each tree. Typically, this area over which the tree is assumed to compete for site resources is represented by a circle whose radius is a
function of tree size. The competitive stress experienced by a given tree is assumed to be a function of the extent to which its competition circle overlaps
those of neighboring trees (Fig. 14-5). Various definitions of the area of influence, the measure of overlap, and the use of weights when summing areas of

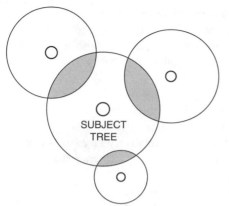

FIGURE 14-5
An illustration of the competition zone overlap used in the definition of point-density measures. *(After Clutter et al., 1983.)*

overlap have led to a large number of point-density expressions, although all are conceptually similar. An example of the area overlap type of point-density measure is the competition index proposed by Gerrard (1969):

$$CI_i = \frac{1}{A_i} \sum_{j=1}^{n} a_j$$

where CI_i = competition index for subject tree i
  A_i = area of competition circle for subject tree i
  n = number of competitors
  a_j = area of overlap of the jth competitor

The basic premise of Gerrard's index is that the competitive stress sustained by a tree is directly proportional to the area of overlap of its competition circle with those of its neighbors and inversely proportional to the area of its own competition circle.

Competition measures based on distance-weighted size ratios involve the sum of the ratios between the size of each competitor to the subject tree, weighted by a function of the distance between the competing trees. The most common measure of tree size is dbh, but other measures (e.g., height, crown size) have also been employed. The index developed by Hegyi (1974), and modified by Daniels (1976) to use point-sampling concepts in the definition of competitors, provides an example of this type of point-density measure:

$$CI_i = \sum_{j=1}^{n} \frac{D_j / D_i}{DIST_{ij}}$$

where CI_i = competition index of the ith subject tree

n = number of competitors (defined by a fixed-radius circle centered at the subject tree or by the number of trees "in" with a fixed BAF sweep with the vertex of the angle centered at the ith tree)

D_j = dbh of the jth competitor

D_i = dbh of the ith subject tree

$DIST_{ij}$ = distance between subject tree i and the jth competitor

From the formulation of this index, it is clear that relatively large trees close to the subject tree are assumed to exert more competitive influence than smaller trees that are farther away.

The third general type of point-density measure that will be presented is based on area available to the subject tree. This approach involves constructing polygons around the subject tree by connecting the perpendicular bisectors of the distance between the subject tree and its competitors (Fig. 14-6). The polygon area for each tree has been termed *area potentially available* (APA). A basic premise underlying APA is that, within limits, larger APA values should result in a higher survival rate and more tree growth (at least in certain dimensions, for example, dbh). If the perpendicular bisectors are placed equidistance between the subject tree and its competitors, the polygons are mutually exclusive and collectively exhaustive of the total area (that is, the individual polygon areas sum to the total area). Various weighting factors, based on tree size, have been used to determine the placement of the bisecting lines between the subject and competitor trees. The asymmetric division of the intertree distances resulting from weighting by tree size (for example, dbh, basal area, total height) may result in open areas. While weighted-distance polygon areas are mutually exclusive, they may not be collectively exhaustive of the stand area.

A number of studies have compared the efficacy of various point-density measures for predicting tree growth (Opie, 1968; Gerrard, 1969; Johnson, 1973; Daniels, 1976; Alemdag, 1978; Noone and Bell, 1980; Martin and Ek, 1984;

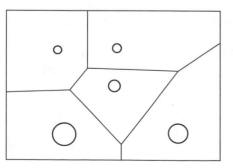

FIGURE 14-6
Polygons constructed by bisecting intertree distances. *(After Daniels et al., 1986.)*

Daniels et al., 1986; Tomé and Burkhart, 1989). The results of these comparisons have been variable, with no index being shown to be universally superior. Some indices seem to be better suited to certain species than others, and within species the performance of a particular index may vary with the stage of stand development and with the management practices followed.

PROBLEMS

14-1 Prepare a brief report on (a) the use of soil characteristics to measure site quality in your locality or (b) the possibilities of using indicator plants of lesser vegetation as a measure of site in your locality.

14-2 Assess the site quality of a forest stand in your locality by determining the site index, using an appropriate site-index relationship.

14-3 Determine site-index values of different species growing in mixture on the same area. What would be the expected difference in cubic volume per unit area for the two species at the index age, assuming that average stem diameter (dbh) is about the same for both species?

14-4 On recent aerial photographs of your locality, locate 10 to 30 circular sample plots that represent a wide density range in terms of crown closure. Then, visit each plot and obtain ground estimates of BA per unit area for the dominant-codominant stems that were visible on the photographs. Plot BA values over crown closure. If a definite trend is evident, fit a regression equation to the plotted points. Explain possible reasons for the pattern of plotted points obtained.

14-5 Given the following data on stand ages A and average height of dominants and codominants H_d:

Age, yr	Height, ft	Age, yr	Height, ft	Age, yr	Height, ft
90	161	27	75	50	89
68	124	24	60	40	82
56	138	23	72	30	69
47	90	20	60	25	61
36	100	100	137	82	111
33	88	82	108	44	72
30	81	60	96	. . .	. . .

a Fit the following simple linear regression to the data:

$$\ln H_d = b_0 + b_1 A^{-1}$$

b From the guide curve fitted in part a, derive a set of anamorphic site-index curves (base age 50 years); plot site curves 80, 90, 100, 110, and 120 ft on graph paper.

14-6 The following data were obtained on *one* ⅕-acre plot in a loblolly pine plantation:

dbh, in.	No. trees tallied	dbh, in.	No. trees tallied
5	7	9	9
6	10	10	5
7	12	11	3
8	24	12	2

Previous research established the following relationships for loblolly pines in the area of interest:

$$\log N = -1.605 \log D + 4.1$$

    where N = number of trees per acre
           D = diameter of tree of average BA, in.

    and CW = 3.56 + 1.61 dbh

    where CW = crown width, ft, for open-grown trees
          dbh = diameter at breast height, in., for open-grown trees

Using the data and previously derived relationships given, compute values for the following measures of stand density: BA per acre, number of trees per acre, stand-density index, CCF.

14-7 An even-aged stand has an average dominant height of 45 ft and 400 trees per acre. Assuming square spacing:

    a Compute the relative spacing.
    b Given that the stand as described is to be thinned to a relative spacing of 0.30, compute the number of stems per acre that would remain following thinning.

REFERENCES

Alban, D. H. 1972. An improved growth intercept method for estimating site index of red pine. *U.S. Forest Serv., North Central Forest Expt. Sta., Res. Paper* NC-80. 7 pp.

Alemdag, I. S. 1978. Evaluation of some competition indexes for the prediction of diameter increment in planted white spruce. *Can. For. Serv., For. Manage. Inst., Inf. Rep.* FMR-X-108. 39 pp.

Alexander, R. R. 1971. Crown competition factor (CCF) for Engelmann spruce in the central Rocky Mountains. *U.S. Forest Serv., Rocky Mt. Forest and Range Expt. Sta., Res. Note* RM-188. 4 pp.

Bailey, R. L., and Clutter, J. L. 1974. Base-age invariant polymorphic site curves. *Forest Sci.* **20:**155–159.

Beck, D. E. 1971a. Growth intercept as an indicator of site index in natural stands of white pine in the Southern Appalachians. *U.S. Forest Serv., Southeast. Forest Expt. Sta., Res. Note* SE-154. 6 pp.

————. 1971b. Height-growth patterns and site index of white pine in the Southern Appalachians. *Forest Sci.* **17**:252–260.

Bickford, C. A., et al. 1957. Stocking, normality, and measurement of stand density. *J. Forestry* **55**:99–104.

Bredenkamp, B. V., and Burkhart, H. E. 1990. An examination of spacing indices for *Eucalyptus grandis. Can. J. For. Res.* **20**:1909–1916.

Burkhart, H. E., and Tennent, R. B. 1977. Site index equations for radiata pine in New Zealand. *N.Z. J. Forestry Sci.* **7**:408–416.

Carmean, W. H. 1972. Site index curves for upland oaks in the Central States. *Forest Sci.* **18**:109–120.

————. 1975. Forest site quality evaluation in the United States. *Adv. Agronomy* **27**:209–258.

Clutter, J. L., Fortson, J. C., Pienaar, L. V., Brister, G. H., and Bailey, R. L. 1983. *Timber management: A quantitative approach.* John Wiley & Sons, New York. 333 pp.

Curtis, R. O. 1970. Stand density measures: An interpretation. *Forest Sci.* **16**:403–414.

————. 1982. A simple index of stand density for Douglas-fir. *Forest Sci.* **28**:92–94.

Daniels, R. F. 1976. Simple competition indices and their correlation with annual loblolly pine tree growth. *Forest Sci.* **22**:454–456.

————, Burkhart, H. E., and Clason, T. R. 1986. A comparison of competition measures for predicting growth of loblolly pine trees. *Can J. For. Res.* **16**:1230–1237.

Doolittle, W. T. 1958. Forest soil-site relationships and species comparisons in the Southern Appalachians. *Soil Sci. Soc. Am. Proc.* **22**:455–458.

Drew, T. J., and Flewelling, J. W. 1977. Some recent Japanese theories of yield-density relationships and their application to Monterey pine plantations. *Forest Sci.* **23**:517–534.

————. 1979. Stand density management: An alternative approach and its application to Douglas-fir plantations. *Forest Sci.* **25**:518–532.

Gerrard, D. J. 1969. Competition quotient: A new measure of the competition affecting individual forest trees. *Mich. Agr. Expt. Sta., Res. Bull.* 20. 30 pp.

Gingrich, S. F. 1967. Measuring and evaluating stocking and stand density in upland hardwood forests in the Central States. *Forest Sci.* **13**:38–53.

Hannah, P. R. 1968. Topography and soil relations for white and black oak in southern Indiana. *U.S. Forest Serv., North Central Forest Expt. Sta., Res. Paper* NC-25. 7 pp.

Hegyi, F. 1974. A simulation model for managing jack-pine stands. In *Growth models for tree and stand simulation.* Edited by J. Fries. Royal College of Forestry, Stockholm, Sweden. pp. 74–90.

Heiberg, S. O., and White, D. P. 1956. A site evaluation concept. *J. Forestry* **54**:7–10.

Johnson, E. W. 1973. Relationship between point density measurements and subsequent growth of southern pines. *Auburn Univ., Ala. Agr. Expt. Sta., Bull.* 447. 109 pp.

Johnson, J. E., Haag, C. L., Bockheim, J. G., and Erdmann, G. G. 1987. Soil-site relationships and soil characteristics associated with even-aged red maple *(Acer rubrum)* stands in Wisconsin and Michigan. *Forest Ecol. and Manage.* **21**:75–89.

Jones, J. R. 1969. Review and comparison of site evaluation methods. *U.S. Forest Serv., Rocky Mt. Forest and Range Expt. Sta., Res. Paper* RM-51. 27 pp.

Krajicek, J. E., Brinkman, K. A., and Gingrich, S. F. 1961. Crown competition—A measure of density. *Forest Sci.* **7**:35–42.

Larson, E. H., and Moehring, D. M. 1972. Site index curves for longleaf pine in East Texas. *Texas A & M Univ., Dept. Forest Sci., Res. Note* No. 1. 3 pp.

Lemin, R. C., Jr., and Burkhart, H. E. 1983. Predicting mortality after thinning in old-field loblolly pine plantations. *So. J. Appl. For.* **7:**20–23.

Martin, G. L., and Ek, A. R. 1984. A comparison of competition measures and growth models for predicting plantation red pine diameter and height growth. *Forest Sci.* **30:**731–743.

Monserud, R. A. 1985. Comparison of Douglas-fir site index and height growth curves in the Pacific Northwest. *Can. J. For. Res.* **15:**673–679.

Nelson, T. C., and Bennett, F. A. 1965. A critical look at the normality concept. *J. Forestry* **63:**107–109.

Noone, C. S., and Bell, J. F. 1980. An evaluation of eight intertree competition indices. *Oregon State Univ., For. Res. Lab., Res. Note* 66. 6 pp.

Oliver, W. W. 1972. Height intercept for estimating site index in young ponderosa pine plantations and natural stands. *U.S. Forest Serv., Pacific Southwest Forest and Range Expt. Sta., Res. Note* PSW-276. 4 pp.

Olson, D. F., Jr., and Della-Bianca, L. 1959. Site index comparisons for several tree species in the Virginia-Carolina Piedmont. *U.S. Forest Serv., Southeast. Forest Expt. Sta., Sta. Paper* 104. 9 pp.

Opie, J. E. 1968. Predictability of individual tree growth using various definitions of competing basal area. *Forest Sci.* **14:**314–323.

Philbrook, J. S., Barrett, J. P., and Leak, W. B. 1973. A stocking guide for eastern white pine. *U.S. Forest Serv., Northeast. Forest Expt. Sta., Res. Note* NE-168. 3 pp.

Reineke, L. H. 1933. Perfecting a stand-density index for even-aged forests. *J. Agr. Res.* **46:**627–638.

Society of American Foresters. 1971. *Terminology of forest science, technology, practice and products.* Society of American Foresters, Washington, D.C. 349 pp.

Sprackling, J. A. 1973. Soil-topographic site index for Engelmann spruce on granitic soils in northern Colorado and southern Wyoming. *U.S. Forest Serv., Rocky Mountain Forest and Range Expt. Sta., Res. Note* RM-239. 4 pp.

Spurr, S. H. 1952. *Forest inventory.* The Ronald Press Company, New York. 476 pp.

———, and Barnes, B. V. 1980. *Forest ecology.* 3d ed. John Wiley & Sons, New York. 687 pp.

Stage, A. R. 1963. A mathematical approach to polymorphic site index curves for grand fir. *Forest Sci.* **9:**167–180.

Strub, M. R., Vasey, R. B., and Burkhart, H. E. 1975. Comparison of diameter growth and crown competition factor in loblolly pine plantations. *Forest Sci.* **21:**427–431.

Tennent, R. B. 1975. Competition quotient in young *Pinus radiata. N.Z. J. Forestry Sci.* **5:**230–234.

Tomé, M., and Burkhart, H. E. 1989. Distance-dependent competition measures for predicting growth of individual trees. *Forest Sci.* **35:**816–831.

West, P. W. 1983. Comparison of stand density measures in even-aged regrowth eucalypt forest of southern Tasmania. *Can. J. For. Res.* **13:**22–31.

Wilson, F. G. 1946. Numerical expression of stocking in terms of height. *J. Forestry* **44:**758–761.

———. 1979. Thinning as an orderly discipline: A graphic spacing schedule for red pine. *J. Forestry* **77:**483–486.

Zeide, B. 1987. Analysis of the 3/2 power law of self-thinning. *Forest Sci.* **33:**517–537.

CHAPTER 15

TREE-GROWTH AND STAND-TABLE PROJECTION

15-1   Increases in Tree Diameter   Tree growth is an intermittent process characterized by changes in stem form and dimension over a period of time. In temperate forests, a growing tree adds a yearly layer of wood just under the bark, from ground level to tip and all around the stem. In cross section, these layers appear as annual rings. Accordingly, tree age can be determined by counting the rings, and the volume of each ring is a measure of the wood added to the central stem that particular year.

Annual rings tend to be wider during the early life of a tree; as age increases, the ring width gradually decreases, resulting in a reduction of annual diameter increment (Fig. 15-1). Even though the *width* of each ring normally decreases as the tree becomes older, this thinner wood layer is added over a larger stem diameter or bole surface. Therefore, the *volume* of wood added annually may be equal to or greater than that of previous years (Gessel et al., 1960). In addition to age, the rate of diameter growth is dependent on the soil moisture availability and the amount of leaf surface functioning in the photosynthetic process. Wider spacing among trees results in more root-growing space and larger crowns, which, in turn, lead to faster diameter growth.

15-2   Increases in Tree Height   Changes in tree height are of prime concern for predicting future stand composition and for selecting the ideal crop trees in pure stands. The typical course of height growth is illustrated by the sigmoid curve in Fig. 15-2. Height growth proceeds slowly until the seedling is well established; this is followed by a period of rapid growth during the next 20 to 30

FIGURE 15-1
Typical pattern of diameter increment for trees in unthinned stands. *(U.S. Forest Service drawing.)*

years, depending on the species and site involved. As a tree begins to attain maturity, height growth gradually tapers off but never completely ceases as long as the tree is living and healthy.

The cumulative growth curve in Fig. 15-2 follows the same general configuration for most functions of tree growth—whether height, diameter, basal area (BA), or cubic volume. Although the exact form of the cumulative growth curve will differ with the variable used and climatic fluctuations, the elongated S-shaped pattern is a characteristic that can be invariably expected. From the foregoing, it can be seen that wood production in the central stem of a tree can be predicted by measuring past rates of diameter and height growth. Indeed, the primary objective of most tree-growth studies is the reliable prediction of future wood yields.

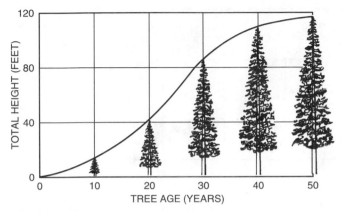

FIGURE 15-2
Cumulative height-growth pattern followed by many coniferous species.

15-3 Periodic and Mean Annual Growth The increase in tree size for 1 year is referred to as the *current annual growth* (or *current annual increment*). Because current growth is difficult to measure for a single year, the average annual growth over a period of 5 to 10 years is commonly substituted instead. The difference in tree size between the start and the end of a growth period, divided by the number of years involved, is properly termed *periodic annual growth* (or *periodic annual increment*). By contrast, the *average,* or *mean, annual growth* (also called *mean annual increment*) is derived by dividing total tree size at any point in time by total age.

Current or periodic annual growth, whether based on volume or other tree size characteristics, increases rapidly, reaches a crest, and then drops off rapidly. In comparison, mean annual growth increases more slowly, attains a maximum at a later age, and falls more gradually. When curves of current and mean annual growth are plotted over tree age, they intersect at the peak of the latter (Fig. 15-3). Similar relationships are exhibited on a stand basis, as well as on a tree basis (Sec. 16-2).

15-4 Past Growth from Complete Stem Analysis The most accurate method of gauging accumulated tree volume growth is by complete stem analysis. Although it is possible to obtain needed measurements and annual ring counts by climbing and boring standing trees, the usual technique requires that sample trees be felled and cut into sections at the end of a designated growth period. Diameter inside bark (dib) at the beginning of the growth period is derived by counting annual rings back to the desired year. The total starting volume of all tree sections is subtracted from current volume to obtain cubic-foot growth.

The exact method followed in making a complete stem analysis, including

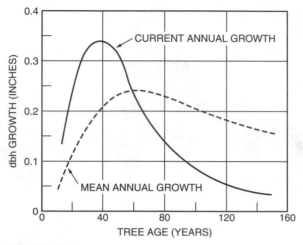

FIGURE 15-3
Graphic comparison of the current annual and mean annual
growth of a tree.

points of stem measurement and intervals between sections, varies according to
tree form and desired precision. Therefore, the procedures outlined in Table
15-1 are intended merely to serve as an illustration of the computations involved. In this example, an 8-year growth period is presumed for a coniferous
tree having a total height of 50 ft at the time of felling.

The tree is severed 1 ft above ground to minimize effects of butt swell; it is
cut into uniform 10-ft lengths, excepting the final 9-ft top section. Present dib is
obtained at each cutting point; for elliptical cross sections, this is derived by averaging minimum and maximum diameters. Next, average cross-section diameters are converted to cross-sectional area in square feet, followed by computations of present cubic volume for each section.

In Table 15-1, stump content is computed as the volume of a cylinder; i.e.,
taper in the first 1-ft section is ignored. *Present* volumes of the four 10-ft sections are derived from Smalian's formula, and content of the top 9-ft section is
computed as the volume of a conoid.

To obtain *previous* stem volume, diameters are measured by counting back
eight annual rings from the present. Cubic volumes for stump and lower stem
sections are calculated as before. For the top section, however, previous length
must be determined by making several trial-and-error cuts from the tip downward—until the first ring preceding the growth period is located. Once the previous top length has been measured, its cubic content is again computed as the
volume of a conoid (or other suitable geometric solid).

The difference in stem volume between the beginning and end of the specified time period represents gross growth. When this value is divided by the num-

TABLE 15-1
SIMPLIFIED STEM-ANALYSIS COMPUTATIONS FOR A TREE-GROWTH PERIOD
OF 8 YEARS

Section height above ground, ft	Average diameter, in.	Cross-sectional area, sq ft	Average cross-sectional area,sq ft		Section length, ft	Section volume, cu ft
			(Stump)	1.163	1	1.163
1	14.6	1.163				
			(Section 1)	1.028	10	10.280
11	12.8	0.894				
			(Section 2)	0.748	10	7.480
21	10.5	0.601				
			(Section 3)	0.507	10	5.070
31	8.7	0.413				
			(Section 4)	0.280	10	2.800
41	5.2	0.147				
			(Top)	0.049	9	0.441
50	0.0	(Conoid)				

Present total height: 50 ft   Present cubic volume: 27.234

Section height above ground, ft	Average diameter, in.	Cross-sectional area, sq ft	Average cross-sectional area, sq ft		Section length, ft	Section volume, cu ft
			(Stump)	0.799	1	0.799
1	12.1	0.799				
			(Section 1)	0.712	10	7.120
11	10.7	0.624				
			(Section 2)	0.533	10	5.330
21	9.0	0.442				
			(Section 3)	0.374	10	3.740
31	7.5	0.307				
			(Section 4)	0.204	10	2.040
41	4.3	0.101				
			(Top)	0.034	2	0.068
43	0.0	(Conoid)				

Previous total height: 43 ft   Previous cubic volume: 19.097
Gross volume increase: 8.137 cu ft
Periodic annual growth: 8.137/8 = 1.017 cu ft per year

ber of years in the period (8 in this example), the result is a measure of periodic annual growth.

Some stem analyses require that sectional cuts be made at both stump and diameter at breast height (dbh) levels, depending on the objectives. In such cases, the stem section below dbh is usually regarded as a cylinder for purposes of deriving cubic volume. Cutting intervals above dbh may also be shortened to 4 ft or less when greater precision is desired. During inclement weather, stem sections about 1 in. thick can be extracted at desired intervals and the actual analysis performed indoors.

15-5   Tree Growth as a Percentage Value   The calculation of tree growth in percentage terms is an expression of the average rate of change in size or volume over a given time period (Belyea, 1959). Because each year's annual ring is added over the cumulative size of the tree stem, tree growth has been most frequently regarded as a compound interest relationship. Despite the apparent logic of the compound interest theory, however, observations of actual volumes in uncut timber stands at three or more points in time indicate that tree growth is sometimes best described by *simple interest rates* (Grosenbaugh, 1958).

Actually, the argument of compound versus simple interest is largely an academic question, because growth percent alone has little practical value in management decision making. A large number of growth percent formulas have been proposed in previous years, many of which are misleading because of the inherent nature of tree growth itself. Because the base dimensions of a tree are constantly increasing, a uniform annual increment results in a progressively lower and lower annual interest rate as the tree gets larger. Thus when the absolute increment remains constant, interest rates can appear astounding for small trees but strictly mediocre for larger ones.

Compound interest formulas are readily available in standard texts on forest finance and valuation. In terms of simple interest, the growth percentage in volume at any age is the current (periodic) annual growth divided by the "base volume" at the beginning of the growth period. Expressed as a formula, annual simple interest rates may be computed by

$$\text{Growth percent} = \frac{V_2 - V_1}{n \times V_1}(100)$$

where V_2 = volume or tree size at end of growth period
V_1 = volume or tree size at start of growth period
n = number of years in growth period

Substituting growth values from Table 15-1, the annual simple interest rate would be

$$\text{Growth percent} = \frac{27.234 - 19.097}{8(19.097)}(100) = 5.3 \text{ percent}$$

15-6   Predictions of Tree Growth   As stated earlier, the principal reason for analyzing the past growth of trees is to establish a pattern for predicting future growth. From the standpoint of practical forest management, growth prediction is usually approached from a *stand* basis rather than in terms of individual trees. However, because tree growth is the integral component of stand growth, the trends of tree size increases are appropriately considered first. Stand-growth prediction is discussed in subsequent sections.

Because the rate of tree growth in diameter, height, form, or volume is heavily dependent on relative age, prediction of future yields from past growth should be limited to short periods of time—usually not more than 5 to 10 years. Otherwise, large errors will result from the assumption that future growth will be equivalent or similar to past growth. As a rule, growth predictions are most reliable during the midlife of a tree, i.e., when size increases can be characterized by the central (near-linear) portion of the cumulative growth curve. When cumulative growth curves are available for the desired species, future growth for short time periods can be approximated by extrapolation of such trends. Curves of periodic or mean annual growth can be utilized in like fashion, but this procedure is not reliable for extended time periods.

15-7   Future Yields from Growth Percentage   This approach is analogous to the foregoing technique, for it presumes that future growth will proceed at the same *rate* as past growth. Even though this may be a reasonable postulation for a 3- to 5-year span, it is not recommended for longer periods. The annual growth rate of 5.3 percent derived in Sec. 15-5 may be used as a simple example. Projecting the present tree volume of 27.234 cu ft ahead by 3 years would result in a theoretical increase of 15.9 percent. Thus future tree volume would be computed as (1.159)(27.234), or 31.564 cu ft.

15-8   Growth Prediction from Diameter and Height Increases   Assume that a tree 14 in. in diameter and 65 ft tall had a dbh of 12 in. and a height of 55 ft 10 years ago. An obvious conclusion is that trees *now* 12 in. in diameter and 55 ft tall (on the same site and in the same relative position in the canopy) will grow 2 in. in diameter and 10 ft in height during the next 10 years. Actual rates of diameter and height increases may be obtained from complete stem analyses, from increment borings of standing trees, or from periodic remeasurement of permanent sample plots.

One simple method of short-term growth prediction for individual trees accommodates changes in height growth by use of a "local" volume table or height/dbh curve for the desired species.[1] Rates of diameter increase are obtained from increment borings at dbh. In the example that follows, the annual cubic-volume growth per tree (outside bark) is based on the number of annual rings in the last $\frac{1}{2}$ in. of tree radius.

---

[1]This procedure may not be suited to long-term growth prediction. Over long periods of time, significant changes may occur in the height/dbh relationship for even-aged stands.

Assume that a tree with a dbh of 12.8 in. has a present merchantable volume of 20.4 cu ft, outside bark, as read from a local volume table. An increment boring at dbh shows four annual rings in the last $1/2$ in. of tree radius; i.e., the tree required 4 years to produce the last full inch of diameter growth. When this tree was exactly 1 in. smaller in diameter, or 11.8-in. dbh, its merchantable volume was 17.2 cu ft—as read from the same "local" volume table. The difference in volume of 3.2 cu ft was thus produced in 4 years, for a periodic annual growth of 3.2/4, or 0.8 cu ft.

For the next 3 to 5 years, it is reasonably safe to assume that cubic volume will continue to increase at the rate of past growth, yielding a merchantable tree volume of 22.8 cu ft within 3 years or 24.4 cu ft after 5 years. As emphasized in Sec. 15-6, tree growth is a near-linear function of age during the middle years of stem development, particularly if the stand is relatively undisturbed by fire, heavy cutting, or unusual changes in density and competition.

Because the relationship of bark thickness to diameter outside bark (dob) changes as a tree grows, predictions more than 5 years ahead should properly be

FIGURE 15-4
Increment borings from stems with eccentric cross sections will display wide differences in growth, depending on the radii selected. Note the acceleration of diameter increment following release from competition. *(U.S. Forest Service photograph.)*

based on wood growth alone, i.e., inside bark measurements. In such instances, future dbh values are developed by computing inside-outside bark ratios for each diameter class involved.

Where continuous forest inventory (CFI) systems are established, the most reliable method of obtaining growth information is by repeated measurements of the same trees on permanent sample plots. The technique of complete stem analysis is also recommended, especially for research studies dealing with patterns and fluctuations in growth cycles. Although the increment borer is a useful inventory tool, this method of growth determination probably ranks below the other two in terms of reliability. Some species exhibit widely divergent patterns of radial growth from one side of the tree to another as viewed on an increment core, depending on live-crown configuration (Fig. 15-4). Many ring-porous hardwoods are extremely difficult to bore with conventional equipment, and certain diffuse-porous species have inconspicuous annual ring delineations. Even though these factors are often beyond the control of the inventory forester, they can nevertheless contribute to erroneous growth estimates.

STAND-TABLE PROJECTION

15-9 Components of Stand Growth The basic elements of stand growth are accretion, mortality, and ingrowth (Gilbert, 1954). *Accretion* is the growth on all trees that were measured at the beginning of the growth period. It includes the growth on trees that were cut during the period plus those trees that died and were utilized. *Mortality* is the volume of trees initially measured that died during a growth period and were not utilized. The volume of those trees that grew into the lowest inventoried diameter class during the growth period is termed *ingrowth*.

Gross growth is a measure of the change in total volume for a given stand. In any given diameter class, it is the change in volume, plus mortality, during the growth period.

Net growth represents the stand-volume increment based on the initial trees after mortality has been deducted. When ingrowth is added to net growth, the result is volume increase, or *production,* i.e., a measure of the net change in volume during a specified growth period. If certain trees were harvested during a growth period, yield volume must also be considered in computing production values.

15-10 Characteristics of Stand-Table Projection This method of growth prediction recognizes the structure of a stand, and growth projections are made according to dbh classes. The method is best suited to uneven-aged, low-density, and immature timber stands. In dense or overmature forests where mortality rates are high, stand-table projection may be of questionable value for providing reliable information on net stand growth.

The procedure ordinarily followed in the stand-table projection method of growth prediction may be briefly summarized as follows:

1 A present stand table showing the number of trees in each dbh class is developed from a conventional inventory.

2 Past periodic growth, by dbh classes, is determined from increment borings or from remeasurements of permanent sample plots. When increment borings are used, growth values must be converted from an inside-bark basis to outside-bark readings.

3 Past diameter growth rates are applied to the present stand table to derive a future stand table showing the predicted number of trees in each dbh class at the end of the growth period. Numbers of trees in each class must then be corrected for expected mortality and predicted ingrowth.

4 Both present and future stand tables are converted to stock tables by use of an appropriate local volume table. Thus for short growth periods, the expected changes in tree height during the growth period are inherently accommodated by diameter increases.

5 Periodic stand growth is obtained as the difference between the total volume of the present stand and that of the future stand.

15-11 Diameter Growth Rates of diameter growth outside bark are best obtained from repeated measurements of permanent sample plots. Consecutive inventories of the same trees provide a direct evaluation of combined wood and bark increment at dbh. As a result, many of the problems encountered in estimating stem growth from increment borings can be avoided.

As an example, Table 15-2 was compiled from remeasurements of dbh outside bark for southern pines and hardwoods distributed throughout the state of Alabama. Tree species and diameters were sampled in proportion to occurrence. All the major southern pines were represented; the principal hardwoods sampled were red and white oaks, hickories, yellow-poplar, sweetgum, and tupelos. The lack of a definite differentiation in growth by diameter classes was attributed to the moderate stand-density levels common in that state (Judson, 1965). Tabulations of this nature, however, are ideally suited to efficient stand-table projections.

When diameter growth is not available in the foregoing form, it is customary to rely on increment borings at dbh instead. Assuming we wish to estimate diameter growth (outside bark) for the last 10 years, estimates for each dbh class might be handled according to the following step-by-step procedure:

1 Measure present dbh to the nearest 0.1 in. and subtract diameter bark thickness to obtain present dib at breast height.

2 From an increment boring, obtain the 10-year wood growth in diameter and subtract from present dib to derive dib at breast height 10 years ago.

3 For each diameter class recognized, plot present diameter bark thickness over present dib at breast height. Draw a smooth, balanced curve (or fit a regres-

TABLE 15-2
ANNUAL DIAMETER GROWTH BY DIAMETER CLASS IN ALABAMA*

dbh class, in.	Pine species			Hardwood species		
	No. of sample trees	Mean growth, in.	Standard deviation, in.	No. of sample trees	Mean growth, in.	Standard deviation, in.
6	522	0.22	0.13	733	0.13	0.10
8	352	0.23	0.13	416	0.13	0.11
10	179	0.24	0.12	255	0.13	0.10
12	88	0.22	0.12	122	0.14	0.10
14	40	0.24	0.14	66	0.14	0.10
16	11	0.26	0.19	40	0.13	0.09
18	10	0.21	0.09	18	0.16	0.10
20+	8	0.18	0.09	12	0.15	0.13
All diameters	1210	0.22	0.13	1662	0.13	0.10

*Diameter growth (outside bark) based on remeasurements of 2872 trees by the U.S. Forest Service. Reprinted from Judson, 1965.

sion equation) through the plotted points. Read off appropriate bark thicknesses for each dib 10 years ago (step 2) and add these values together to arrive at an estimate of dbh (outside bark) 10 years ago.

4 Subtract dbh (outside bark) 10 years ago from present dob to derive the estimated growth in diameter during the stated time period. If future growth is presumed to equal past growth, this information may be applied directly in a stand-table projection.

15-12 Stand Mortality and Ingrowth The reliability of stand-table projections leans heavily on the derivation of realistic estimates of mortality and ingrowth. As with diameter increment, such information is preferably obtained from consecutive reinventories of permanent sample plots; in reality, there is no other sound procedure for making these predictions. Mortality rates are desired for each dbh recognized in the stand table, because the natural demise of smaller stems is usually much greater than for larger diameters. Only when growth predictions are made for very short time periods (perhaps 3 years or less) can mortality be regarded as a negligible factor.

For growth predictions of 5 to 10 years, ingrowth is usually accounted for by having the present stand table include several diameter classes below the minimum dbh desired in the future stand table. As an illustration, if 10-year growth predictions are planned for trees 10-in. dbh and larger, the initial stand table might include all stems that might logically grow into the 10-in.-dbh class during the interim, e.g., those stems presently 6 in. or more in diameter.

15-13   A Sample Stand Projection   For purposes of illustration, it will be assumed that the information on pine species in Table 15-2 represents a 20-acre stand for which a 10-year volume-growth prediction is desired for stems in the 10-in.-dbh class and larger. Present and future board-foot volumes are to be derived from a local volume table based on the Scribner log rule. The present stand table, including adjustments for mortality and applicable decadal growth rates, appears as shown in Table 15-3. The 6-in. and 8-in. trees are included in the present stand table to accommodate ingrowth into larger-diameter classes.

Because mortality has been deducted from the present stand in Table 15-3, the next step is the application of diameter growth rates in deriving a future stand table. The upward movement of trees into larger-dbh classes is proportional to the ratio of growth to the chosen diameter-class interval

$$\text{Growth-index ratio } = \frac{g}{i}$$

where g is the diameter growth in inches and i is the diameter-class interval in inches.

Using the 6-in.-dbh class from Table 15-3 as an example,

$$\text{Growth-index ratio } = \frac{2.2}{2.0} = 1.10$$

The interpretation of a growth-index ratio of 1.10 is that 100 percent of the trees move up one dbh class, and 0.10, or 10 percent, of these advance two classes. Thus, of the 313 trees expected to survive in the 6-in. class, 90 percent (282 trees) move up to the 8-in. class, and 10 percent (31 trees) move all the

TABLE 15-3
PRESENT STAND TABLE, MORTALITY, AND EXPECTED 10-YEAR
DIAMETER GROWTH FOR A 20-ACRE PINE STAND IN ALABAMA

dbh class, in.	Present stand, no. of stems	Expected mortality, percent	Expected survival, no. of stems	10-year dbh growth, in.
6	522	40	313	2.2
8	352	35	229	2.3
10	179	25	134	2.4
12	88	20	70	2.2
14	40	15	34	2.4
16	11	10	10	2.6
18	10	10	9	2.1
20+	8	20	6	1.8
Total	1210		805	

TABLE 15-4
APPLICATION OF GROWTH-INDEX RATIOS IN DERIVING A FUTURE STAND
TABLE FOR A 20-ACRE PINE STAND IN ALABAMA

dbh class, in.	Present stand surviving, no. of stems	Growth-index ratio, g/i	No. of stems moving up, by dbh classes			Future stand table, no. of stems
			No change	1 class	2 classes	
6	313	1.10	0	282	31	0
8	229	1.15	0	195	34	282
10	134	1.20	0	107	27	226
12	70	1.10	0	63	7	141
14	34	1.20	0	27	7	90
16	10	1.30	0	7	3	34
18	9	1.05	0	8	1	14
20	6	0.90	1	5	0	12
22	0	· · · ·	0	0	0	6
Total	805	· · · ·	1	694	110	805

way to the 10-in. class. None will remain in the 6-in. class in this instance. If the growth-index ratio had been less than unity, for example, 0.80, 80 percent of the trees would move up one class interval, and 20 percent would remain in the present dbh class. For the dbh classes in Table 15-3, growth-index ratios and the future stand table are shown in Table 15-4.

Once the future stand table has been derived, present and future volumes (stock tables) can be obtained from an appropriate local volume table (Table 15-5).

TABLE 15-5
PREDICTED 10-YEAR VOLUME PRODUCTION OF A 20-ACRE PINE STAND
IN ALABAMA

dbh class, in.	Present stand table, no. of stems	Future stand table, no. of stems	Scribner volume per tree, bd ft	Present stock table, bd ft	Future stock table, bd ft	Volume production, bd ft
6	313	0				
8	229	282				
10	134	226	42	5,628	9,492	3,864
12	70	141	86	6,020	12,126	6,106
14	34	90	136	4,624	12,240	7,616
16	10	34	201	2,010	6,834	4,824
18	9	14	280	2,520	3,920	1,400
20	6	12	369	2,214	4,428	2,214
22	0	6	481	0	2,886	2,886
Total	805	805	· · · ·	23,016	51,926	28,910

Volume production is computed for each dbh class as the difference between present and future volumes. For this hypothetical stand, the predicted net volume growth for the 10-year period is 28,910 bd ft, or 1445 bd ft per acre. On an *annual* basis, the predicted growth per acre is estimated as 144.5 bd ft.

PROBLEMS

15-1 Prepare curves of periodic and mean annual growth for an important timber species growing in your locality. Does the culmination of mean annual growth roughly coincide with the accepted rotation age for that species? Give reasons for differences, if any.

15-2 Make a complete stem analysis of a tree that is 20 to 40 years old. Using a growth period of 5 to 10 years, compute (a) present cubic volume, (b) periodic annual growth, and (c) predicted future volume 5 to 10 years hence.

15-3 Using simple interest rates, compute the growth percent of an even-aged stand in your locality. Explain possible reasons why simple interest rates may give erroneous estimates of growth percent in even-aged stands.

15-4 Predict periodic annual growth of an uneven-aged stand in your locality by applying the stand-table projection method.

15-5 Suppose a tree with dbh equal to 8.00 in. grows with a constant basal-area increment of 0.0345 sq ft per year over the next 5 years. (a) Compute the diameter increment for the tree just described for each year of the 5-year period. (b) With a constant basal-area increment, is the diameter increment constant, increasing, or decreasing?

15-6 Following are growth-index ratios by dbh class.

dbh class, in.	Growth-index ratio, g/i	dbh class, in.	Growth-index ratio, g/i
6	1.45	14	1.00
8	1.30	16	0.90
10	1.20	18	0.85
12	1.10	20	0.80

a Apply the growth-index ratio values given to the present stand table shown in Table 15-4 to develop a future stand table.

b Use the future stand table developed in part a in conjunction with the local volume table given in Table 15-5 to derive a future stock table and volume-production values by dbh class.

REFERENCES

Barrett, J. P., and Allen, P. H. 1966. Predicting yield of extensively managed white pine stands. *Univ. New Hampshire Agr. Expt. Sta. Tech. Bull.* 108. 15 pp.

Belyea, H. C. 1959. Two new formulae for predicting growth per cent. *J. Forestry* **57**:104–107.

Cameron, R. J., and Lea, R. 1980. Band dendrometers or diameter tapes? *J. Forestry* **78:**277–278.

Ffolliott, P. F. 1965. Determining growth of ponderosa pine in Arizona by stand projection. *U.S. Forest Serv., Rocky Mt. Forest and Range Expt. Sta., Res. Note* RM-52. 4 pp.

Gessel, S. P., Turnbull, K. J., and Tremblay, F. T. 1960. *How to fertilize trees and measure response.* National Plant Food Institute, Washington, D.C. 67 pp.

Gilbert, A. M. 1954. What is this thing called growth? *U.S. Forest Serv., Northeast. Forest Expt. Sta. Paper* 71. 5 pp.

Grosenbaugh, L. R. 1958. Allowable cut as a new function of growth and diagnostic tallies. *J. Forestry* **56:**727–730.

Hall, O. F. 1959. The contribution of remeasured sample plots to the precision of growth estimates. *J. Forestry* **57:**807–811.

Herman, F. R., DeMars, D. J., and Woollard, R. F. 1975. Field and computer techniques for stem analysis of coniferous forest trees. *U.S. Forest Serv., Pacific Northwest Forest and Range Expt. Sta., Res. Paper* PNW-194. 51 pp.

Judson, G. M. 1965. Tree diameter growth in Alabama. *U.S. Forest Serv., Southern Forest Expt. Sta., Res. Note* SO-17. 3 pp.

Kirby, C. L. 1962. The growth and yield of white spruce-aspen stands in Saskatchewan. *Forestry Branch Dept. Nat. Resources Prov. Saskatchewan Tech. Bull.* 4. 58 pp.

Liming, F. G. 1957. Homemade dendrometers. *J. Forestry* **55:**575–577.

Meyer, H. A. 1952. Structure, growth, and drain in balanced, uneven-aged forests. *J. Forestry* **50:**85–92.

Moser, J. W., Tubbs, C. H., and Jacobs, R. D. 1979. Evaluation of a growth projection system for uneven-aged northern hardwoods. *J. Forestry* **77:**421–423.

Smith, R. B., Hornbeck, J. W., Federer, C. A., and Krusic, P. J., Jr. 1990. Regionally averaged diameter growth in New England forests. *U.S. Forest Serv., Northeast. Forest Expt. Sta., Res. Paper* NE-637. 26 pp.

Zedaker, S. M., Burkhart, H. E., and Stage, A. R. 1987. General principles and patterns of conifer growth and yield. Pp. 203–241 in *Forest vegetation management for conifer production,* J. D. Walstad and P. J. Kuch (eds.), John Wiley & Sons, New York.

CHAPTER 16

GROWTH AND YIELD MODELS

16-1 Introduction Forest management decisions are predicated on information about both current and future resource conditions. Inventories taken at one instant in time provide information on current volumes and related statistics. Forests are dynamic biological systems that are continuously changing, and it is necessary to project these changes to obtain relevant information for prudent decision making.

Stand dynamics (i.e., the growth, mortality, reproduction, and associated changes in the stand) can be predicted through direct or indirect methods. Direct methods, such as the stand-table projection technique discussed in Chap. 15, involve field observations in existing stands. Past growth and mortality trends are used to infer future trends in the stands observed.

There are many situations in which direct observation of forest growth and mortality are not feasible, however. Diameter growth, mortality, and ingrowth relationships developed through stand-table projection techniques are not reliable for long periods of time. Furthermore, managers often wish to evaluate a broad range of treatment alternatives. Inferences from past growth are limited to the conditions under which that growth occurred. Also, the costs of direct observation are sometimes prohibitive. Consequently, foresters often rely on indirect methods of predicting stand dynamics—i.e., growth, mortality, and related quantities of a stand are inferred from the study of other stands. These inferences are made through the use of tables, equations, or computer simulation models. Techniques for forecasting stand dynamics are collectively referred to as growth and yield models.

Growth and yield forecasts may be required for a short-term or long-term basis, for the overall stand volume or volume by product and size classes. With the wide variety of existing stand conditions and the diverse objectives and needs of users of growth and yield models, it is not surprising that numerous approaches have been proposed. These approaches range from models that provide only a specified aggregate stand volume to models with information about individual trees. Regardless of the structural complexity and amount of output detail provided, all growth and yield models have a common purpose: to produce estimates of stand characteristics [such as the volume, basal area (BA), and number of trees per unit area] at specified points in time. Representative examples of growth and yield models are discussed in this chapter. These examples are indicative of points along a continuum of modeling complexity. Many of the growth and yield models that are used operationally in forestry are mixtures of the model types described here. Descriptions of growth and yield models for even- and uneven-aged stands are provided.

16-2 Growth and Yield Relationships   Before proceeding further, it is important to define the terms growth and yield. *Growth* is the increase (increment) over a given period of time. *Yield* is the total amount available for harvest at a given time. Thus yield can be regarded as the summation of the annual increments. To be meaningful, growth and yield values must be qualified with regard to the part of the tree and the portion of the stand being considered. Further, one must be certain of the unit of measure being used, and, for growth, of the time period involved.

The factors most closely related to growth and yield of forest stands are (1) the point in time in stand development, (2) the site quality, and (3) the degree to which the site is occupied. For even-aged stands, these factors can be expressed quantitatively through the variables of stand age, site index, and stand density, respectively. The measure of stand density most commonly used in growth and yield models for natural stands has been BA per unit area, whereas most models for planted stands have employed number of trees per unit area. For a given site index and initial stand-density level, volume per unit area (yield) plotted over stand age results in a sigmoid curve (Fig. 16-1). The growth curve (often referred to as current annual growth or current annual increment) increases up to the inflection point of the yield curve and decreases thereafter. Another important quantity is the mean annual growth or increment, defined as the yield at any given age divided by the total number of years (age) required to achieve that yield. Rotation age is sometimes set as the age of maximum mean annual increment because, for a given parcel of land, that is the harvest age which will maximize total wood production from a perpetual series of rotations. The rotation age actually selected, however, is also dependent on trends in stumpage values, tree size specifications for various products, and other management considerations. One will note that, in Fig. 16-1, the current annual growth curve crosses the

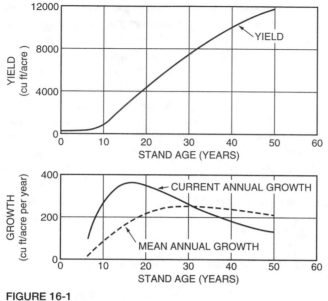

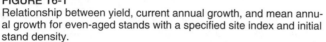

FIGURE 16-1
Relationship between yield, current annual growth, and mean annual growth for even-aged stands with a specified site index and initial
stand density.

mean annual growth curve at its highest value. Section 16-3 contains a description of the equations used to construct Fig. 16-1 and the mathematical relationships between these curves.

16-3   Mathematical Relationships between Growth and Yield   The
yield curve in Fig. 16-1 was generated from the equation

$$Y = e^{10-32A^{-1}}$$

where Y is the yield in cubic feet per acre and A is the stand age in years. Equations of this type are often fitted by linear regression techniques after performing
logarithmic transformations. That is, the equation $\ln Y = a + bA^{-1}$ can be fitted as
a simple linear regression, where $\ln Y$ denotes the natural logarithm of Y. Because yield is the summation of the annual increments, or, to state it another
way, growth is the rate of change in the yield function, the growth curve (current
annual growth) can be derived through methods of differential calculus. Taking
the first derivative of yield with respect to age *(dY/dA)* gives the current annual
growth G equation

$$G = (e^{10-32A^{-1}})(32A^{-2})$$

To compute the age at which growth is maximized, one takes the derivative of growth with respect to age *(dG/dA)* or the second derivative of yield with respect to age *(d^2Y/dA^2)*, sets the quantity equal to zero, and solves for *A*:

$$dG/dA = (e^{10-32A^{-1}})[-2(32)A^{-3}] + (32A^{-2})(e^{10-32A^{-1}})(32A^{-2}) = 0$$

Simplifying and solving results in

$$A = 16 \text{ years}$$

To obtain the age of maximum mean annual growth or increment (denoted by MAI here), one divides the yield function by age (this gives the MAI function by definition), takes the first derivative with respect to age, sets the result equal to zero, and solves for age:

$$\text{MAI} = (e^{10-32A^{-1}})(A^{-1})$$
$$d\text{MAI}/dA = (e^{10-32A^{-1}})(-A^{-2}) + (A^{-1})(e^{10-32A^{-1}})(32A^{-2}) = 0$$

The solution for *A* is

$$A = 32 \text{ years}$$

GROWTH AND YIELD MODELS FOR EVEN-AGED STANDS

16-4 Normal Yield Tables Yield prediction began in the United States with the development of normal yield tables for natural stands. Temporary plots were deliberately located in fully stocked or "normal" density portions of a series of stands of varying ages representing various site qualities. These plot observations of volume per unit area were then sorted into site-quality classes, and volume values were plotted over age. A volume-age curve was then drawn through the points for each site-quality class by using graphical techniques. Values were read from the curve for selected site-quality classes and ages to compile a normal yield table. Table 16-1 is an example of a normal yield table for Douglas-fir in the Pacific Northwest. It should be noted that many normal yield tables contain auxiliary information, such as BA, number of trees per unit area, and diameter distributions, as well as volume per unit area.

Normal yield tables were constructed in an era when only two variables could be included readily by graphical techniques. Thus analysts eliminated the variable of density by holding it constant at fully stocked or "normal" levels. With modern computing technology and analytical techniques, there is no longer any need to restrict the number of variables considered in growth and yield analyses.

Normal yield tables were generally regarded as a model of an ideal, fully

TABLE 16-1
NORMAL YIELD TABLE FOR DOUGLAS-FIR IN THE PACIFIC
NORTHWEST, BOARD FEET PER ACRE, SCRIBNER RULE*

Age, years	Site index, ft (base age 100)				
	80	110	140	170	200
30	0	0	300	2,600	8,000
40	0	200	4,500	11,900	24,400
50	30	3,300	12,400	27,400	44,100
60	1,100	8,100	23,800	42,800	62,000
70	2,400	14,000	35,200	57,200	78,200
80	4,400	20,100	45,700	70,000	92,500
90	6,900	26,000	55,000	81,000	104,800
100	9,600	31,400	62,800	90,400	115,100
110	12,200	36,300	69,400	98,300	123,700
120	14,700	40,700	75,000	105,100	131,100
130	17,000	44,700	80,000	111,000	137,700
140	19,200	48,300	84,500	116,300	143,500
150	21,300	51,600	88,600	121,200	148,700
160	23,300	54,600	92,400	125,700	153,500

*Volume of all trees 11.6-in. dbh and larger. Assumed stump height is 2 ft; minimum top diameter is 8 in.; trimming allowance is 0.3 ft for each 16-ft log. (From McArdle et al., 1961.)

stocked forest to be strived for in management. Today, few foresters believe that the stands shown in normal yield tables constitute a rational management goal. However, for some timber types, the only yield tables available are normal yield tables, and it is important for foresters to be acquainted with the methods used to construct these tables in order to apply them when necessary.

When constructing normal yield tables, one assumes that the temporary plots used have always been fully stocked. A series of fully stocked stands of various ages in a given site-quality class is taken to represent stages in a single growth curve. These assumptions and procedures are questionable, because most stands that are fully stocked at a given time have been overstocked or understocked at some previous time in their development. Consequently, the stand progressions implied in normal yield tables are not likely to be found in nature.

In addition, the definition of normal, or fully stocked, is subjective, and it is not likely that a normal stand would be recognized if it existed. The utility of normal yield tables is severely limited, because no reliable methods are available for predicting yields of nonnormal or understocked stands. The usual procedure has been to compute the ratio of the BA of the stand of interest to the BA shown for the normal yield table and to apply this ratio to the volume. This procedure has not, however, proved to be very satisfactory, because of difficulties in attempting to project changes in this ratio through time.

16-5 Empirical Yield Tables   An empirical yield table is similar to a normal yield table except that it supposedly applies to "average" rather than full, or normal, stocking. Thus the problem of defining normal stocking is eliminated, but an empirical yield table applies only to the average density levels found on the sample plots used. Consequently, difficulties in the application of these tables are similar to those for normal yield tables—adjustments must be made for deviations from the "average" density of the yield-table plots when applying empirical tables to other stands. Empirical yield tables provide few advantages over normal yield tables; the principal idea behind their construction was that the resultant tables should more closely approximate realizable yields under operational forest management than would the values from normal yield tables.

The term *empirical yield table* is somewhat of a misnomer, because all yield tables are ultimately empirical, for they are based on plot observations from a specified forest population. However, the term is fairly well established in forestry terminology as a means of identifying yield tables that apply to "average" stand density. Modern growth and yield modeling techniques do not rely on either "normal" or "average" density concepts, but, rather, include density as a dynamic part of the stand-projection system. Such growth and yield models are commonly termed *variable-density* tables (or equations).

16-6 Variable-Density Growth and Yield Equations   A multiple regression approach to yield estimation, which takes stand density into account, was first applied by MacKinney and Chaiken (1939). Their prediction model for natural stands of loblolly pine was

$$\log Y = b_0 + b_1 A^{-1} + b_2 S + b_3 \log \text{SDI} + b_4 C$$

where $\log Y$ = logarithm of yield (total cu ft per acre of loblolly pine)
$\qquad A^{-1}$ = reciprocal of stand age
$\qquad S$ = site index
$\quad \log \text{SDI}$ = logarithm of stand-density index
$\qquad C$ = composition index (BA per acre of loblolly pine divided by total stand BA)

The measure of density used was Reineke's stand-density index (Sec. 14-15) and a "composition index" was included, because not all of the sample plots were pure loblolly pine. This milestone study in quantitative analysis for growth and yield estimation is akin to methods still being used. Since MacKinney and Chaiken's work, many investigators have used multiple regression techniques to predict growth and/or yield for total stands or for some merchantable portion of stands. Stand-level variables, such as age, site index, BA, or number of trees per acre, are used to predict some specified aggregate stand volume. No information on volume distribution by size class is provided; thus resultant equations from this approach are sometimes referred to as *whole-stand models.*

The variable forms used in subsequent analyses have generally been similar to those employed by MacKinney and Chaiken. Logarithmic transformation of yield is generally made prior to equation fitting to conform to the assumptions customarily made in linear regression analysis. Furthermore, the use of the logarithm of yield as the dependent variable is a convenient way to mathematically express the interaction of the independent variables in their effect on yield. In other words, a unit change in site index, for example, has a differential effect on yield, depending on the level of the other independent variables (age and stand density). This differential effect—sometimes called *interaction*—would not be the case if the dependent variable were yield and interaction terms were not explicitly included.

In most yield analyses, stand age has been expressed as a reciprocal to allow for the "leveling off" (asymptotic) effect of yield with increasing age. Site index is not often transformed prior to fitting, but sometimes logarithmic or reciprocal transformations are employed. In some models, height of the dominant stand has been used in conjunction with age and the variable site index eliminated. (Note that if any two of the three variables—age, site index, height of the dominant stand—are known, the third can be determined.) Use of height rather than site index has the advantage that it is a measured rather than a predicted variable and, thus, more nearly satisfies the assumptions of regression analysis. The measure of stand density is commonly subjected to logarithmic transformation—particularly in models employing BA—but the exact form in which density is included is quite variable, especially for plantation models which utilize number of trees per unit area as a predictor variable.

Early work did not attempt to relate growth analyses to yield analyses, although the biological relationships can be readily expressed mathematically. Buckman (1962) and Clutter (1963) were the first researchers in the United States to explicitly recognize the mathematical relationships between growth and yield in their analyses. Clutter derived compatible growth and yield models for loblolly pine by ensuring that the algebraic form of the yield model could be obtained by mathematical integration of the growth model. Subsequently, Sullivan and Clutter (1972) extended Clutter's models by estimating yield and cumulative growth as a function of initial stand age, initial BA, site index, and future age. When the future age equals the current age (i.e., when the projection period is zero years), the projection model is reduced to a conventional yield model. Thus it is simultaneously a yield model for current conditions and a projection, or growth, model for future conditions.

As an example of compatible growth and yield models, the following equations (adapted from Sullivan and Clutter) were published by Beck and Della-

Bianca (1972) for thinned stands of yellow-poplar:

$$\ln Y_2 = 5.36437 - 101.16296(S^{-1}) - 22.00048(A_2^{-1})$$
$$+ 0.97116(A_1/A_2)(\ln \text{BA}_1) + 3.71796(1 - A_1/A_2) \qquad (1)$$
$$+0.01619(S)(1 - A_1/A_2)$$

where Y_i = stand volume per acre at age A_i
 S = site index, ft (base age 50)
 BA_i = basal area, sq ft per acre, at age A_i
 A_i = stand age at time i (A_1 = present stand age, A_2 = projected stand age)
 ln = natural logarithm

Note that when $A_2 = A_1 = A$, i.e., the projection period is zero years, and $\text{BA}_1 = \text{BA}_2 = \text{BA}$, then equation 1 reduces to the conventional yield model

$$\ln Y = 5.36437 - 101.16296(S^{-1}) - 22.00048(A^{-1}) \qquad (2)$$
$$+ 0.97116(\ln \text{BA})$$

Projected BA per acre is computed by

$$\ln \text{BA}_2 = (A_1/A_2)(\ln \text{BA}_1) + 3.82837(1 - A_1/A_2) \qquad (3)$$
$$+ 0.01667(S)(1 - A_1/A_2)$$

Table 16-2, a variable-density yield table, was prepared by substituting appropriate values for current age, site index, and BA into equation 2. Equations 1 through 3 can also be used to project stand growth. For example, one might assume that a stand growing on site-index-110 land is 40 years old and has 90 sq ft of BA per acre. From equation 2, the current volume is estimated as

$$\ln Y = 5.36437 - 101.16296(110^{-1}) - 22.00048(40^{-1}) + 0.97116(\ln 90)$$
$$\ln Y = 8.2647$$
$$Y = 3884 \text{ cu ft per acre}$$

To estimate the yield 10 years hence, i.e., at age 50, one substitutes $A_1 = 40$, $A_2 = 50$, $S = 110$, and $\text{BA}_1 = 90$ into equation 1 and solves, resulting in

$$\ln Y_2 = 5.36437 - 101.16296(110^{-1}) - 22.00048(50^{-1})$$
$$+ 0.97116(40/50)(\ln 90) + 3.71796(1 - 40/50)$$
$$+ 0.01619(110)(1 - 40/50)$$
$$\ln Y_2 = 8.6005$$
$$Y_2 = 5434 \text{ cu ft per acre}$$

Alternatively, one can project the BA to age 50 by using equation 3. Here, one would substitute the projected basal area (BA_2) for BA, the projected age A_2

TABLE 16-2
VARIABLE-DENSITY YIELD TABLE FOR THINNED STANDS OF YELLOW-POPLAR,
TOTAL CUBIC-FOOT VOLUME PER ACRE*

Age, years	Site index, ft (base age 50)	Basal area, sq ft per acre					
		50	70	90	110	130	150
	90	1032	1431				
	100	1155	1602				
20	110	1266	1756				
	120	1367	1896	2420			
	130	1459	2023	2582			
	90	1490	2065	2636			
	100	1667	2311	2950	3584		
30	110	1827	2533	3234	3930	4622	
	120	1973	2735	3491	4242	4990	
	130	2105	2918	3725	4527	5324	6118
	90	1789	2481	3166	3848		
	100	2002	2776	3543	4306	5064	
40	110	2195	3043	3884	4720	5552	6379
	120	2370	3286	4194	5096	5994	6888
	130	2529	3506	4475	5438	6395	7349
	90	1997	2769	3535	4295	5052	
	100	2235	3099	3955	4806	5653	6496
50	110	2450	3397	4336	5269	6197	7121
	120	2645	3668	4682	5689	6691	7688
	130	2822	3913	4995	6070	7139	8204
	90	2149	2980	3804	4622	5436	
	100	2405	3334	4256	5172	6083	6990
60	110	2637	3656	4666	5670	6669	7663
	120	2847	3947	5038	6122	7200	8273
	130	3037	4211	5375	6532	7682	8828
	90	2265	3140	4008	4870	5728	6582
	100	2534	3514	4485	5450	6410	7366
70	110	2778	3852	4917	5975	7027	8075
	120	3000	4159	5309	6451	7587	8718
	130	3201	4438	5664	6883	8095	9302
	90	2356	3266	4169	5066	5958	6846
	100	2636	3655	4665	5668	6667	7661
80	110	2890	4006	5114	6214	7309	8399
	120	3120	4326	5521	6709	7891	9068
	130	3329	4615	5891	7159	8420	9675

*Yields are for wood and bark of the total stem for all trees 4.5-in. dbh and larger. (From Beck and Della-Bianca, 1972.)

for A, and the site index S (note that site index does not change over time) into equation 2 and solve for yield:

$$\ln BA_2 = (40/50)(\ln 90) + 3.82837(1 - 40/50)$$
$$+ 0.01667(110)(1 - 40/50)$$
$$\ln BA_2 = 4.7323$$
$$BA_2 = 113.56 \text{ sq ft per acre}$$
$$\ln Y = 5.36437 - 101.16296(110^{-1}) - 22.00048(50^{-1})$$
$$+ 0.97116(\ln 113.56)$$
$$\ln Y = 8.6005$$
$$Y = 5434 \text{ cu ft per acre}$$

Identical results are obtained by both options for projecting growth and yield because numerical consistency was ensured in the process used to estimate the coefficients in a set of analytically compatible models.

16-7   Size-Class Distribution Models   The distribution of volume by size classes, as well as the overall volume, is needed as input to many forest management decisions. A variety of approaches providing the distribution of volume by size classes (generally dbh classes) has been taken in the development of growth and yield models. One widely applied technique for even-aged stands is a diameter-distribution modeling procedure. In this approach, the number of trees per unit area in each diameter class is estimated through the use of a mathematical function which provides the relative frequency of trees by diameters. Mean total tree heights are predicted for trees of given diameters in stands of specified characteristics (i.e., of specified age, site index, and stand density). Volume per diameter class is calculated by substituting the predicted mean tree heights and the diameter-class midpoints into tree volume or taper equations. Yield estimates are obtained by summing the volumes in the diameter classes of interest. Although only total stand values (e.g., age, site index, and number of trees per unit area) are needed as input, detailed stand distributional information is obtainable as output.

A typical dbh distribution for pure, even-aged stands is shown by the histogram in Fig. 16-2. As a rule, these distributions have a single peak (i.e., they are unimodal) and are slightly skewed. Curves can be fitted to such diameter distributions by a variety of mathematical functions, but the two most popular functions used in past forest yield studies are the beta and the Weibull functions. Both of these functions are unimodal but highly flexible in the shapes that they can assume. Consequently, they have been found very satisfactory for describing the relative frequencies by dbh class in unthinned stands where the underlying dbh distribution is generally within the range of shapes that the mathematical function can approximate.

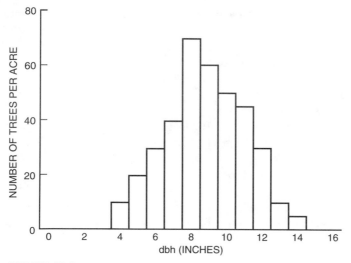

FIGURE 16-2
A typical dbh distribution for pure, even-aged stands.

Table 16-3 is a variable-density yield table for unthinned slash pine plantations that is based on the diameter-distribution analysis technique just described. It will serve as an example of how this technique is applied to forest yield prediction.

Diameter distributions shown in Table 16-3 were generated with the Weibull function. This function has three parameters, commonly denoted as *a, b,* and *c.* The *a* parameter is the "location" parameter; it indicates the lower end of the diameter distribution. "Spread" in the diameter distribution is controlled by the *b* parameter, while the "shape" of the distribution is determined by *c.* Regression equations are used to relate the parameters in the Weibull function to the stand attributes. For a given set of stand attributes, the Weibull function completely characterizes the diameter distribution.

To further illustrate this diameter-distribution modeling technique, one might consider these Table 16-3 entries: 500 trees per acre planted, age 15, site index 40. These stand characteristics are used to estimate the parameters *a, b,* and *c* in the Weibull function which is then applied to generate the number of trees by dbh class. Under these stand conditions, the relative frequency for the 5-in.-dbh class, for example, is 0.2697. Multiplying 0.2697 times the number of surviving trees at age 15 (356) gives 96 trees per acre in the 5-in. class. Substituting a dbh value of 5 and the stand attributes (age 15, site index 40, trees surviving 356) into a height-diameter relationship results in a predicted total tree height of 25 ft. The dbh and total-height values are substituted into a tree volume equation to estimate the average volume per tree in the 5-in. class. For this illustration, each tree in the 5-in.-dbh class has a volume of

$$V = -1.045389 + 0.002706 \text{ dbh}^2 H$$
$$= -1.045389 + 0.002706(5)^2(25)$$
$$= 0.646 \text{ cu ft}$$

This volume per tree times the number of trees per acre (0.646×96) results in an estimated volume per acre of 62 cu ft. All other values in Table 16-3 are computed similarly. The volumes per acre are summed for the appropriate dbh classes to obtain an estimate of the yield per acre of the desired portion of the stand. With the entire stand table (numbers of trees per unit area) available, this technique is obviously flexible for generating stock tables (volume or weight per unit area) in any desired units and for any portion of the stand. Section 16-8 gives the equations used to develop Table 16-3 and shows, in detail, their use.

Other types of size-class distribution models have been developed for even-aged stands. The diameter-distribution analysis procedure illustrated here has been more widely applied than other alternatives, however, and it typifies the size-class distribution approach.

16-8 Example of Computations for Size-Class Distribution Model Equations and procedures used to generate Table 16-3, a variable-density yield table based on a diameter-distribution analysis technique, are demonstrated here for the age 15 entries with 500 trees per acre planted on site-index-40 land.

1 Estimate the number of trees per acre surviving at age 15 (equation from Coile and Schumacher, 1964).

$$\log T_s = \log T_p + (0.023949 - 0.012505 \log T_p)A$$

where T_s = number of trees per acre surviving
T_p = number of trees per acre planted
A = age (number of years since planting)
$\log$ = logarithm to base 10

For our example,

$$\log T_s = \log 500 + (0.023949 - 0.012505 \log 500)15$$
$$= 2.5519$$
$$T_s = 356$$

2 Compute the height of dominants and codominants at age 15 for site index 40 from a selected site-index equation (from Farrar, 1973).

$$\log H_d = \log S - 8.80405(A^{-1} - A_i^{-1}) + 22.7952(A^{-2} - A_i^{-2})$$

TABLE 16-3
VARIABLE-DENSITY YIELD TABLE, SHOWING INFORMATION BY DBH CLASSES, FOR UNTHINNED SLASH PINE PLANTATIONS*

Site index 40 (base age 25)

Age, years	dbh, in.	Trees planted per acre: 500			Trees planted per acre: 750			Trees planted per acre: 1000		
		Trees per acre	Ave. ht., ft	Cu ft per acre	Trees per acre	Ave. ht., ft	Cu ft per acre	Trees per acre	Ave. ht., ft	Cu ft per acre
15	1	2	5	...	10	5	...	28	6	...
	2	21	13	...	58	15	...	105	16	...
	3	62	19	...	119	20	...	172	21	...
	4	98	23	...	142	24	...	167	24	...
	5	96	25	62	105	26	74	103	27	80
	6	56	28	94	47	28	79	40	28	67
	7	18	29	50	12	29	33	10	29	27
	8	3	30	12	2	30	8	1	30	4
		356		218	495		194	626		178
25	1	1	5	...	3	5	...	7	5	...
	2	8	11	...	19	13	...	33	14	...
	3	22	19	...	43	21	...	63	22	...
	4	41	26	57	64	28	81	84	28	...
	5	55	31	132	73	32	169	88	32	98
	6	56	35	191	69	36	203	75	36	184
	7	48	38	182	51	38	182	53	38	211
	8	31	40	122	31	40	122	31	40	182
	9	15	42	65	15	42	65	15	42	122
	10	6	44	13	6	44	27	6	43	63
	11	1	45		2	45		2	44	26
	12	0	...		0	...		1	45	16
		284		762	376		849	458		902

1	1	5	.	2	5	.	4	5	.
2	6	9	.	12	10	.	19	11	.
3	15	18	.	25	19	.	35	20	.
4	25	26	.	37	27	.	46	28	.
5	33	32	36	44	33	52	52	34	65
6	37	37	94	42	37	107	50	38	132
7	36	41	158	40	41	175	43	42	194
8	29	44	190	32	44	210	33	45	222
9	21	47	194	22	46	198	23	47	212
10	13	49	158	14	49	170	15	49	183
11	7	51	109	8	50	122	8	51	125
12	3	53	58	4	52	76	4	52	76
13	1	54	23	2	53	46	2	53	46
14	0	.	.	1	54	27	1	54	27
	227		1020	285		1183	335		1282

Site index 60 (base age 25)

2	1	14	.	3	17	.	9	18	.
3	7	23	.	22	26	.	45	27	.
4	29	29	.	66	32	.	112	33	.
5	67	34	84	123	36	170	175	37	255
6	101	38	268	144	40	410	167	40	476
7	94	40	400	98	42	443	91	42	411
8	46	42	286	34	44	223	24	44	157
9	10	44	85	5	46	45	3	46	27
10	1	46	11	0	.	.	0	.	.
	356		1134	495		1291	626		1326

331

TABLE 16-3
(Continued)

Age, years	dbh, in.	Trees planted per acre: 500			Trees planted per acre: 750			Trees planted per acre: 1000		
		Trees per acre	Ave. ht., ft	Cu ft per acre	Trees per acre	Ave. ht., ft	Cu ft per acre	Trees per acre	Ave. ht., ft	Cu ft per acre
					Site index 60 (base age 25)					
25	2	0	...	...	1	15	...	2	17	...
	3	2	25	...	6	27	...	12	29	...
	4	9	34	...	19	37	...	32	38	...
	5	23	42	41	41	44	79	61	46	126
	6	41	48	148	66	50	252	88	51	345
	7	59	53	352	83	54	507	99	55	618
	8	62	57	547	76	58	683	83	59	761
	9	50	60	605	52	61	640	83	59	640
	10	27	63	432	24	63	384	52	61	358
	11	9	65	182	7	66	143	22	64	123
	12	2	67	50	1	67	25	6	66	25
		284		2357	376		2713	1	67	2996
								458		
35	2	0	...	...	...	...	...	1	13	...
	3	2	23	...	3	25	...	6	26	...
	4	6	34	...	11	36	...	16	38	...
	5	13	43	24	22	45	43	30	47	64
	6	24	51	94	35	53	144	46	54	193
	7	35	57	227	47	59	318	57	60	393
	8	40	63	394	52	64	521	59	65	602
	9	41	67	559	47	68	651	51	69	718
	10	33	71	599	35	72	645	36	72	663

11	20	74	463	20	75	470	20	75	470
12	9	77	260	9	77	260	9	77	260
13	3	79	105	3	80	106	3	80	106
14	1	82	42	1	82	42	1	82	42
	227		2767	285		3200	335		3511

Site index 80 (base age 25)

15									
2	0	·	·	0	·	·	1	20	·
3	1	27	·	3	·	·	7	32	·
4	6	36	·	16	31	·	33	40	·
5	21	42	37	51	39	·	89	46	183
6	54	47	190	106	45	101	160	51	627
7	94	51	537	147	50	405	181	54	1106
8	103	54	855	118	54	898	116	57	1023
9	61	57	698	47	56	1021	35	59	416
10	15	59	223	7	59	558	4	61	61
11	1	61	18	0	61	108	0	·	·
	356		2558	495		3091	626		3416

25									
3	0	·	·	1	32	·	2	34	·
4	2	42	·	4	45	·	8	47	·
5	6	52	14	14	55	37	23	56	63
6	17	60	81	32	63	162	49	64	254
7	34	67	266	58	69	470	81	70	667
8	55	73	637	81	74	953	103	75	1230
9	66	77	1044	86	78	1380	96	79	1561
10	58	81	1210	62	82	1310	63	82	1332
11	33	84	873	29	85	776	26	85	696
12	11	87	361	8	88	265	6	88	199
13	2	90	80	1	90	40	1	90	40
	284		4566	376		5393	458		6042

TABLE 16-3
(Continued)

Site index 80 (base age 25)

Age, years	dbh, in.	Trees planted per acre: 500			Trees planted per acre: 750			Trees planted per acre: 1000		
		Trees per acre	Ave. ht., ft	Cu ft per acre	Trees per acre	Ave. ht., ft	Cu ft per acre	Trees per acre	Ave. ht., ft	Cu ft per acre
35	3	0	....	....	0	....	....	1	32	....
	4	1	42	....	2	45	....	4	47	....
	5	4	55	10	7	58	20	11	59	32
	6	10	66	53	16	68	89	23	69	130
	7	19	74	166	30	76	270	40	78	371
	8	31	82	407	44	83	586	56	84	756
	9	42	88	766	57	89	1052	65	90	1214
	10	45	93	1085	54	94	1317	60	95	1479
	11	38	98	1179	41	99	1286	43	99	1348
	12	24	102	928	23	102	890	22	103	859
	13	10	106	474	9	106	426	8	106	379
	14	3	109	170	2	109	113	2	109	113
		227		5238	285		6049	335		6681

*Age is number of years since planting; cubic-foot volumes are outside bark for a 4-in. top limit (ob) for all trees in the 5-in.-dbh class and larger. Diameter- and height-distribution equations from Dell et al., 1979; tree volume equation from Bennett et al., 1959.

where H_d = average height of dominants and codominants, ft
$\qquad$ S = site index, ft (base age 25)
$\qquad$ A_i = base age for site index

In this illustration,

$$\log H_d = \log 40 - 8.80405(15^{-1} - 25^{-1}) + 22.7952(15^{-2} - 25^{-2})$$
$$= 1.4321$$
$$H_d = 27 \text{ ft}$$

3 Using stand characteristics specified and computed in steps 1 and 2, estimate the Weibull distribution parameters and apply this estimated distribution for computing the numbers of trees in each dbh class (equations from Dell et al., 1979).

$$a = 1.3986 + 2.9217 \log H_d - 1.8477 \log A - 1.1126 \log T_s$$
$$= 1.3986 + 2.9217 \log 27 - 1.8477 \log 15 - 1.1126 \log 356$$
$$a = 0.56881$$

$$b = 2.5800 + 10.138 \log H_d - 2.5005 \log A - 3.6275 \log T_s - a$$
$$= 2.5800 + 10.138 \log 27 - 2.5005 \log 15 - 3.6275 \log 356 - 0.56881$$
$$b = 4.32615$$

$$c = 9.1471 + 6.5959 \log H_d - 7.6706 \log A - 2.4479 \log T_s$$
$$= 9.1471 + 6.5959 \log 27 - 7.6706 \log 15 - 2.4479 \log 356$$
$$c = 3.32121$$

The parameters a, b, and c completely specify a Weibull distribution which is used to generate the relative frequencies of trees by dbh class for the given overall stand characteristics. To compute relative proportions of trees by dbh class, substitute the upper and lower limits of the class into the cumulative-distribution function. Subtracting the cumulative distribution up to the lower limit of the class from the upper limit gives the proportion of trees in that class. The Weibull cumulative-distribution function is

$$F(x) = 1 - e^{-[(x-a)/b]^c}$$

where $a \leq x < \infty$.

Continuing the present example and demonstrating for the 5-in.-dbh class:

$$F(5.5) = 1 - e^{-[(5.5-0.56881)/4.32615]^{3.32121}}$$
$$= 0.7866$$
$$F(4.5) = 1 - e^{-[(4.5-0.56881)/4.32615]^{3.32121}}$$
$$= 0.5169$$

Proportion in 5-in. class $= 0.7866 - 0.5169 = 0.2697$

Multiplying the proportion in the 5-in. class times the number of trees surviving gives the number of trees in that class:

$$(0.2697)(356) = 96$$

Analogous procedures are followed for all other dbh classes. The process is continued until a predicted frequency of less than one tree per unit area results. Note that the sum of the proportions for all dbh classes must equal 1, and thus the sum of the number of trees in all dbh classes will equal the total number surviving.

4 Use the midpoint of each dbh class and the stand attributes to predict the mean total tree height for each class (from Dell et al., 1979).

$$\log (H_d/H_i) = -0.050341 + (D_i^{-1} - D_{max}^{-1})[3.1868$$
$$+ 0.000015708(A)(T_s) + 0.0114942(T_s)A^{-1}$$
$$- 2.0981 \log (T_s/A) + 1.4034 \log (H_d/A)]$$

where H_i = total height, ft, for trees D_i in. in dbh
D_{max} = midpoint of largest-diameter class containing at least one tree per acre, as defined by Weibull distribution and number of trees surviving

Following through for the 5-in. class (*note: D_{max} = 8 in. from Table 16-3):

$$\log (27/H_i) = -0.050341 + (5^{-1} - 8^{-1})[3.1868$$
$$+ 0.000015708(15)(356) + 0.0114942(356)(15^{-1})$$
$$-2.0981 \log (356/15) + 1.4034 \log (27/15)]$$
$$= 0.0259$$
$$H_i = 25 \text{ ft}$$

5 Substitute the midpoint diameter and the predicted total tree height into a tree volume or taper equation to estimate the average volume per tree in each dbh class. Using a tree volume equation from Bennett et al. (1959) and the values for the 5-in. class in our example results in

$$V = -1.045389 + 0.002706 \text{ dbh}^2H$$
$$= -1.045389 + 0.002706(5)^2(25)$$
$$= 0.646 \text{ cu ft}$$

Analogous procedures are used to compute volume per tree for the other diameter classes of interest.

6 Multiply the volume per tree times the number of trees per acre in the dbh classes of interest to obtain volume per acre for each dbh class. For the 5-in.-dbh class,

$$(0.646 \text{ cu ft per tree})(96 \text{ trees per acre}) = 62 \text{ cu ft per acre}$$

7 Sum the diameter-class values of interest to obtain an estimate of stand volume per acre.

Using equations shown in this illustration and values from Table 16-3 where appropriate (for age 15, trees per acre planted 500, and site index 40), the cubic-foot volumes per acre by dbh classes and the overall merchantable volumes are

dbh	Volume per tree	Trees per acre	Volume per acre, cu ft
5	0.646	96	62
6	1.682	56	94
7	2.800	18	50
8	4.150	3	12
Total			218

16-9 Individual-Tree Models Approaches to predicting stand growth and yield which use individual trees as the basic unit are referred to as *individual-tree models*. The components of tree growth in these models are commonly linked together through a computer program which simulates the growth of each tree and then aggregates these to provide estimates of stand growth and yield. Models based on individual-tree growth provide detailed information about stand dynamics and structure, including the distribution of stand volume by size classes.

Individual-tree models may be divided into two classes, distance independent and distance dependent, depending on whether or not individual tree locations are required tree attributes. Distance-independent models project tree growth either individually or by size classes, usually as a function of present size and stand-level variables (e.g., age, site index, and BA per unit area). It is not necessary to know individual-tree locations when applying these models. Typically, distance-independent models consist of three basic components: (1) a diameter-growth component, (2) a height-growth component (or a height-diameter relationship to predict heights from dbh values), and (3) a mortality component. Mortality may be stochastically generated (i.e., determined through a random process), or it may be predicted as a function of growth rate.

Distance-dependent models vary in detail but are quite similar in overall concept and structure. Initial stand conditions are input or generated, and each tree is assigned a coordinate location. The growth of each tree is simulated as a function of its attributes, the site quality, and a measure of competition from neighbors. The competition index varies from model to model but in general is a function of the size of the subject tree and the size of and distance to competitors.

Tree growth is commonly adjusted by a random component representing genetic and/or microsite variability. Survival is controlled either stochastically or deterministically as a function of competition and/or individual-tree attributes. Yield estimates are obtained by summing the individual-tree volumes (computed from tree volume or taper equations) and multiplying by appropriate expansion factors.

A stand simulator for loblolly pine plantations, PTAEDA2, is, in many aspects, typical of distance-dependent, individual-tree models. The PTAEDA2 model (Burkhart et al., 1987) consists of two main subsystems—one dealing with the generation of an initial precompetitive stand and another with the growth and dynamics of that stand. Management subroutines have been added to this framework to simulate varying hardwood competition levels, fertilization, and thinning. Input/output routines make the model operable and also add flexibility.

A number of options are available for creating rectangular spatial patterns in PTAEDA2. Users may specify the distance between trees and between rows in a conventional manner (e.g., 6 × 8 ft, 6 × 12 ft) allowing the program to compute the planted number of trees. Alternatively, the number of trees may be specified along with the ratio of planting distance to row width (e.g., 3:4, 1:2). If this ratio is omitted, square spacing is assumed. From this information, a simulation plot is generated and coordinate values are assigned to each of the planting locations. The juvenile stand is then advanced to an age of 8 years where intraspecific competition is assumed to begin. At this point, predicted juvenile mortality is assigned at random. Individual-tree dimensions are then generated for the residual stand. Dbh is generated from a two-parameter Weibull distribution; the parameters of the Weibull distribution are estimated as functions of stand age, number of trees surviving, and average height of dominants and codominants at that age. Height is predicted for each tree from an equation involving dbh, average height of dominants and codominants, trees surviving, and age. Crown ratio for each tree is then calculated as a function of its total height, dbh, and age.

After assigning dimensions to each tree, the competition effect of neighboring trees is calculated for each individual tree as

$$CI_i = \sum_{j=1}^{n} \frac{dbh_j/dbh_i}{DIST_{ij}}$$

where CI_i = competition index of ith subject tree

n = number of competitors "in" with BAF 10 sq ft per acre sweep centered at ith tree

dbh_j = dbh of jth competitor

dbh_i = dbh of ith subject tree

$DIST_{ij}$ = distance between subject tree i and jth competitor

After generation of the precompetitive stand, competition is evaluated and simulated trees are grown individually on an annual basis. In general, growth in height and diameter is assumed to follow some theoretical growth potential. An adjustment or reduction factor is applied to this potential increment based on a tree's competitive status and vigor, and a random component is then added representing microsite and/or genetic variability.

The potential height increment for each tree is the change in average height of the dominant and codominant trees, obtained as the first difference with respect to age of a site-index equation. A tree may grow more or less than this potential, depending on its individual attributes.

Crown ratio is considered to be an expression of a tree's photosynthetic potential. It is used in conjunction with competition index to compute an adjustment factor for height growth. The adjustment factor times the potential height growth (determined from a site-index equation) gives the estimated actual height growth for an individual tree with a given crown ratio and competition index.

The maximum dbh attainable for an individual tree of given height and age is considered to be equal to that of loblolly pines grown in the open. An equation describing this relationship, developed from open-grown tree data, is

$$dbh_0 = b_0 + b_1 H + b_2 A$$

where dbh_0 = dbh of open-grown trees
H = total tree height
A = age

The first difference of this equation with respect to age represents maximum potential diameter increment

$$PDIN = b_1 HIN + b_2$$

where PDIN is the potential diameter increment and HIN is the observed height increment. This potential diameter increment is adjusted by a reduction factor that is a function of the tree's competition index and crown ratio, a measure of photosynthetic potential. A normally distributed random component is added to diameter-growth determinations.

The probability that a tree remains alive in a given year is assumed to be a function of its competitive stress (CI) and individual vigor as measured by photosynthetic potential (expressed by crown ratio, CR, in this equation). The "probability of survival" equation is

$$PLIVE = b_1 CR^{b_2} e^{-b_3 CI^{b_4}}$$

where PLIVE is the probability that a tree remains alive. In PTAEDA2, survival probability is calculated for each live tree each year and used to determine annual mortality. The calculated PLIVE is compared with a uniform random variate between zero and one. If PLIVE is less than this generated number, the tree is considered to have died.

These components, along with subroutines to simulate the effects of various levels of hardwood competition, thinning, and fertilization, were linked together in a computer program to simulate individual-tree growth and stand development. Figure 16-3 is a schematic diagram showing relationships between tree and stand components in this growth and yield model for loblolly pine plantations.

Amateis et al. (1989) developed individual-tree diameter increment and survival equations for loblolly pine plantations by substituting a distance-inde-

FIGURE 16-3
Schematic diagram showing relationships between tree and stand components for an individual-tree, distance-dependent model of loblolly pine plantations. (From *Burkhart et al., 1987*).

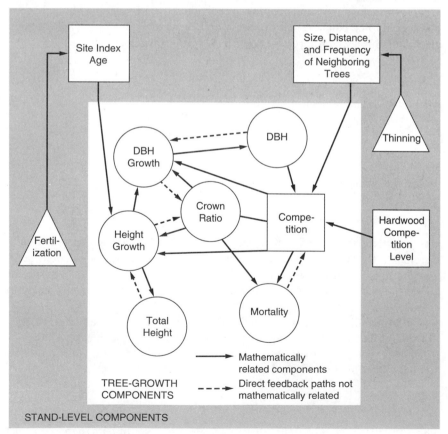

pendent measure of competition into the basic functional forms used in the PTAEDA2 model. The measure of competition used involves the ratio of quadratic mean dbh of the stand (computed from basal area and numbers of trees per unit area) and individual-tree dbh. Hence, the relative competitive position of each tree in the stand is expressed, but knowledge of tree locations is not required.

GROWTH AND YIELD MODELS FOR UNEVEN-AGED STANDS

16-10 Difficulties in Modeling Uneven-Aged Stands Uneven-aged stands are composed of trees that differ markedly in age. Consequently, age is not a usable variable for growth- and yield-prediction purposes. Also, site-quality assessment by site-index methods is questionable because of initial suppression of advance reproduction, especially for tolerant species. Furthermore, site index is an age-dependent variable. Growth and yield models without age as a variable, and without site index for assessing site quality, have been developed for uneven-aged stands. Modeling techniques for uneven-aged stands may also, in theory, be applied to even-aged conditions. However, in most situations, models involving age (such as those described in Secs. 16-4 through 16-9) have been applied to even-aged stands. Sections 16-11 through 16-13 describe approaches and provide examples of growth and yield models that have been developed specifically for uneven-aged conditions—i.e., conditions where age is not a usable variable.

16-11 Growth and Yield Equations Moser and Hall (1969) developed a volume growth-rate equation for uneven-aged stands of mixed northern hardwoods. Solution of their growth-rate equation provides a yield function expressed in terms of elapsed time from a given initial condition. The variable time (in lieu of stand age) was introduced into the yield function by assigning a relative time t_0 at some identified point in the stand's development with initial condition Y_0. The resulting yield equation, expressed as a function of time and initial volume and BA, is

$$Y = [(Y_0)(8.3348\text{BA}_0^{-1.3175})]$$
$$\times [0.9348 - (0.9348 - 1.0203\text{BA}_0^{-0.0125})e^{-0.0062t}]^{-105.4}$$

where Y_0 = initial volume, cu ft per acre
$\quad$ BA_0 = initial basal area, sq ft per acre
$\quad\quad$ t = elapsed time interval, years from initial conditions
$\quad\quad$ Y = predicted volume, cu ft, t years after observation of initial conditions
$\quad\quad\quad$ Y_0 and BA_0 at time t_0

If one assumes that a stand with a volume of 1500 cu ft per acre and BA of 60 sq ft per acre is observed, then the predicted yield 10 years hence would be

$$Y = [(1500)(8.3348 \ 60^{-1.3175})]$$
$$\times [0.9348 - (0.9348 - 1.0203 \ 60^{-0.0125})e^{-0.0062(10)}]^{-105.4}$$
$$= 1885 \text{ cu ft per acre}$$

Table 16-4 is a yield table constructed from solution of Moser and Hall's equation for selected initial conditions and elapsed time intervals. It provides an example of an approach to yield prediction for uneven-aged stands.

16-12 Size-Class Distribution Models Diameter distributions in regular, uneven-aged stands are inverse J-shapes (Fig. 16-4). Relative frequency curves, such as the Weibull function, described in Sec. 16-7, can assume this inverse J-shape and can, thus, be used to model diameter distributions in uneven-aged stands. Modifications are necessary, however, to express the parameters of the diameter distribution in terms of variables other than stand age. One alternative, for example, would be to express the parameters as functions of some initial value and the elapsed time from that initial value. Such a procedure would be somewhat analogous to the methods outlined in Sec. 16-11 for overall stand volume.

Another approach to size-class distribution modeling in uneven-aged stands is the use of the stand-table projection models developed by Ek (1974). He pre-

TABLE 16-4
YIELD TABLE, SHOWING FINAL VOLUME (CU FT PER ACRE) AS A FUNCTION OF INITIAL CONDITIONS AND ELAPSED TIME, FOR UNEVEN-AGED STANDS OF MIXED NORTHERN HARDWOOD

Initial volume, cu ft per acre	Elapsed time, years	Initial basal area, sq ft per acre		
		60	80	100
1500	10	1885	1844	1813
	15	2101	2033	1982
	20	2333	2235	2162
1750	10	2200	2152	2115
	15	2451	2372	2313
	20	2722	2608	2522
2000	10	2514	2459	2417
	15	2801	2711	2643
	20	3111	2981	2883

Source: Moser and Hall, 1969.

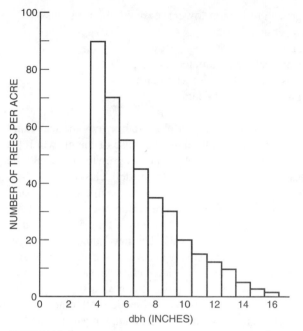

FIGURE 16-4
Typical dbh distribution for regular, uneven-aged stands.

sented equations to predict periodic ingrowth, mortality, and survivor growth, by 2-in.-diameter classes, in northern hardwood stands. Net 5-year change (i.e., the change from t_0, the time of initial measurement, to t_1, the time of final measurement, where t_1 is 5 years after t_0) in the number of trees in a diameter class Δn was defined as

$$\Delta n = \text{stand ingrowth} - \text{mortality} - \text{upgrowth} + \text{ingrowth}$$

The component equations of this generalized stand-table projection model are

$$n_{is} = 15.123 N^{0.38753} e^{(-0.32908\text{BA}^{1.58011} N^{-1})} \tag{1}$$
$$n_m = 0.03443 n[(\text{BA}/N)/(\text{ba}/n)]^{0.54748} \tag{2}$$
$$n_u = 0.01070 n^{0.81433} S[(\text{ba}/n)/(\text{BA}/N)]^{0.14611} e^{-0.00160\text{BA}} \tag{3}$$

where n_{is} = stand ingrowth = merchantable trees at t_1 that were nonmerchantable at t_0

n_m = mortality = trees present in a diameter class at t_0 but dead at t_1

n_u = upgrowth = trees present in a diameter class at t_0 but growing into next larger diameter class at t_1

n_i = ingrowth = upgrowth from next lower measured diameter class
N = number of trees per acre in stand
BA = stand basal area, sq ft per acre
n = number of trees in specified diameter class
ba = basal area of trees in specified diameter class
S = site index (height, ft, at age 50 years)

Equations 1 through 3 can be used to project observed or hypothetical stand tables which can then be converted to stock tables. Because of the growth rates involved for the northern hardwoods example, stand ingrowth n_{is} would be added only to the smallest merchantable class (6-in.-dbh class) when 2-in. groupings are used. Diameter-class ingrowth n_i is equal to the upgrowth n_u computed from the next lower diameter class (e.g., upgrowth computed for the 8-in. class is ingrowth to the 10-in. class). For projections longer than 5 years, a new stand table must be constructed at 5-year intervals. The new stand table is prepared by adding ingrowth trees from smaller size classes to the number of survivors in a class that did not move to the next larger class. Basal areas can be computed by using the class midpoint diameters. The new stand table then serves as the initial conditions for the next 5-year projection. Volumes can be computed by applying an appropriate size-class volume equation, such as

$$v = b_1 \mathrm{ba}^{b_2} S^{b_3} \mathrm{BA}^{b_4}$$

where v is merchantable volume in a specified diameter class, the b_i's are constants to be estimated from data, and the other variables remain as previously defined.

16-13 Individual-Tree Models Growth and yield models which use individual trees as the basic unit have been developed for mixed species, uneven-aged as well as pure, even-aged stands. An example is FOREST, a computer model published by Ek and Monserud (1974) for simulating the growth and reproduction of even- or uneven-aged mixed species stands.

Usual input for FOREST, a distance-dependent model, is a set of tree coordinates and associated tree characteristics (e.g., height, diameter, age, clear bole length, and species). Tree coordinates and tree characteristics may also be generated by the program. Each tree is then "grown" for a number of projection periods based on potential growth functions, modified by an index of competition. The competition index is based on the assessment of relative tree size, crowding, and shade tolerance. Mortality is obtained when the probability of survival for a stem falls below a threshold value, which is dependent on the competitive status of a tree. In any "year" of the simulation, optional reproduction routines may be called to allow for regeneration by seed and sprout production of the overstory. Silvicultural treatments, including site alteration, cutting, or pruning operations, may also be specified for implementation as the stand develops. Output of the model is in the form of periodic stand tables with yield and mortality for various products.

APPLYING GROWTH AND YIELD MODELS

16-14 Choosing an Appropriate Growth and Yield Model Growth and yield models provide input to forest management decisions regarding individual stands, forests, and broad regions. The projection period and the level of stand detail required may vary in each case. In choosing appropriate growth and yield models, foresters must be concerned with the reliability of estimates, the flexibility to reproduce desired management alternatives, the ability to provide sufficient detail for decision making, and the efficiency for providing this information.

Although advantages and disadvantages cannot be ascribed to different modeling approaches except in the context of specific uses, general characteristics of the various alternatives can be briefly described. Equations for predicting overall stand values can generally be applied with existing inventory data and are computationally efficient. However, such equations do not provide size-class information needed to evaluate various utilization options and usually cannot be used to analyze a wide range of stand treatments.

Size-class distribution models, such as the diameter-distribution models described in Sec. 16-7, require only overall stand values as input but provide detailed size-class information as output. Thus alternative utilization options can be evaluated. Computationally, these models are somewhat more expensive to apply than equations for overall stand values, and they are generally not flexible enough to evaluate a broad range of stand treatments because of the required assumptions about the shape of the underlying diameter distribution.

Individual-tree models provide maximum detail and flexibility for evaluating alternative utilization options and stand treatments. They are, however, more expensive to develop, require a more detailed data base to implement, and are much more expensive to apply, requiring sophisticated computing equipment and greater execution time for comparable stand estimates than the overall stand or size-class distribution models.

16-15 A Word of Caution When applying growth and yield models, one assumes a relatively homogeneous stand with regard to independent variables (e.g., age, site index, BA) used to predict stand values. If there is significant variation in variables such as site or stand density for a given area, the area must be stratified into reasonably homogeneous stands and predictions made separately for each of these stands to ensure accurate results.

Growth and yield predictions apply to net area; all nonproductive areas must be deducted when making estimates on an area basis.

In most growth and yield tables, no allowance is made for logging breakage or other losses during harvest, and it is implicitly assumed that all material meeting minimum merchantability standards will be utilized. Adjustments must often be made in predicted values from growth and yield models to approximate volumes that are likely to be realized under local harvesting and utilization conditions.

PROBLEMS

16-1 For a commercially important species in your area, conduct a literature search to determine the extent and nature of growth and yield information that is available. Prepare a report on the kinds of growth and yield models developed and the relationships shown by these models for the species chosen for study.

16-2 Using the yield values shown in Table 16-1, determine the age of culmination of mean annual growth for each site index. Are the ages of culmination consistent from a biological standpoint?

16-3 Apply linear regression techniques to fit the following model to yield values for site indexes 80, 140, and 200 shown in Table 16-1:

$$\ln Y = b_0 + b_1 A^{-1}$$

where $\ln Y$ is the natural logarithm of yield and A is the stand age. Using the fitted relationships for yield, derive equations for current annual growth and mean annual growth for each site index. Solve for the age of maximum current annual and maximum mean annual growth. Are the growth relationships consistent with expectations from a biological standpoint?

16-4 Using the growth and yield equations for yellow-poplar in Sec. 16-6:

    a Estimate the current volume for a stand of age 30, BA per acre of 60 sq ft, and site index of 105 feet.

    b Compute the periodic (10-year) growth of the stand described in part a from age 30 to 40 years.

    c Calculate the mean annual growth at age 40.

    d Convert answers in parts a, b, and c to metric units.

16-5 Plot histograms of the dbh distributions given in Table 16-3 for site indexes 40, 60, and 80 at age 25 with 750 trees per acre planted; for ages 15, 25, and 35 at site index 60 with 750 trees per acre planted; and for 500, 750, and 1000 trees per acre planted with site index 60 and age 25. Are changes in the dbh distribution logically related, from a biological standpoint, to plantation age, site index, and number of trees planted?

16-6 Compute entries similar to those shown in Table 16-3 for the 5-, 6-, and 7-in.-dbh classes in a stand of age 25, site index 70, and 900 trees per acre planted by applying the equations given in Sec. 16-8.

16-7 Apply the equation for uneven-aged stands shown in Sec. 16-11 to develop a yield tabulation similar to Table 16-4 for the following conditions:

Initial volumes   2400, 2600 cu ft per acre
Elapsed times   5, 10 years
Initial basal areas   90, 110 sq ft per acre

16-8 Given the following initial stand conditions:

Site index $S = 65$ ft (base age 50)
Basal area BA = 69 sq ft per acre
Number of trees per acre $N = 200$

and the following initial stand table:

dbh class, in.	Number of trees per acre
6	100
8	50
10	30
12	20

Use the stand-table projection equations from Sec. 16-12 to estimate a stand table 5 years hence and 10 years hence.

REFERENCES

Amateis, R. L., Burkhart, H. E., and Walsh, T. A. 1989. Diameter increment and survival equations for loblolly pine trees growing in thinned and unthinned plantations on cutover, site-prepared lands. *So. J. Appl. For.* **13**:170–174.

Bailey, R. L., and Dell, T. R. 1973. Quantifying diameter distributions with the Weibull function. *Forest Sci.* **19**:97–104.

Barrett, J. P., Alimi, R. J., and McCarthy, K. T. 1976. Growth of white pine in New Hampshire. *J. Forestry* **74**:450–452.

Beck, D. E., and Della-Bianca, L. 1972. Growth and yield of thinned yellow-poplar. *U.S. Forest Serv., Southeast. Forest Expt. Sta., Res. Paper* SE-101. 20 pp.

Belcher, D. W., Holdaway, M. R., and Brand, G. J. 1982. A description of STEMS—The stand and tree evaluation and modeling system. *U.S. Forest Serv., North Central Forest Expt. Sta., Gen. Tech. Report* NC-79. 18 pp.

Bennett, F. A., and Clutter, J. L. 1968. Multiple-product yield estimates for unthinned slash pine plantations—pulpwood, sawtimber, gum. *U.S. Forest Serv., Southeast. Forest Expt. Sta., Res. Paper* SE-35. 21 pp.

———, McGee, C. E., and Clutter, J. L. 1959. Yield of old-field slash pine plantations. *U.S. Forest Serv., Southeast. Forest Expt. Sta., Paper* 107. 19 pp.

Bowling, E. H., Burkhart, H. E., Burk, T. E., and Beck, D. E. 1989. A stand-level, multi-species growth model for Appalachian hardwoods. *Can. J. For. Res.* **19**:405–412.

Buckman, R. E. 1962. Growth and yield of red pine in Minnesota. *U.S. Dept. of Agr., Forest Serv. Tech. Bull.* 1272. 50 pp.

Burkhart, H. E., and Sprinz, P. T. 1984. Compatible cubic volume and basal area projection equations for thinned old-field loblolly pine plantations. *Forest Sci.* **30**:86–93.

———, Farrar, K. D., Amateis, R. L., and Daniels, R. F. 1987. Simulation of individual tree growth and stand development in loblolly pine plantations on cutover, site-prepared areas. *Virginia Poly. Inst. and State Univ. Pub.* FWS-1-87. 47 pp.

Clutter, J. L. 1963. Compatible growth and yield models for loblolly pine. *Forest Sci.* **9**:354–371.

———, Fortson, J. C., Pienaar, L. V., Brister, G. H., and Bailey, R. L. 1983. *Timber management: A quantitative approach.* John Wiley & Sons, New York. 333 pp.

Coile, T. S., and Schumacher, F. X. 1964. *Soil-site relations, stand structure, and yields of slash and loblolly pine plantations in the Southern United States.* T. S. Coile, Inc., Durham, N.C. 296 pp.

Curtis, R. O., Clendenen, G. W., and DeMars, D. J. 1981. A new stand simulator for coast Douglas-fir: DFSIM user's guide. *U.S. Forest Serv., Pacific Northwest Forest and Range Expt. Sta., Gen. Tech. Report* PNW-128. 79 pp.

Daniels, R. F., and Burkhart, H. E. 1975. Simulation of individual tree growth and stand development in managed loblolly pine plantations. *Virginia Poly. Inst. and State Univ., Pub.* FWS-5-75. 69 pp.

———, and Burkhart, H. E. 1988. An integrated system of forest stand models. *For. Ecol. and Manage.* **23**:159–177.

Dell, T. R., Feduccia, D. P., Campbell, T. E., Mann, W. F., Jr., and Polmer, B. H. 1979. Yields of unthinned slash pine plantations on cutover sites in the West Gulf region. *U.S. Forest Serv., Southern Forest Expt. Sta., Res. Paper* SO-147. 84 pp.

Ek, A. R. 1974. Nonlinear models for stand table projection in northern hardwood stands. *Can. J. For. Res.* **4**:23–27.

———, and Monserud, R. A. 1974. FOREST: A computer model for simulating the growth and reproduction of mixed species forest stands. *Univ. of Wisconsin, Res. Paper* R2635. 13 pp. plus appendices.

Farrar, R. M., Jr. 1973. Southern pine site index equations. *J. Forestry* **71**:696–697.

Hafley, W. L., and Buford, M. A. 1985. A bivariate model for growth and yield prediction. *Forest Sci.* **31**:237–247.

Hann, D. W., and Larsen, D. R. 1991. Diameter growth equations for fourteen tree species in southwest Oregon. *Oregon State Univ., Res. Bull.* 69. 18 pp.

Hilt, D. E. 1985. OAKSIM: An individual-tree growth and yield simulator for managed, even-aged, upland oak stands. *U.S. Forest Serv., Northeast. Forest Expt. Sta., Res. Paper* NE-562. 21 pp.

Hyink, D. M., and Moser, J. W., Jr. 1983. A generalized framework for projecting forest yield and stand structure using diameter distributions. *Forest Sci.* **29**:85–95.

Knoebel, B. R., Burkhart, H. E., and Beck, D. E. 1986. A growth and yield model for thinned stands of yellow-poplar. *Forest Sci. Monograph* 27. 62 pp.

Lenhart, J. D. 1988. Diameter-distribution yield-prediction system for unthinned loblolly and slash pine plantations on non-old-fields in East Texas. *So. J. Appl. For.* **12**:239–242.

Lynch, T. B., and Moser, J. W., Jr. 1986. A growth model for mixed species stands. *Forest Sci.* **32**:697–706.

MacKinney, A. L., and Chaiken, L. E. 1939. Volume, yield, and growth of loblolly pine in the mid-Atlantic coastal region. *U.S. Forest Serv., Appalachian Forest Expt. Sta., Tech. Note* 33. 30 pp.

Matney, T. G., and Sullivan, A. D. 1982. Compatible stand and stock tables for thinned and unthinned loblolly pine stands. *Forest Sci.* **28**:161–171.

McArdle, R. E., Meyer, W. H., and Bruce, D. 1961. The yield of Douglas fir in the Pacific Northwest. *U.S. Dept. of Agr., Forest Service, Tech. Bull.* 201 (rev.). 74 pp.

Mitchell, K. J. 1975. Dynamics and simulated yield of Douglas-fir. *Forest Sci. Monograph* 17. 39 pp.

Moser, J. W., Jr., and Hall, O. F. 1969. Deriving growth and yield functions for uneven-aged forest stands. *Forest Sci.* **15**:183–188.

Pienaar, L. V., and Turnbull, K. J. 1973. The Chapman-Richards generalization of von Bertalanffy's growth model for basal area growth and yield in even-aged stands. *Forest Sci.* **19**:2–22.

Schreuder, H. T., Hafley, W. L., and Bennett, F. A. 1979. Yield prediction for unthinned natural slash pine stands. *Forest Sci.* **25**:25–30.

Smalley, G. W., and Bailey, R. L. 1974. Yield tables and stand structure for loblolly pine plantations in Tennessee, Alabama, and Georgia highlands. *U.S. Forest Serv., Southern Forest Expt. Sta., Res. Paper* SO-96. 81 pp.

Sullivan, A. D., and Clutter, J. L. 1972. A simultaneous growth and yield model for loblolly pine. *Forest Sci.* **18**:76–86.

Wycoff, W. R., Crookston, N. L., and Stage, A. R. 1982. User's guide to the Stand Prognosis Model. *U.S. Forest Serv., Intermount. Forest and Range Expt. Sta., Gen. Tech. Report* INT-133. 112 pp.

Zedaker, S. M., Burkhart, H. E., and Stage, A. R. 1987. General principles and patterns of conifer growth and yield. Pp. 203–241 in *Forest vegetation management for conifer production,* J. D. Walstad and P. J. Kuch (eds.), John Wiley & Sons, New York.

CHAPTER 17

ASSESSING RANGELAND, WILDLIFE, WATER, AND RECREATIONAL RESOURCES

17-1 Purpose of Chapter A detailed treatment of measurement and inventory of forage, wildlife, water, and recreation resources is beyond the scope of an introductory book on forest measurements. Typically, measurement of these resources is discussed in detail in textbooks that focus on management of a specific forest land resource. Nevertheless, an introduction to assessment of these resources is sometimes provided in measurements courses aimed primarily at quantifying the tree overstory. Integrated inventories that consider measurement of several resources are often conducted, and many of the sampling, measurement, and prediction principles discussed in this volume apply to resources other than timber. This chapter provides a brief introduction and overview of assessing rangeland, wildlife, water, and recreation resources. Additional detail can be obtained in the references cited at the end of the chapter.

MEASURING RANGELAND RESOURCES

17-2 Forage Resources *Rangelands* may be defined as those lands supporting vegetation suitable for grazing. The use of these lands is not limited to livestock and wildlife; rangelands may have concurrent utility for watershed protection, recreation, timber-growing, mining, or other activities. The range manager's objective is to obtain maximum coordinated use of these lands on a sustained yield basis.

Included in this section are range-measurement techniques utilized by resource managers to measure characteristics of grass, grasslike, forb, and shrub vegetation. Such information is used in developing range management plans and in evaluating the results of implemented management practices.

For estimating livestock and wildlife grazing capacity, information may be needed on forage production, patterns of forage utilization, or changes in vegetative communities resulting from grazing use or specific management practices. Measurements of vegetative cover may be desired for relating management influences on water infiltration and runoff. It may also be necessary to document changes in understory vegetation that result from timber-cutting practices, since different cutting systems influence fire danger, timber regeneration, wildlife habitat, and soil erosion. And finally, vegetative measurements might be needed to determine the suitability of an area for outdoor recreational use—or to determine the effect of recreational use on changes in plant cover and composition. Thus at certain points in time, all resource managers are likely to be interested in measurements of the herbaceous and shrubby plants that are characteristic of rangelands.

17-3 Planning Range Measurements   Marking the boundaries of the planning unit or study area on maps and aerial photographs is the first step in deciding on the sampling procedures and measurements to be made. The size of the planning unit and the variability within the unit determine, to a great extent, the procedures and techniques that should be used. The specific data required, time available for measurements, accuracy desired, financing available, and abilities of resource personnel should also be considered prior to the collection of field data.

The range manager seldom encounters a planning unit that is sufficiently homogeneous for sampling without some degree of stratification. As a rule, this stratification is accomplished by ground surveillance, by visual reconnaissance from low-flying aircraft, or by detailed study of aerial photographs. The recognition and delineation of various strata makes it feasible to employ more efficient means of sampling, e.g., stratified random sampling (Sec. 8-8).

Range sites sometimes occur in such variable patterns that it is not practical to map each site separately. The mapping unit then becomes a complex of sites. The proportion of each site within the mapping unit may be determined by dot grids (Sec. 3-11). Sampling in the field may be accomplished by sites, with data interpreted and applied to the mapping unit on the basis of the percentage of the sites within the complex mapping unit.

By using acetate overlays on aerial photos, preliminary boundary location, range-site delineation, and general planning of the measurements can be quickly accomplished in the office. Preliminary site delineations can then be field-checked and adjusted on the overlays where necessary.

17-4 Sampling Considerations   The *time of sampling* range vegetation is important because range plant communities are composed of plants which reach their peak development at different seasons, or at different periods within a season. There are "cool-season plants" that tend to exhibit maximum growth in early spring and fall and "warm-season plants" that grow mainly in the summer, provided moisture conditions are favorable during these periods.

Near the end of the major growing season is often the best time for many vegetation surveys because species are best recognized at this time, and total herbage and browse production are near maximum. The resource manager should realize, however, that biases may exist from sampling at this time because an important species that reached peak development earlier may have largely passed from the scene. Utilization studies may be made during the grazing period to determine seasonal patterns of use, but if only a single utilization check is planned, it is best done toward the end of a grazing period and prior to a new plant growth period.

With respect to the *sampling design,* most of the techniques described earlier in this book apply equally to range inventories (Chap. 8). The resource manager must decide on the technique that will provide the needed information and measure a sufficient number of sample units to achieve the desired level of sampling precision. Stratified random sampling will often provide a cost-effective inventory system for extensive surveys when good-quality aerial photographs are available. Where systematic designs are specified, the resource manager should be aware of the pitfalls of treating such samples as if they were randomly selected (Sec. 8-7).

Sample units for range measurements may be circular, square, or rectangular plots. As with most forms of sampling, the plot size selected should be the smallest possible for convenience and efficiency of sampling, yet large enough so that the sampling variation between plots is not extreme.

The plot size may be chosen for the attribute being measured on the basis of previous data from the same or a similar range site, or by a preliminary study accomplished to determine the best plot size and number to achieve a specified level of sampling precision. For the latter approach, data will be needed on the *time* required to measure different-sized plots and on the influence of plot *sizes* on the coefficient of variation. These relationships, along with formulas for calculating sampling intensity, are discussed in Chap. 8.

17-5 Determining Grazing Capacity   Livestock grazing capacity is usually expressed in terms of animal unit months (AUM). An animal unit is the equivalent of 1000 lb of animal live weight, or a cow and a calf. The dry-weight forage required to provide for one animal unit month on rangeland is approximately 900 to 1050 lb.

Animal unit conversion factors may be used to convert the amount of forage required per AUM to forage requirements for a particular class of livestock or species of wildlife. Some conversion factors are mule deer, 0.2; ewe and lamb,

0.2; white-tail deer, 0.14; grown horse, 1.25 (Range Term Glossary Committee, 1964).

Grazing capacity is the number of grazing animals that can be maintained on a range without depleting the range resource. The basic calculations to determine grazing capacity from herbage- and browse-weight measurements may be summarized by the following relationship:

$$\text{Grazing capacity per unit area } = \frac{\sum \left(\begin{matrix} \text{dry wt. per unit area} \\ \text{for each plant species} \end{matrix} \times \text{species use factor} \right)}{\text{animal unit requirement}}$$

As an example, the following herbage and browse weights, along with corresponding use factors, may be assumed:

Range plant species	Weight per acre, lb	Use factor, percent
Blue grama	370	60
Burroweed	56	0
Poverty threeawn	224	20

If we further assume that the animal unit requirement will be approximately 900 lb/AUM, then the grazing capacity would be estimated as

$$\frac{(370 \times 0.60 + 56 \times 0.0 + 224 \times 0.20)}{900} = \frac{267}{900} = 0.3 \text{ AUM per acre}$$

Therefore one animal unit could graze on 1 acre for 0.3 month; i.e., 3.3 acres would be required to yield forage for 1 AUM. Forty acres would be needed to carry one animal unit for a year.

17-6 Clipped-Plot Technique Clipped plots may be used to determine weight of herbage and browse on range sites. The procedure is to locate plots of known area and clip herbaceous plants as near to ground level as practical, and to clip the current year's production on browse plants. Individual plant species may be clipped separately and weighed in the field, if their growth characteristics allow for easy separation. Then, samples of clipped species may be collected and dried to determine percentages of dry weights. If too much time is required in the field to separate the species, or if plants are mechanically clipped by power clippers, total herbage clipped on plots may be saved for sorting and weighing in the laboratory.

In the past, square plots that are 3.1 ft on a side (9.6 sq ft) have been popular for herbage clipping, because the plot yield in grams is converted to pounds per acre when multiplied by 10. When using the metric system, 1-m^2 plots (that is, 1 m on a side) are popular. With this plot size, the herbage weight in grams, multiplied by 10, will be converted to yield in kilograms per hectare. Various plot sizes may be used for different kinds of vegetation, of course.

Plots used to determine *total* herbage and browse weight may be clipped on range sites prior to grazing or on plots protected from grazing by wire cages. There are a number of sources of error which may be associated with clipped plots unless precautions are taken during clipping. Care must be taken to clip accurately at the plot boundary; otherwise data will be biased upward or downward. The inclusion of previous-year dry-matter production along with current growth will result in an overestimation of annual production. And carelessness in the height of clipping can introduce bias because a high proportion of the weight of some plants (e.g., bunch grasses) is near the base.

Since clipped-plot samples are expensive and time-consuming, they are largely employed in conjunction with other estimation techniques. Ocular estimates of vegetation weight, in combination with clipping and weighing, are sometimes used in range analysis. The estimator is trained to recognize plant-weight units of 5 or 10 g, or some other convenient unit, and then proceeds to estimate the number of these units for each plant species on sample plots. Ocular estimates have proved most successful when few species are present on the plots.

Clipped plots and ocular estimates can be combined in a system of double sampling (Chap. 8). By this technique, ocular estimates are made for all sample plots, from which a subsample (perhaps 1 plot in 10) is clipped and weighed. The relationship between estimated and clipped plots is established by regression analysis, and all estimated data are then adjusted by using the fitted regression. The application of regression techniques to relate an easily measured attribute to herbage or browse weight is a common approach for determining plant-weight production on a range site.

17-7 Range-Utilization Estimates The utilization of forage plants is commonly expressed as the percentage of the current year's herbage or browse weight that is removed by grazing animals. There are many different techniques for determining utilization, but the simplest method is to select a *key species* (a palatable and abundant plant) and estimate the percentage use on this species along transect lines through the range sites.

One use of the utilization percentage is for calculating a grazing capacity. If a range is stocked for 60 days with 40 animal units (80 AUM), the proper use factor for a key species is 60 percent, and the current use is estimated at 20 percent, grazing capacity may be calculated by this relationship:

$$\frac{80 \text{ AUM}}{0.20 \text{ current use}} = \frac{x \text{ AUM}}{0.60 \text{ proper use}}$$

The grazing capacity is thus calculated as 240 AUM. One-third of this amount (80 AUM) is currently being used; therefore, there are 160 AUM remaining to be utilized.

Utilization studies are also needed to determine *where* a range is being grazed. The key species selected for this type of evaluation should be widely distributed and palatable and should be a species for which use can be easily estimated. The pattern of usage should be mapped, with notations on areas that are overgrazed or undergrazed. Armed with this information, the range manager can plan management improvements, e.g., water development, fencing, herding, or trails to obtain more uniform use on a range.

Another benefit derived from utilization studies is that of learning which plants are used during different seasons of the year. This information is useful in evaluating the effects of grazing on plant vigor and in planning grazing levels and systems that will maintain the vigor of preferred herbage species.

Ocular estimates of utilization are often employed. Resource managers make such estimates by clipping plants to various levels, estimating the weights removed, and then verifying the actual utilization by clipping and weighing remaining plant parts. Such estimates may be made over general areas, on sample plots of fixed area, or on individual plants along transect lines.

Another technique employs *paired* sample plots, plants, or plant parts to estimate utilization. One of the pair is clipped *before* grazing and the other *after* grazing; the difference, expressed as a percentage of the ungrazed weight, is the utilization. For pasture studies, the forage weight differences may be obtained from caged versus open (grazed) plots, but such techniques are too expensive and require too many sample units to be cost-effective for most range inventories.

Measurement of *twig length* before and after grazing is a technique used in some browse-utilization studies. In this case, use is not a percentage of total plant weight, but the method does provide an estimate of the *relative* degree of plant utilization. The greatest difficulties with this approach are those associated with twig growth between measurements and the problem of obtaining an unbiased and adequate sample.

The *number of grazed twigs* on browse plants may also be used to estimate relative plant utilization. With this method, individual plants and branches from each plant should be selected at random. Ten twigs (or some other convenient number), beginning at the tip of the branch, are then observed. The percentage of total twigs that are grazed provides an estimate of relative utilization for the

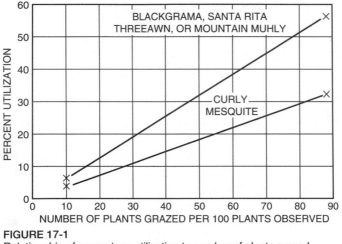

FIGURE 17-1
Relationship of percentage utilization to number of plants grazed.
(Courtesy Arizona Inter-Agency Range Committee.)

plant species sampled. Averages for large numbers of plants may be required to obtain a reliable estimate for an entire site.

Regression techniques are useful for relating relatively "easy-to-measure" characteristics to plant utilization. Relationships between percent of utilization and average height of grazed plants or number of plants grazed can be developed for specific range sites; then, for future studies, use estimates are greatly simplified (Fig. 17-1).

One of the major problems encountered on utilization surveys is that of evaluating use during a period when regrowth may occur; this is the main reason for planning field observations near the end of the growing season for key species.

17-8 Range Condition and Trend It has long been recognized that overgrazing of certain plant communities can be documented by changes in the species composition of those communities (Sampson, 1917). It has also been shown that grazing capacity is influenced by the stage of plant succession on a range area, with higher grazing capacities usually associated with stages of succession closest to herbaceous climax vegetation.

Range condition is the status or stage of succession that a plant community expresses as compared with the potential or climax vegetation possible for the site. Thus the condition may be classed as excellent, good, fair, or poor, depending on present characteristics of the range site. Impacts of grazing and other influences on a range site are reflected by changes in species composition, plant cover, numbers of plants, relative vigor of plants, and soil erosion.

Trend is the direction of change—whether stable, toward the potential for the site, or away from the potential for the site. A trend toward poorer range conditions might be caused by such factors as excessive grazing, drought, fire, plant diseases, or mechanical disturbances of the soil. The range manager must be able to determine the causes of site deterioration. A poor range condition is usually an indication that improvements in management are needed to restore the site to its maximum productivity and use. Although range condition and trend evaluations tend to emphasize grazing influences, these concepts may also be extended to measure impacts of any origin on the range vegetation.

Plant cover, density, and frequency are the attributes of vegetation that are most frequently measured to determine condition and trend on range sites in response to grazing or other impacts.

Plant crown cover is the proportion (or percentage) of the ground surface under live aerial parts of plants. *Plant basal cover* is the percentage of the ground covered by plant bases, and *total cover* may include the combined aerial parts of plants, mulch, and rocks. Cover may be determined from ocular estimates of sample plots, by charting or mapping vegetation on plots, or it may be estimated from low-altitude aerial photographs.

The proportion of cover may also be estimated along taped transects, or "point frames" may be employed (Fig. 17-2). A frame holding 10 vertical or in-

FIGURE 17-2
A point frame for sampling range vegetation.

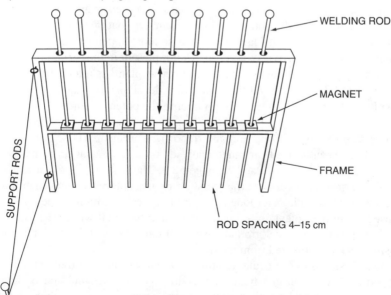

clined pins is a standard measuring device; the pins are lowered through the vegetation, and the percentage of "hits" provides an estimate of cover.

Density is defined as the number of plants or specific plant parts per unit area of ground surface, e.g., number of mesquite plants per hectare. The counting of individual plants on sample plots of known area is a simple means of deriving density estimates. For sod-forming plants, however, it is difficult to decide on which plant units to count. And in some vegetation areas, the number of plants per unit area is so large that counting is not practical. Density does not give an indication of the size of individuals unless counts are made and recorded by size classes.

Frequency is the number of plots on which a species occurs divided by the total number of plots sampled. Frequency data are used to detect changes in plant abundance and distribution on a range site over time, or to identify differences in species responses to varying management practices. Estimates of frequency are simple and objective; the resource manager must merely identify the species and record its presence if it is found on the sample plot.

Selection of the proper plot size is extremely important for estimating frequency, and more than one plot size may be needed for varying plant species and plant distributions. Frequency data are easily obtained, but numerous sample plots must often be evaluated before reliable estimates can be derived.

The relative proportion of each plant species on a given range site is termed the *species composition.* It may be expressed in terms of weight, cover, or density (Range Term Glossary Committee, 1964).

Species are grouped into three categories based on their response to grazing. Plants that are present in the potential plant community but decrease with heavy grazing are called *decreasers.* Plants that become more abundant, at least initially, with heavy use are *increasers,* and species not present in the potential vegetation but which invade the site as cover of the native plants is reduced are called *invaders* (Fig. 17-3). Range *condition guides* are developed by gathering species composition data from sites which are producing the potential for that particular locale.

A three-step method of condition and trend measurements (U.S. Forest Service, 1965) has been utilized for range analyses on lands managed by the U.S. Forest Service and Bureau of Land Management. The first step is to establish a cluster of two or three permanent transects. At regular intervals along each transect, a 2-cm- (3/4-in.-) diameter loop is lowered vertically from the tape marking the transect. A tally of plant "hits," rock, litter, and bare soil within the loops is made. Then, plant cover, total ground cover including rocks and litter, and plant composition can be estimated from the data.

The second step is to use ocular appraisals and a summary of transect data to classify current condition and trend of the site. Vegetation condition is determined by "scores" based on plant composition, cover, and plant vigor. A soil condition rating is also determined.

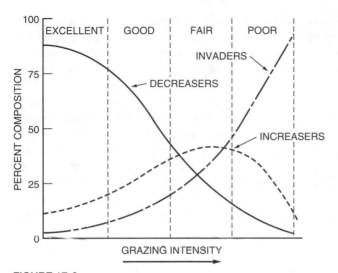

FIGURE 17-3
As grazing intensity increases, range condition deteriorates and
the proportion of decreasers, increasers, and invaders changes.
(From *Stoddart et al., 1975.*)

The third step is to take a close-up photo and a photo of the general view of
the transect for future reference. These are permanent study transects, and when
data are collected for the same site at a future date, a new estimate of condition
is attained, trend is indicated, and adjustments in management can be made accordingly.

MEASURING WILDLIFE RESOURCES

17-9   Animal Populations and Habitat   Management of the wildlife resource implies an effort to attain a degree of balance between the food and cover
available and the animal populations that are favored. Thus the inventory problems are twofold:

1 Estimating, through periodic sampling techniques, numbers of animals,
composition, trend, and the natural range of various wildlife populations

2 Determining the food and cover requirements of different species, and
evaluating the adequacy of various habitat units for supporting wildlife populations.

In many respects, the problems associated with wildlife measurements are
similar to those described in evaluating rangeland resources for domesticated
livestock. This is particularly true with respect to wildlife habitat assessments,

i.e., the determination of carrying capacity, forage production, forage utilization, or condition and trend. With wild animals, however, the exact population size is rarely, if ever, known. Therefore census techniques, along with estimates of animal productivity and population trend, are also of special importance.

17-10 Population Estimates The various techniques for measuring populations of vertebrates may be grouped into three classes: *direct* census, *indirect* census, and mark-recapture techniques. A direct census implies the counting of the animals or birds themselves; such counts may be made on drives, by visual surveillance from aircraft, from aerial photographs or thermal imagery, and by ground observations on flushing strips.

For an indirect census, observations other than animal counts are recorded, and the population estimate is derived from this indirect evidence of animal presence. Included here are bird-call counts, track counts, and pellet-group counts.

Mark-recapture methods are based on a measured change in a population. For example, a known number of game birds may be captured, banded, and returned to the population. When these birds are seen, recaptured, or taken by hunters at some later point in time, the ratio of the total banded birds to those recaptured can be used for estimating the population size.

The drive-census method requires a line of observers or "beaters" moving steadily through an area to flush out the desired species. Recorders are stationed at the opposite boundaries and along the edges of the tract to count the animals pushed out. If vegetation is sparse and the line of beaters is widely spaced, each person records all birds or animals that go back through the line on their right side. A fair amount of noise is often desirable, and regular calling back and forth among beaters (to be sure that all observations are recorded) will aid in obtaining a reliable count.

The drive census is one of the oldest methods in use, and where sufficient manpower is available, it will provide reliable population estimates. The fact that 25 to 50 persons may be needed to cover an area of 1 mi^2 tends to limit the application of the method to level terrain that is already surrounded by clearings such as firebreaks or logging roads. Under the limited circumstances where drives are economically feasible, the data derived serve as a useful control or base for evaluating alternative census procedures.

Visual observation of wildlife from low-flying aircraft is an effective census technique in areas where animals tend to congregate and where overstory cover is relatively sparse. The method is mainly applicable to large mammal counts in open plains country and for migratory waterfowl wintering on open coastal waters.

A common procedure involves low-level flights in a grid pattern with a pilot and observer. Helicopters, though more expensive, may be effectively used to count desert bighorn sheep in rough, low mountains or to enumerate elk that are concentrated on winter ranges with sparse tree cover. In some states, visual surveys are also used to census moose and to count breeding waterfowl on systematically spaced aerial transects.

Helicopter counts are apt to be more reliable than surveys from fixed-wing aircraft, and the slower flights may enable observers to obtain a listing of species sighted by sex and age classes. The basic census procedures are simple for good observers and pilots, especially where the exact flight paths and areas to be covered are carefully mapped out in advance.

Wildlife species that can be counted on visual surveys are also susceptible to enumeration on aerial photographs. Large-scale photography has been successfully used to count waterfowl along the Pacific Coast flyway and to determine trends in gull populations by combining visual estimates of numbers with photography of nesting areas and associated flocks. Such trend estimates do not provide a complete picture of the population, but they may be useful in detecting large annual changes, i.e., variations of 25 percent or more. For inventories of waterfowl and large mammals, photographic scales of 1:3000 and larger are recommended; color or infrared color films should be specified to provide maximum contrast between the wildlife species and associated backgrounds or native habitats.

The *strip-flushing census method* may be used for species that will hold to cover until an observer approaches, and then fly or run away when flushed. Ruffed grouse, woodcock, and even white-tailed deer may be enumerated by this technique.

The procedure is similar to a strip system of timber cruising sometimes used by foresters. The observer walks along parallel lines or transects through a tract and records the perpendicular *flushing distance* each time an animal is sighted. In practice, the technique tends to be limited to fairly level and open terrain where walking is very easy and where good visibility conditions prevail.

Indirect census methods are commonly used for smaller, short-lived animals and for big game that is relatively inaccessible because of weather, terrain, or cover. Such surveys may be less precise than a direct census, but they can provide valuable information on population trends.

Call counts are used to assess population trends for game birds such as the mourning dove. When doves migrate northward in spring, the males establish territories and attempt to attract a female by calling (cooing). In many states, a series of 20-mile routes have been established in dove habitats for checking during the calling season. Observers begin their routes just before daylight, stopping at specified intervals to count all dove calls heard during a 3-min interval. Doves seen perched or flying across the road between stops are recorded separately.

Most male birds that proclaim territories by calling can be counted by various modifications of this method. Male call counts alone do not constitute a complete census, but they may be used in conjunction with sex-ratio data to provide population estimates and trends.

Track counts is a census method that is best adapted to land areas that are criss-crossed with a grid pattern of nonsurfaced (i.e., easily imprinted) roads and trails. Checking these roads a few hours after heavy rains appears to give the best results. The technique has received limited endorsement, because counts are often highly variable, especially for low population densities.

The behavioral pattern and activity cycle of the species should be well known to observers before this census procedure is attempted. Track counts are most useful for assessing populations when a species is migrating or when herds are moving from one seasonal range to another.

The *pellet-group-count method* is based on the assumption that periodic accumulations of animal defecations are related to population density. For example, studies have shown that mule deer and elk leave about 13 clusters of fecal pellets on the ground every 24 hours. Thus the number of pellet groups on a range unit may be converted to one deer or elk day for each 13 groups found. To convert deer days to numbers of animals, it is necessary to know how long the deer has been on the area (which can be determined for migratory herds), or the survey must be cleared of pellet groups and then counted *after* a known time interval, e.g., 1 month.

The *Lincoln index* is a commonly used ratio method of population estimation based on the banding or marking of animals. A number of animals are captured, marked or banded, and then released back into the population. When animals from the population are later recaptured, the ratio of total banded animals to the banded individuals caught can be used to estimate the size of the population *(N):*

$$N = \frac{\text{total no. banded}}{\text{no. of banded caught}} \times \text{total no. caught}$$

The accuracy of the method depends upon an adequate sample size, and the banded or unbanded individuals must have an equal chance of being recaptured. Recapturing animals with food bait may result in a biased sample if "trap-happy" animals return for the bait more frequently than individuals who have not previously found it.

17-11 Habitat Measurement Populations of wildlife tend to differ in numbers because of variations in hunting pressures, species characteristics, adaptability, and the quality of available habitats. The essential habitat components are water, food, and cover (i.e., protection from inclement weather and predators, including man).

Different types of plant cover used for nest sites, resting, bed grounds and other purposes will also vary with the wildlife species, its size, and mobility. It is necessary to evaluate the cover requirements for each species, since adequate cover for the cottontail will be inadequate for deer. Of equal importance to the elements of water, food, and cover is the *spatial arrangement* of these components. Without the proper arrangement of the essential habitat elements, there will be a poor population or none at all. Most wildlife need *all* these elements within their normal daily range of movement. And if the basic elements periodically reoccur in several locales over a landscape, a highly productive population is much more likely to develop.

Procedures previously described for estimating forage yields, use, condition, and trend of rangelands will apply equally to wildlife habitats. Measurements of the edible herbage utilized from shrubs or trees are based on twig lengths before and after browsing. Carrying capacities for various habitats are computed by the same procedure used for determining livestock grazing capacity. In fact, in some instances, a given range may be utilized by either domestic or semiwild animals. Animal unit requirements will differ for each species, of course.

An important source of wildlife food that is of minimal concern to domestic livestock is plant fruit, including acorns, nuts, and seeds. Measurements of fruit production are difficult because of variations in time of ripening and the length of time they are held on the producing plant.

Efforts to estimate fruit yields have been generally confined to acorns, nuts, or large fleshy fruits. Sampling may be based on fixed-area plots by setting funnel-type traps to collect dropped fruits or by counting fruit on sample branches of the woody plant. With trap samples, the number of fruits per trap is expanded to the number for the total crown area; four to six traps per tree are often used for estimating acorn yields.

When numbers of fruits on sample branches are used for estimation purposes, tree totals may then be expanded by determining yields from various classes of tree size, form, and species in stands of known composition and density. Area sampling consists of collecting data from small plots distributed over the entire stand. Open, unprotected plots on the ground may be useful if samples can be taken frequently during the fruiting season. Otherwise, traps must be used to protect against losses due to deterioration and animal consumption.

One measurable quantity that is related to the *arrangement* of the basic habitat elements is the amount of *edge* present on a given range. Edge may be defined as the total linear measure of the borderline between two distinct vegetation types; i.e., it must be distinct in the eyes of the wildlife species under consideration. Some animals find these borders between classes of vegetation attractive because they often provide food supplies near nesting grounds, cover for travel, or a clear escape route near a bed ground.

The amount of edge is subject to periodic change because of events such as forest fires and timber harvesting. A square opening of 20 acres surrounded by forest has less than one-third of the border or edge provided by ten scattered 2-acre openings. This fact can be an important consideration in planning clearcut logging in conjunction with wildlife habitat requirements.

There are many techniques for measuring edge on a selected range (e.g., on type maps or aerial photographs), but the land manager must exercise keen judgment in deciding *which edges* will be of interest and importance to a given species of wildlife. For example, a deer might not consider a difference between two grass types to be significant, whereas a grass-bush border would probably constitute an important edge.

The concept of edge is particularly important for species that are characterized by a limited home range and a small radius of daily mobility.

MEASURING WATER RESOURCES

17-12 Importance of Water Water is important to resource managers in its liquid, solid, and vapor forms. It is one of our most dynamic natural resources, and at times it may completely dominate or limit the use and management of other resources on the land. During periods of heavy rainfall, runoff, or snow cover, for example, water may limit access to the land, timber harvesting, grazing, and other operational functions.

In both liquid and solid forms, water influences the production rates of timber, forage, and wildlife. In the form of snow, water provides winter recreational opportunities, and liquid water on and from wildlands affects fishing, swimming, boating, and other water-based recreational activities. In many localities, streamflow runoff derived from liquid and solid precipitation constitutes an important water-supply source for municipal, industrial, and agricultural applications downstream. Runoff may also provide significant sources of power to aid in satisfying energy needs. Excessive runoff may cause upstream and downstream damage as a result of erosion, flooding, or sedimentation. Thus, the quantity, timing, and quality of water produced from wildlands are of critical importance; either too much or too little water can be as limiting as water of impaired quality.

In vapor form, water is a factor in determining evaporation and transpiration rates from wildlands. And its contribution to atmospheric humidity may exert a significant influence on the forest-fire hazard because of the effect on fuel moisture contents.

Interactions between water and the utilization of other resources on the land can also be important. As an illustration, the harvesting of trees may increase streamflow yields, at least temporarily. At the same time, poor access-road engineering on watersheds may increase rates of erosion and sedimentation. It is therefore apparent that not only does water affect other resources, but the use and management of those resources affects water quantity and quality. To assess the significance of water both on and off the land, the resource manager should be able to inventory water in its various forms, just as is done for timber, rangelands, and wildlife resources.

17-13   Factors Affecting Runoff   The flow of a stream is essentially controlled by two factors, one depending on the physical characteristics of the watershed and the other upon weather and climatic characteristics (e.g., precipitation) that directly affect the watershed. Land managers concerned with water resources must have an appreciation of the mensurational aspects of both sets of factors to achieve desired water resource objectives and to properly evaluate available management opportunities.

A watershed is an area of internal drainage, the size and shape of which is determined by surface topography. A watershed is completely encircled by a divide or ridge line. Precipitation falling on one side of the divide drains toward the outlet or mouth of the watershed on that side of the divide; precipitation falling on the other side of the divide drains toward outlets of other watersheds. The resource manager uses the watershed as his basic land unit for planning and management purposes; in this sense, it is roughly comparable to the forester's working circle or the range manager's grazing allotment.

In essence, the watershed is a limited unit of the earth's surface within which climatic conditions can be assessed; it also represents a system where a balance can be struck in terms of the inflow and outflow of moisture. Various sizes of watersheds may be recognized and delineated, depending on the objectives of users. Small, experimental watersheds may encompass only a few hectares in area, while large watersheds may be composed of entire river basins.

17-14   Physical Characteristics of a Watershed   Perhaps the easiest characteristic of a watershed to measure is its areal extent. Such evaluations are commonly made by using a planimeter or dot grids to measure the delineated area on planimetric maps, topographic maps, or aerial photographs. Area is an important consideration because the total volume of water carried by a stream is directly related to watershed area.

Another important consideration is the effect of watershed area on peak flows. As with total water yields, higher peak flows are usually associated with larger areas. However, when such outputs are expressed in terms of *flow per unit of watershed area,* it is the smaller watersheds that characteristically have the greater rates of flow.

The outline form or shape of a watershed can sometimes have a marked effect on streamflow patterns. For example, a long, narrow basin would be expected to have attenuated flood-discharge periods, whereas basins with round or oval shapes are expected to produce sharply peaked flood discharges. Although watershed shape is a difficult parameter to quantify, various indices have been developed for comparing the configurations of different basins. One such index, based on the degree of roundness or circularity of a watershed, may be computed by

$$\text{Shape index} = \frac{0.28 \times \text{watershed perimeter (units)}}{\sqrt{\text{watershed area (units}^2)}}$$

When a watershed is circular in shape, the index value will be approximately 1. The closer a shape-index value is to unity, the greater the likelihood that precipitation will be quickly concentrated in the main stream channel, possibly resulting in high peak flows. Watersheds that are noncircular in shape will have index values greater than unity.

The slopes of various land surfaces within a watershed, usually expressed in percentage terms, can greatly influence the velocity and associated erosive power of overland flow. In addition, slope is related to infiltration, evapotranspiration, soil moisture, and the groundwater contribution to streamflow.

For small areas, watershed slope can be estimated by computing the average slope from several on-the-ground measurements obtained with an Abney level or clinometer. For larger areas, or for evaluating a number of different watersheds, slope estimates can be computed from topographic maps issued by the U.S. Geological Survey. One procedure consists in randomly selecting locations within each watershed and measuring slopes directly on the contour maps; the slope for the entire watershed can then be approximated by computing an arithmetic average of these values. Watershed slope can also be estimated by the following relationship, based on determining the area between contour lines within a watershed:

$$\text{Slope (\%)} = \frac{c \times l}{a}(100)$$

where c = contour interval (ft or m)
l = total length of contours (ft or m)
a = area of watershed (ft^2 or m^2)

For those occasional situations where slopes are relatively uniform over an entire watershed, an estimate of average slope may be derived by the relationship

$$\text{Slope (\%)} = \frac{e}{d}(100)$$

where e = elevational difference between the highest and lowest points on the
        watershed (ft or m)
d = horizontal distance between high and low elevations (ft or m)

The mean elevation and the variations in elevation of a watershed are important factors with respect to temperature and precipitation patterns, especially in
mountainous topography. Temperature patterns, in turn, are associated with
evaporative losses and with the timing of periods of snowpack accumulation and
melt. Precipitation patterns, such as annual amounts or the proportion of annual
precipitation that falls as snow, may also be related to elevation and are important in assessing the total water flow from a watershed.

Transpiration and evaporation losses on a watershed, factors that affect the
amount of water available for streamflow, are influenced by the general orientation or aspect of the basin. Also, the accumulation and melting of snow is related
to the orientation of a watershed. For example, if the orientation is southerly,
successive snowfalls may soon melt and infiltrate into the ground or produce
runoff. For those watersheds with a northerly orientation, individual snowfalls
may accumulate throughout the winter and melt late in the spring, producing
high flow rates.

The orientation of a watershed is normally expressed in degrees azimuth or in
terms of the major compass headings (i.e., N, NE, E, etc.); the designation indicates the direction that the watershed "faces." The direction of flow for the main
stream channel can also be used to indicate the general orientation of a watershed.

The pattern or arrangement of natural streams on a watershed is an important
physical characteristic of any drainage basin for two primary reasons. First of
all, it affects the efficiency of the drainage system and thus its hydrographic
characteristics. Secondly, the drainage provides the land manager with a knowledge of soil and surface conditions existing on the watershed; more specifically,
the erosive forces of stream channels are related to and restricted by the type of
materials from which the channels are carved. Most drainage patterns can be
classed as dendritic, i.e., an irregular branching of tributaries, often with no pre-
dominant direction or orientation.

A quantitative approach to classifying streams in a basin can consist in systematically *ordering* the network of branches and tributary streams. Each non-
branching tributary, regardless of whether it enters the main stream channel or
its branches, is designated a first-order stream. Streams receiving only non-
branching tributaries are termed second order; third-order streams are formed
by the junction of two second-order streams, and so on (Fig. 17-4). The order
number of the main stream at the bottom of the watershed indicates the extent
of branching and is an indication of the size and extent of the drainage network.

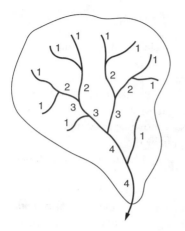

FIGURE 17-4
Horton's system of designating stream
orders.

Another method of quantifying the drainage network of a watershed consists in determining drainage density by the following relationship:

$$\text{Drainage density} = \frac{l}{a}$$

where l = total length of perennial and intermittent streams on a watershed (units)

a = watershed area (units^2)

Drainage density is an expression of the closeness of spacing of stream channels on a watershed. In general, low drainage densities are favored in regions of highly permeable subsoils, dense vegetative cover, and low relief.

17-15 Measurement of Water Quantity Of the various types of hydrologic information, one of the most important to a watershed manager is streamflow data. Such data can provide the manager with information on daily, seasonal, and annual runoff volumes, as well as peak and low flows. Streamflow from natural watersheds is basically a result of precipitation. This relationship is greatly modified, however, by factors such as weather, soils, vegetation, and topography. For some portions of the United States, where watershed characteristics do not differ greatly from one area to another, the streamflow measured on one watershed may be used to index flows from nearby or adjacent watersheds. Regression methods can be used for establishing these relationships. In other areas, however, the diverse nature of natural watersheds and the large number of factors affecting streamflow will often prohibit such extrapolations. Because of this situation, direct measurements of streamflow may represent the only method of accurately determining runoff from a particular watershed.

To obtain an estimate of the quantity of streamflow from a watershed, a measurement of discharge is necessary. Discharge Q, or rate of flow, is the volume of water that passes a particular location per unit of time. Two types of information are required for discharge estimates: the cross-sectional area a of the channel and the mean velocity v for this cross-sectional area. Discharge can then be computed by

$$Q = a \times v$$

Since the discharge of a stream is the product of its cross-sectional area and mean velocity, accurate measurements of both quantities are necessary. Units of discharge are cubic meters per second or liters per second.

Perhaps the simplest way of estimating discharge (and one of the least accurate) is to observe how far a floating object, tossed into the stream, travels in a given length of time. Dividing this distance by the time interval provides a rough estimate of the velocity of the water. Because the velocity at the surface is greater than the mean velocity of the stream, a reduction factor is necessary to obtain an estimate of mean velocity. This factor is commonly assumed to be about 85 percent. Using this corrected velocity and a measurement of the cross-sectional area of the stream, the discharge can be computed from the preceding formula.

Another method of measuring streamflow is based on the following relationship:

$$Q = \frac{1.49a \times r^{2/3} \times s^{1/2}}{b}$$

where Q = discharge (ft³/sec)
$\quad a$ = cross-sectional area (ft²)
$\quad r$ = hydraulic radius (ft)
$\quad s$ = slope of channel (ft/ft)
$\quad b$ = roughness coefficient

(If m and sec are used, the constant 1.49 is omitted.)

The hydraulic radius is computed by dividing the cross-sectional area a of the stream by the wetted perimeter. The roughness coefficient b must be estimated from conditions of the channel and may vary from approximately 0.030 to 0.060 for natural channels; an average value for natural streams is about 0.035. The main source of error in applying this formula (often referred to as *Manning's equation*) to natural channels results from estimating the roughness coefficient. An error of 0.001 represents about a 3 percent error in discharge. This method, also called the *slope area method,* can perhaps best be used for estimating the discharge of peak flows in natural channels where sufficient high-water marks can be determined.

A more common and accurate method of estimating stream discharge than either of the foregoing procedures is to use a stream-current meter. A current meter is an instrument used to measure the velocity of flowing water by means of a rotating element. When placed in a flowing stream, the number of revolutions per unit of time is related to the velocity of the water. To make these velocity determinations, the current meter is either mounted on a hand-held rod or suspended from a cable. Pressure-sensing devices which give a direct reading of stream velocity are also available.

The velocity of a stream varies from point to point in a given cross section, and a number of velocity measurements with a current meter are necessary to obtain a reliable estimate of discharge. Because of streambed roughness, channel configuration, and turbulence, velocity profiles of natural streams are subject to considerable variation. To properly weight the various velocities found in a given stream, the stream is divided into several vertical sections, and the mean velocity and area are determined separately for each section. The discharge Q for the entire stream can then be obtained by summing the product of area a and velocity v for each section:

$$Q = a_1 v_1 + a_2 v_2 + \cdots + a_n v_n$$

where n is the number of sections. The greater the number of sections, the closer the approximation to the true discharge. Under ideal conditions, 10 may be adequate, but an evaluation of 15 to 20 sections is desirable. The actual number taken depends primarily on the size of the stream channel and the amount of turbulence.

The velocity and depth of each vertical section can be measured from cable cars, boats, and bridges, or simply by wading. For depths greater than 0.5 m, two measurements are made for each section: at 0.2 and 0.8 of the depth of the water. These two velocities are averaged to obtain the mean velocity for that section. For depths less than 0.5 m, the current meter is set at 0.6 of the depth as measured from the water surface. For deep streams, velocity measurements are made at relatively close intervals, and the actual velocity profile is estimated; however, this method is both costly and time-consuming.

Once the discharge rate has been determined, the volume of flow can be calculated for any specified time period by

$$\text{Volume of runoff} = \text{discharge} \times \text{time}$$

A measurement of discharge is applicable only to those stream conditions and flow levels existing at the time of measurement. Yet, flow levels in natural streams usually change with time, and a relationship of discharge to some other variable is desirable. In practice, the "stage" or depth of water above a given datum at a specified stream cross section is used. When a sufficient number of stages and their associated discharges have been measured, a stream-rating curve can be constructed. This relationship, along with a record of the stage of a stream, can be used to estimate discharge for those periods when the stage is not constant (Fig. 17-5).

Rating curves are affected mainly by channel characteristics and thus must be developed individually for all gauging stations utilizing natural channels. Stage measurements may be obtained from systematic readings of the water-level surface on a gauge or graduated rod or from automatic water-stage recorders.

On smaller streams, the channel configuration may be modified to alter the flow characteristics along a particular stretch of channel. This is done to obtain more accurate measurements of streamflow. Two general categories of structures, viz., weirs and flumes, may be utilized, depending upon the characteristics of the stream channel.

Weirs are among the oldest and most reliable types of structures that can be used to measure the flow of water in small streams (Fig. 17-6). A weir is usually a simple overflow structure built across an open channel to create an upstream

FIGURE 17-5
Rating curve for a natural stream channel (above) and a streamflow hydrograph developed from a rating curve and record of stage (below).

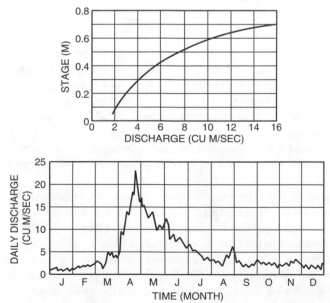

FIGURE 17-6
This trapezoidal weir at the Coweeta Hydrologic Laboratory in North Carolina is used to measure streamflow continuously. *(U.S. Forest Service Photograph.)*

pool. The discharge over the crest of the weir is determined by the vertical distance between the crest of the weir and the water surface in the upstream pool; this height is usually referred to as the *head*. Depending upon the shape of the opening, weirs may be identified as V notch, rectangular, or trapezoidal.

Weirs can be used most effectively whenever there is a fall of 1 ft or more over the crest. An advantage of standard weirs is that equations exist to estimate streamflow directly without the necessity of special calibration. Weirs should not be used on streams that have considerable bedload or suspended sediment load.

Flumes are somewhat similar to weirs, and numerous types are employed, depending upon channel and stream characteristics. In contrast to weirs, flumes do not create an upstream ponding effect but instead provide a smooth length of channel that allows the stream to pass through freely. In addition, the velocity of flow is usually sufficiently high so that any suspended sediment or bedload will not be deposited. The more common types of flumes have been precalibrated, although some field measurements may be necessary to check the calibration.

Streamflow measuring devices such as weirs and flumes are usually selected to meet the specific needs for each location. Flow occurs on some watersheds only as the result of an occasional large storm. Other watersheds may have streams that flow continuously. In most cases, it is necessary to measure streamflow at low, medium, and high rates as accurately as possible. Whenever possible, equipment with a precalibrated stage-discharge relationship should be used to gauge the entire flow.

17-16 Measurement of Water Quality   Water-quality samples are obtained so that the various kinds and amounts of substances present in water can be evaluated. A sampling program is often necessary because of the difficulties in attempting to continuously monitor the water quality of a lake or stream. Knowing when, why, and where to collect samples is basic to obtaining good water-quality measurements. This requires a knowledge of not only the system being sampled but also the expected time and space distribution patterns of the variables being sampled and their behavior in solution. Nonrepresentative sampling is perhaps the major source of error in obtaining water-quality information.

One of the goals of a water-quality investigation may be to provide information from which the general composition of water can be precisely determined. Other investigations may concentrate only on specific water-quality problems. The distribution of materials in an aquatic environment is influenced by the source of the material, mobility, phase (solid, dissolved, or gaseous), and type of system (stream, lake, reservoir). Sampling sites should be both accessible and representative of major sections of a stream or lake system.

For some streams, a "grab sample" obtained with a clean glass or plastic container may provide a satisfactory sample for preliminary analysis. This method of sample collection usually assumes that stream turbulence causes adequate mixing and that the sample is representative of the entire stream cross section. A single grab sample, however, should be regarded as representative of the discharge only at the time of sampling.

Grab samples collected periodically or continuous water sampling instrumentation (e.g., coshoton wheel), can provide a reasonable indication of water quality for many streams. Systematically sampling a number of locations about the same time is of value for reconnaissance purposes. If at all possible, flow measurements should also be available at sampling sites. If streamflow and water quality are closely related, a record of stream discharge provides a convenient means for attempting an extrapolation of the chemical record.

Once obtained, the water samples may have to be treated immediately to protect against degradation of the contents. For example, provisions for freezing or otherwise preserving the sample must be available, if water samples are to be used later for organic analysis. A possible alternative is to measure certain variables in the field.

If a water sample is to be retained for later analysis in the laboratory, the sample container should be marked or tagged so that its identity is not lost. Information such as time, date, water temperature, place where the sample was obtained (e.g., middle of lake, edge of stream, etc.), and streamflow may be necessary to correctly interpret the results of a water-quality analysis.

MEASURING RECREATIONAL RESOURCES

17-17   The Problem   The last three decades have seen a heightened interest in forest-based recreational activities in the United States. Many public lands, e.g., national and state parks and some national forests, are managed *primarily* for recreational benefits, and the public increasingly looks to forest lands for recreational opportunities. During the same period, environmental legislation has mandated that recreation (and related aesthetic and ecological) values be considered more fully in forest management decisions. For these reasons, recreational resources can be as important to the land manager as the more tangible values of wood and water and forage. Consequently, managers need some objective means of describing, comparing, assessing, or measuring recreational resources and the benefits or user satisfactions derived therefrom.

While the timber productivity of a forest can be described in terms of volume per unit of land and grazing productivity can be expressed in animal unit months, it is more difficult to define and quantify the direct productivity of recreational-aesthetic resources. It is not sufficient to simply enumerate recreational visits, overnight campground stays, or fishing boats launched over a given period of time. These measures may reflect the popularity of a resource, and therefore may be important for guiding management activities. However, they do not measure the capacity or potential for recreation, as sites are often crowded beyond some level of ecological or social capacity.

Rather, resource managers and recreational specialists recognize that the products of recreational-aesthetic resources are the *satisfactions or benefits derived* from the recreational experience; in other words, the nature and quality of the experience better reflects what is produced than the total number of persons utilizing a resource. However, recreational experiences are only partially determined by resource conditions and facilities; what the recreation participant brings to the site in the way of equipment, knowledge, skills, companions, and so forth also plays a large role. The challenge for resource managers has been to devise methods to operationally inventory forest lands in a way that will provide valid measurements of those resource conditions that substantially shape the quality of recreational experiences.

Although much work remains to be done, considerable progress has been made in inventorying recreational resources and relating resource conditions to recreational experiences. From these efforts, workable techniques have been devised for estimating visitor use on existing recreational sites and for assessing the potential of particular areas to support various recreational pursuits. It is with these procedures that the ensuing discussions are primarily concerned.

17-18   Visitor Use of Recreational Facilities   The output of recreational management is commonly measured in terms of visitor use, i.e., visitor-days.[1] This information is valuable for scheduling visitor information programs and maintenance operations and for planning additional recreational sites or facilities. Use data are also of assistance in the prediction of the rates at which facilities depreciate. When the intensity of use on a site is compared with its capacity, patterns of use or activity changes over time can be ascertained.

Recreational use is the result of the interaction of the *supply* of opportunities provided and the *demand* of the public for such opportunities. Thus, in economic terms, use may be regarded as the *quantity* of recreation that is consumed.

An obvious means of measuring use at a recreational site is to obtain a *complete registration* of visitors entering or leaving an area. To employ this method effectively requires absolute control over access to the site, and a registration station must be manned around the clock. This approach is seldom used on a year-round basis, and it is usually limited to facilities where fees or permits are needed for entry. Many parks maintain entry checkpoints or registration stations during peak-use seasons, but they are rarely operated on a 24-hour basis because of the expense involved. Where such systems are feasible, they do provide an opportunity to gain detailed visitor information, e.g., number of persons per vehicle or family unit, length of visit, types of activities pursued by different age groups, and so on.

For estimating visitor use on dispersed or undeveloped recreational sites such as wilderness areas, various systems of *self-registration* have been employed. One major problem is that many people do not voluntarily register their presence; thus this method has not worked well in areas where there are numerous routes for ingress and egress. It is sometimes feasible to designate a "calibration period" whereby the proportion of visitors who do register can be estimated to determine whether nonregistrants differ from registrants in terms of their activity and use characteristics.

For both estimating and regulating visitor use on dispersed sites, there appears to be considerable merit in a permit or license-fee system. By this approach, registration data can be obtained in the same manner as currently supplied by applicants for hunting and fishing licenses.

Because of the problem of counting or registering *all* visitors at recreational sites, some form of sampling may be employed to estimate visitor use. This approach, of course, creates a new problem—which persons and what proportion of the total number of visitors should be counted, registered, or interviewed? As

---

[1]A visitor-day is usually defined as 12 visitor-hours; i.e., 2 visitor-days equal 1 calendar day.

with any public opinion poll, a *representative sample* is the objective, but this is not easily achieved with mobile populations of varying age groups, differing activity patterns, and divergent recreational preferences.

If the only information desired is the total visitor use during a particular season, a number of sample days might be selected, followed by tallies of visitors present on those dates. To account for weekly cycles of visitor use, a sampling design must be developed to select "measuring dates" having a representative proportion of weekdays, holidays, and weekends; otherwise any extrapolations from such estimates may be severely biased.

A variation of this technique is to stop all persons leaving the site on certain dates to determine how long they have been at the recreational facility; this approach permits an estimate of use in terms of visitor-days. One difficulty of applying such estimates is that they are valid only for the current season; the sampling must be repeated during subsequent years to update or revise use evaluations.

One method of minimizing the annual problem of sampling use is to find an indicator that is more cost-effective for predicting use than direct counts of people themselves. This approach is analogous to the indirect census techniques previously outlined for wildlife populations. Indicator measurements that have been employed include a diversity of inventory data such as traffic counts, water consumption, number of boat launchings per unit of time, weight of garbage collected, and quantity of sewage effluent.

The essential characteristic of an indicator variable is that it should be highly correlated with the kind or pattern of recreational use to be predicted. The basic procedural steps that have been followed on many surveys are:

1 Select the season of year for which predictions of visitor use are desired.

2 Choose an indicator (e.g., traffic counts) that presumably rises and falls in the same pattern as the recreational use to be predicted. Obtain indicator data continuously—throughout the season.

3 Randomly select several sample days (perhaps 10 to 30 days during the selected season) with equal representation of weekdays, holidays, and weekends. On the sample days, measure the recreational use of interest, e.g., by interviews, camper registration, turnstile counts, etc.

4 By regression analysis, attempt to establish a relationship between the indicator data and recreational use, based on paired measurements for the sample days. Calculate the precision of the prediction equation.

5 For future seasons (assuming that use patterns are similar), continue to obtain indicator data only. Use the regression equation to predict recreational use.

The indicator most frequently used at developed, unsupervised recreational sites is an axle count obtained from an automatic traffic counter. Not only is this technique much used, but it is also much abused. The popularity of such devices may be attributed to two characteristics: they are relatively cheap, requiring little maintenance, and they produce numbers. Unless they are intelligently located, however, the data that result can be quite misleading.

In a study of recreational sites in the Appalachian Mountains, James and Ripley (1963) found strong correlations between visitor use and axle counts. In one instance, a linear equation of the form $Y = a + bX$ was fitted to the relationship with this result:

$$Y \text{ (total visits)} = 110 + 0.3X \text{ (axle count)}$$

For some variables, of course, the relationship between an indicator variable and visitor use may be curvilinear. The sampling precision for this type of use-prediction equation may be less than that obtained in predicting other products of resource management. For example, many use surveys (based on 10 to 12 sampling days per site) produce sampling errors of ± 25 percent of the estimated variable at a probability level of 0.68.

Where water is supplied to developed recreational sites through a metered system, consumption may be closely related to hours of recreational use. In a study of a recreational site in Arizona, for example, it was found that water consumption was more highly correlated with the number of visitor-days than were axle counts. A comparison of the two indicator estimates is given in Table 17-1.

Where water meters are pretested to ensure that they accurately record low-water flow, they may have several advantages over such indicators as pneumatic traffic counters. Even though they may be more expensive to install, a single meter may suffice for certain sites where two or more traffic counters might be

TABLE 17-1
ESTIMATES OF RECREATIONAL USE ON AN ARIZONA CAMPGROUND*

Activity	Based on water meter		Based on axle count	
	Estimate	Sampling error†	Estimate	Sampling error†
	Visitor-days	Percent	Visitor-days	Percent
Camping	27,420	5.7	27,276	7.5
Spectator	485	21.3	408	24.2
Viewing scenery	303	18.7	266	19.9
Misc. activities	272	—	171	—
Total use	28,480	5.3	28,121	7.3

*From James and Tyre (1967).
†Probability level of 0.68.

needed. Also, water meters require little maintenance, they are less susceptible to vandalism, and they are unaffected by snow and ice. And finally, meters provide added information on site water requirements, including the need for pumps or sewage disposal facilities.

17-19 Assessing Potential Recreational Sites The preceding section was concerned with measurements of visitor use, i.e., the *output* of recreational planning and management. Here we are concerned with the inventory of natural resources with respect to their potential as *inputs* in the production of recreational use on developed or undeveloped areas.

Many of the measurement techniques discussed in other parts of this book are applicable to inventories of potential recreational sites. For example, the recreational planner will often require information on vegetative types, timber volumes, water resources, and wildlife populations. Although the assessment of these values as recreational input factors will differ from the assessment from other standpoints, the physical inventory requirements may be quite similar. Existing techniques are readily available for the physical inventory of recreational sites; the real problem is that of establishing inventory criteria, i.e., standards based on what potential visitors consider a recreational resource.

As a rule, inventory criteria employed in the past have been based on a combination of physical characteristics and administrative considerations. For example, the criteria for locating potential campgrounds would include some *physical* requirements regarding the amount of slope, the kind of vegetation, and the availability of water supplies. There would also be *administrative* criteria concerning the minimum site area and the maximum distance from population centers.

The inventory itself may be conducted in a manner similar to that of preparing a land-use or a vegetation type map. Standards are established to (1) discover potential sites, (2) designate selected areas for development, and (3) rank the proposed sites by desirability classes or assign them to various management categories. The sites that meet the original standards may then be shown on a map—a presentation that makes up the finished product of the inventory. Except for statistical summaries of site areas, carrying capacities, and demographic information, few calculations are involved.

Much of the recent work on systems for inventorying recreational resources has focused on the concept of a spectrum of recreational opportunities. Recreation opportunities can be expressed in terms of three principal components: the activities, the setting, and the experience (U.S. Forest Service, 1982). Mixes or combinations of activities, settings, and probable experience opportunities have been arranged along a spectrum, or continuum, called the *recreation opportunity spectrum* (ROS). The ROS approach is now commonly applied by agencies responsible for management of public lands.

PROBLEMS

17-1 Using aerial photographs of a local range area, locate the boundaries of a pasture and delineate the range sites. Then determine the area of each site with dot grids or a planimeter.

17-2 Design and participate in a drive census or a strip-flushing census for an important game species in your locality. Compare results with similar surveys conducted by public agencies. Describe the main problems encountered during your census, and suggest possible ways to improve the efficiency of the method.

17-3 Design and conduct some form of indirect census for a wildlife species, e.g., bird-call counts, track counts, or pellet-group counts. Prepare a written report on your findings.

17-4 What rare, unusual, or endangered species of wildlife occur in your locality? What steps are (or should be) taken to ensure their survival?

17-5 In a particular wildlife population, 50 animals were banded. During a period of trapping, 10 banded and 30 unbanded individuals were captured. Use the Lincoln index method to estimate the size of the population.

17-6 A particular watershed has a perimeter of 2950 m and an area of 64 ha. Compute the shape index for this watershed. What can you conclude about the shape of this watershed?

17-7 Obtain topographic maps of two watersheds in different portions of the United States. Delineate the watershed boundaries and classify streams on each watershed by stream order. Plot the number of streams versus stream order for each watershed on the same graph. Compute the drainage density for each watershed.

17-8 Assume that you have selected several sample days for interviewing families at a local campground or trailer park. If you wish to interview six family groups per day, *how* should these groups be selected? List several criteria to be observed that might improve the chances for an unbiased and representative sample.

REFERENCES

Arizona Inter-Agency Range Committee. 1972. *Proper use and management of grazing land.* 48 pp.

Black, P. E. 1990. *Watershed hydrology.* Prentice-Hall, Englewood Cliffs, N.J. 408 pp.

Brooks, K. N., Ffolliott, P. F., Gregersen, H. M., and Thames, J. L. 1991. *Hydrology and the management of watersheds.* Iowa State University Press, Ames, Iowa. 392 pp.

Brown, D. 1954. *Methods of surveying and measuring vegetation.* Bull. 42, Commonwealth Bureau of Pastures and Field Crops, Commonwealth Agricultural Bureau, Farnham Royal, Bucks, England. 223 pp.

Brown, G. W. 1989. *Forestry and water quality.* 2d ed. O.S.U. Bookstores, Inc., Corvallis, Oreg. 142 pp.

Buhyoff, G. J., Williams, S. B., and Klemperer, W. D. 1981. Gravity model formulation for an extensive national parkway site. *Environ. Manage.* **5**:253–262.

Burnham, K. P., Anderson, D. R., and Laake, J. L. 1980. Estimation of density from line transect sampling of biological populations. *Wildlife Monograph* No. 72. 202 pp.

Campbell, F. L. 1970. Participant observation in outdoor recreation. *J. Leisure Research* **2**:226–236.

Caughley, G. 1977. *Analysis of vertebrate populations*. John Wiley & Sons, New York. 234 pp.

Cook, C. W., and Stubbendieck, J. (eds.). 1986. *Range research: Basic problems and techniques*. Society for Range Management, Denver, Colo. 317 pp.

Cooperrider, A. Y., Boyd, R. J., and Stuart, H. R. (eds.). 1986. *Inventory and monitoring of wildlife habitat*. U.S.D.I. Bur. Land Management Service Center, Denver, Colo. 858 pp.

Driver, B. L., Brown, P. J., Stankey, G. H., and Gregoire, T. G. 1987. The ROS planning systems: Evaluation, basic concepts, and research needed. *Leisure Sci.* **9**:201–212.

Ffolliott, P. F., and Worley, D. P. 1965. An inventory system for multiple use evaluations. *U.S. Forest Serv., Rocky Mt. Forest and Range Expt. Sta., Res. Paper* RM-17. 15 pp.

Forest-Range Task Force. 1972. The nation's range resources—A forest-range environmental study. *Forest Resource Report* No. 19, U.S. Forest Service, Washington, D.C. 147 pp.

Golden, M. S., Tuttle, C. L., Kush, J. S., and Bradley, III, J. M. 1984. Forestry activities and water quality in Alabama: Effects, recommended practices, and an erosion-classification system. *Auburn Univ., Ala. Agr. Expt. Sta., Bull.* 555. 87 pp.

Hendee, J. C., and Lucas, R. C. 1973. Mandatory wilderness permits: A necessary management tool. *J. Forestry* **71**:206–209.

Hewlett, J. D. 1982. *Principles of forest hydrology*. University of Georgia Press, Athens, Ga. 183 pp.

———, and Douglass, J. E. 1968. Blending forest uses. *U.S. Forest Serv., Southeast. Forest Expt. Sta., Res. Paper* SE-37. 15 pp.

Holechek, J. L., Pieper, R. D., and Herbel, C. H. 1989. *Range management*. Prentice-Hall, Englewood-Cliffs, N.J. 501 pp.

James, G. A., and Henley, R. K. 1968. Sampling procedures for estimating mass and dispersed types of recreation use on large areas. *U.S. Forest Serv., Southeast. Forest Expt. Sta., Res. Paper* SE-31. 15 pp.

———, and Ripley, T. H. 1963. Instructions for using traffic counters to estimate recreation visits and use. *U.S. Forest Serv., Southeast. Forest Expt. Sta., Res. Paper* SE-3. 12 pp.

———, and Tyre, G. L. 1967. Use of water-meter records to estimate recreation visits and use on developed sites. *U.S. Forest Serv., Southeast. Forest Expt. Sta., Res. Note* SE-73. 3 pp.

Lee, R. 1980. *Forest hydrology*. Columbia University Press, New York. 349 pp.

Pollock, K. H., Nichols, J. D., Brownie, C., and Hines, J. E. 1990. Statistical inference for capture-recapture experiments. *Wildlife Monograph* No. 107. 97 pp.

Ralph, C. J., and Scott, J. M. (eds.). 1981. Estimating the numbers of terrestrial birds. *Studies in Avian Biology* No. 6. Cooper Ornithological Soc. 630 pp.

Range Term Glossary Committee. 1964. *A glossary of terms used in range management*. American Society of Range Management, Portland, Oreg. 32 pp.

Sampson, A. W. 1917. Succession as a factor in range management. *J. Forestry* **15**:593–596.

Satterlund, D. R. 1972. *Wildland watershed management*. John Wiley & Sons, New York. 370 pp.

Schemnitz, S. D. (ed.). 1980. *The wildlife management techniques manual*. The Wildlife Society, Washington, D.C. 686 pp.

Schroeder, H. W. 1983. Measuring visual features of recreational landscapes. Pp. 189–202 in *Recreation planning and management,* S. R. Lieber and D. R. Fesenmaier (eds.), Venture Publications, State College, Pa.

Seber, G. A. F. 1982. *The estimation of animal abundance and related parameters.* 2d ed. Charles Griffin & Co. Ltd., London, England. 653 pp.

Stoddart, L. A., Smith, A. D., and Box, T. W. 1975. *Range management,* 3d ed. McGraw-Hill Book Company, New York. 532 pp.

U.S. Forest Service. 1965. Range analysis field guide—Southwestern region. (mimeo. paper.) 153 pp.

———. 1982. *ROS user's guide.* Washington, D.C. 38 pp.

Wenger, K. F. (ed.). 1984. *Forestry handbook.* 2d ed. John Wiley & Sons, New York. 1335 pp.

Yoho, N. S. 1980. Forest management and sediment production in the South: A review. *So. J. Appl. For.* **4:**27–36.

APPENDIX TABLES

APPENDIX TABLE 1
SELECTED CONVERSIONS: ENGLISH TO METRIC UNITS

Length		
Multiply	by	to obtain
Inches	25.40	millimeters (mm)
Inches	2.54	centimeters (cm)
Feet	30.480	centimeters (cm)
Feet	0.3048	meters (m)
Yards	0.9144	meters (m)
Chains (Gunter's)	20.1168	meters (m)
U.S. statute miles	1.6093	kilometers (km)
Nautical miles	1.852	kilometers (km)

Area		
Square inches	645.16	square millimeters (mm^2)
Square inches	6.4516	square centimeters (cm^2)
Square feet	929.03	square centimeters (cm^2)
Square feet	0.0929	square meters (m^2)
Square yards	0.8361	square meters (m^2)
Square chains	404.6856	square meters (m^2)
Acres	0.4047	hectares (ha)
Square miles	2.5899	square kilometers (km^2)

Volume		
Cubic inches	16.387	cubic centimeters (cm^3)
Cubic feet	0.02832	cubic meters (m^3)
Cubic yards	0.7646	cubic meters (m^3)

Special conversions		
Square feet per acre	0.2296	square meters per hectare (m^2/ha)
Cubic feet per acre	0.06997	cubic meters per hectare (m^3/ha)
Cubic feet per second	101.941	cubic meters per hour (m^3/h)
Feet per second	1.097	kilometers per hour (km/h)
Gallons per acre	11.2336	liters per hectare (l/ha)
Gallons per minute	0.0757	liters per second (l/s)
Pounds per acre	1.1208	kilograms per hectare (kg/ha)
Pounds per cubic foot	16.0185	kilograms per cubic meter (kg/m^3)
Number (e.g., stems) per acre	2.471	number per hectare (no./ha)

APPENDIX TABLE 2
SCALE CONVERSIONS FOR MAPS AND VERTICAL PHOTOGRAPHS

Ratio scale	Feet per inch	Inches per 1000 feet	Inches per mile	Miles per inch	Meters per inch	Acres per square inch	Square inches per acre	Square miles per square inch
1: 500	41.667	24.00	126.72	0.008	12.700	0.0399	25.091	0.00006
1: 600	50.00	20.00	105.60	.009	15.240	.0574	17.424	.00009
1: 1,000	83.333	12.00	63.36	.016	25.400	.1594	6.273	.00025
1: 1,200	100.00	10.00	52.80	.019	30.480	.2296	4.356	.00036
1: 1,500	125.00	8.00	42.24	.024	38.100	.3587	2.788	.00056
1: 2,000	166.667	6.00	31.68	.032	50.800	.6377	1.568	.00100
1: 2,400	200.00	5.00	26.40	.038	60.960	.9183	1.089	.0014
1: 2,500	208.333	4.80	25.344	.039	63.500	.9964	1.004	.0016
1: 3,000	250.00	4.00	21.12	.047	76.200	1.4348	.697	.0022
1: 3,600	300.00	3.333	17.60	.057	91.440	2.0661	.484	.0032
1: 4,000	333.333	3.00	15.84	.063	101.600	2.5508	.392	.0040
1: 4,800	400.00	2.50	13.20	.076	121.920	3.6731	.272	.0057
1: 5,000	416.667	2.40	12.672	.079	127.000	3.9856	.251	.0062
1: 6,000	500.00	2.00	10.56	.095	152.400	5.7392	.174	.0090
1: 7,000	583.333	1.714	9.051	.110	177.800	7.8117	.128	.0122
1: 7,200	600.00	1.667	8.80	.114	182.880	8.2645	.121	.0129
1: 7,920	660.00	1.515	8.00	.125	201.168	10.00	.100	.0156
1: 8,000	666.667	1.500	7.92	.126	203.200	10.203	.098	.0159
1: 8,400	700.00	1.429	7.543	.133	213.360	11.249	.089	.0176
1: 9,000	750.00	1.333	7.041	.142	228.600	12.913	.077	.0202
1: 9,600	800.00	1.250	6.60	.152	243.840	14.692	.068	.0230
1: 10,000	833.333	1.200	6.336	.158	254.000	15.942	.063	.0249
1: 10,800	900.00	1.111	5.867	.170	274.321	18.595	.054	.0291
1: 12,000	1,000.00	1.0	5.280	.189	304.801	22.957	.044	.0359
1: 13,200	1,100.00	.909	4.800	.208	335.281	27.778	.036	.0434
1: 14,400	1,200.00	.833	4.400	.227	365.761	33.058	.030	.0517

1: 15,000	1,250.00	.80	4.224	.237	381.001	35.870	.028	.0560
1: 15,600	1,300.00	.769	4.062	.246	396.241	38.797	.026	.0606
1: 15,840	1,320.00	.758	4.00	.250	402.337	40.000	.025	.0625
1: 16,000	1,333.333	.750	3.96	.253	406.400	40.812	.024	.0638
1: 16,800	1,400.00	.714	3.771	.265	426.721	44.995	.022	.0703
1: 18,000	1,500.00	.667	3.52	.284	457.201	51.653	.019	.0807
1: 19,200	1,600.00	.625	3.30	.303	487.681	58.770	.017	.0918
1: 20,000	1,666.667	.60	3.168	.316	508.002	63.769	.016	.0996
1: 20,400	1,700.00	.588	3.106	.322	518.161	66.345	.015	.1037
1: 21,120	1,760.00	.568	3.00	.333	536.449	71.111	.014	.1111
1: 21,600	1,800.00	.556	2.933	.341	548.641	74.380	.013	.1162
1: 22,800	1,900.00	.526	2.779	.360	579.121	82.874	.012	.1295
1: 24,000	2,000.00	.50	2.640	.379	609.601	91.827	.011	.1435
1: 25,000	2,083.333	.480	2.534	.395	635.001	99.639	.010	.1557
1: 31,680	2,640.00	.379	2.000	.500	804.674	160.000	.006	.2500
1: 48,000	4,000.00	.250	1.320	.758	1,219.202	367.309	.003	.5739
1: 62,500	5,208.333	.192	1.014	.986	1,587.503	622.744	.0016	.9730
1: 63,360	5,280.00	.189	1.000	1.000	1,609.347	640.00	.0016	1.0000
1: 96,000	8,000.00	.125	.660	1.515	2,438.405	1,469.24	.0007	2.2957
1: 125,000	10,416.667	.096	.507	1.973	3,175.006	2,490.98	.0004	3.8922
1: 126,720	10,560.00	.095	.500	2.00	3,218.694	2,560.00	.0004	4.00
1: 250,000	20,833.333	.048	.253	3.946	6,350.012	9,963.907	.0001	15.5686
1: 253,440	21,120.00	.047	.250	4.00	6,437.389	10,244.202	.0001	16.00
1: 500,000	41,666.667	.024	.127	7.891	12,700.025	39,855.627	.000025	62.2744
1: 1,000,000	83,333.333	.012	.063	15.783	25,400.050	159,422.507	.0000062	249.0977

Source: Aerial-Photo Interpretation in Classifying and Mapping Soils, Agriculture Handbook 294, Soil Conservation Service, U.S. Department of Agriculture, Washington, D.C.,

APPENDIX TABLE 3
SCRIBNER DECIMAL C LOG RULE FOR LOGS 6 TO 32 FT IN LENGTH

Diameter, in.	Length, ft													
	6	8	10	12	14	16	18	20	22	24	26	28	30	32
	Contents, bd ft in tens													
6	0.5	0.5	1	1	1	2	2	2	3	3	3	4	4	5
7	0.5	1	1	2	2	3	3	3	4	4	4	5	5	6
8	1	1	2	2	2	3	3	3	4	4	5	6	6	7
9	1	2	3	3	3	4	4	4	5	6	6	7	8	9
10	2	3	3	3	4	6	6	7	8	9	9	10	11	12
11	2	3	4	4	5	7	8	8	9	10	11	12	13	14
12	3	4	5	6	7	8	9	10	11	12	13	14	15	16
13	4	5	6	7	8	10	11	12	13	15	16	17	18	19
14	4	6	7	9	10	11	13	14	16	17	19	20	21	23
15	5	7	9	11	12	14	16	18	20	21	23	25	27	28
16	6	8	10	12	14	16	18	20	22	24	26	28	30	32
17	7	9	12	14	16	18	21	23	25	28	30	32	35	37
18	8	11	13	16	19	21	24	27	29	32	35	37	40	43
19	9	12	15	18	21	24	27	30	33	36	39	42	45	48
20	11	14	17	21	24	28	31	35	38	42	45	49	52	56
21	12	15	19	23	27	30	34	38	42	46	49	53	57	61
22	13	17	21	25	29	33	38	42	46	50	54	58	63	67
23	14	19	23	28	33	38	42	47	52	57	61	66	71	75
24	15	21	25	30	35	40	45	50	55	61	66	71	76	81
25	17	23	29	34	40	46	52	57	63	69	75	80	86	92
26	19	25	31	37	44	50	56	62	69	75	82	88	94	100
27	21	27	34	41	48	55	62	68	75	82	89	96	103	110
28	22	29	36	44	51	58	65	73	80	87	95	102	109	116
29	23	31	38	46	53	61	68	76	84	91	99	107	114	122
30	25	33	41	49	57	66	74	82	90	99	107	115	123	131
31	27	36	44	53	62	71	80	89	98	106	115	124	133	142
32	28	37	46	55	64	74	83	92	101	110	120	129	138	147
33	29	39	49	59	69	78	88	98	108	118	127	137	147	157
34	30	40	50	60	70	80	90	100	110	120	130	140	150	160
35	33	44	55	66	77	88	98	109	120	131	142	153	164	175
36	35	46	58	69	81	92	104	115	127	138	150	161	173	185
37	39	51	64	77	90	103	116	129	142	154	167	180	193	206
38	40	54	67	80	93	107	120	133	147	160	174	187	200	214
39	42	56	70	84	98	112	126	140	154	168	182	196	210	224
40	45	60	75	90	105	120	135	150	166	181	196	211	226	241
41	48	64	79	95	111	127	143	159	175	191	207	223	238	254
42	50	67	84	101	117	134	151	168	185	201	218	235	252	269
43	52	70	87	105	122	140	157	174	192	209	227	244	262	279
44	56	74	93	111	129	148	166	185	204	222	241	259	278	296
45	57	76	95	114	133	152	171	190	209	228	247	266	286	304
46	59	79	99	119	139	159	178	198	218	238	258	278	297	317
47	62	83	104	124	145	166	186	207	228	248	269	290	310	331
48	65	86	108	130	151	173	194	216	238	260	281	302	324	346
49	67	90	112	135	157	180	202	225	247	270	292	314	337	359
50	70	94	117	140	164	187	211	234	257	281	304	328	351	374

APPENDIX TABLE 3 *(Continued)*

Diameter, in.	\multicolumn{14}{c}{Length, ft}													
	6	8	10	12	14	16	18	20	22	24	26	28	30	32
	\multicolumn{14}{c}{Contents, bd ft in tens}													
51	73	97	122	146	170	195	219	243	268	292	315	341	365	389
52	76	101	127	152	177	202	228	253	278	304	329	354	380	405
53	79	105	132	158	184	210	237	263	289	316	341	368	395	421
54	82	109	137	164	191	218	246	273	300	328	355	382	410	437
55	85	113	142	170	198	227	255	283	312	340	368	397	425	453
56	88	118	147	176	206	235	264	294	323	353	382	411	441	470
57	91	122	152	183	213	244	274	304	335	365	396	426	457	487
58	95	126	158	189	221	252	284	315	347	379	410	442	473	505
59	98	131	163	196	229	261	294	327	359	392	425	457	490	523
60	101	135	169	203	237	270	304	338	372	406	439	473	507	541
61	105	140	175	210	245	280	315	350	385	420	455	490	525	560
62	108	145	181	217	253	289	325	362	398	434	470	506	542	579
63	112	149	187	224	261	299	336	373	411	448	485	523	560	597
64	116	154	193	232	270	309	348	387	425	464	503	541	580	619
65	119	159	199	239	279	319	358	398	438	478	518	558	597	637
66	123	164	206	247	288	329	370	412	453	494	535	576	617	659
67	127	170	212	254	297	339	381	423	466	508	550	593	635	677
68	131	175	219	262	306	350	393	437	480	524	568	611	655	699
69	135	180	226	271	316	361	406	452	497	542	587	632	677	723
70	139	186	232	279	325	372	419	465	512	558	605	651	698	744
71	144	192	240	287	335	383	430	478	526	574	622	670	717	765
72	148	197	247	296	345	395	444	493	543	592	641	691	740	789
73	152	203	254	305	356	406	457	508	559	610	661	712	762	813
74	157	209	261	314	366	418	471	523	576	628	680	733	785	837
75	161	215	269	323	377	430	484	538	592	646	700	754	807	861
76	166	221	277	332	387	443	498	553	609	664	719	775	830	885
77	171	228	285	341	398	455	511	568	625	682	739	796	852	909
78	176	234	293	351	410	468	527	585	644	702	761	819	878	936
79	180	240	301	361	421	481	541	602	662	722	782	842	902	963
80	185	247	309	371	432	494	556	618	680	742	804	866	927	989
81	190	254	317	381	444	508	572	635	699	762	826	889	953	1016
82	196	261	326	391	456	521	586	652	717	782	847	912	977	1043
83	201	268	335	401	468	535	601	668	735	802	869	936	1002	1069
84	206	275	343	412	481	549	618	687	755	824	893	961	1030	1099
85	210	281	351	421	491	561	631	702	772	842	912	982	1052	1123
86	215	287	359	431	503	575	646	718	790	862	934	1006	1077	1149
87	221	295	368	442	516	589	663	737	810	884	958	1031	1105	1179
88	226	301	377	452	527	603	678	753	829	904	979	1055	1130	1205
89	231	308	385	462	539	616	693	770	847	924	1001	1078	1155	1232
90	236	315	393	472	551	629	708	787	865	944	1023	1101	1180	1259
91	241	322	402	483	563	644	725	805	886	966	1047	1127	1208	1288
92	246	329	411	493	575	657	739	822	904	986	1068	1150	1232	1315
93	251	335	419	503	587	671	754	838	922	1006	1090	1174	1257	1341
94	257	343	428	514	600	685	771	857	942	1028	1114	1199	1285	1371
95	262	350	437	525	612	700	788	875	963	1050	1138	1225	1313	1400

APPENDIX TABLE 3 *(Continued)*

Diameter, in.	Length, ft													
	6	8	10	12	14	16	18	20	22	24	26	28	30	32
	Contents, bd ft in tens													
96	268	357	446	536	625	715	804	893	983	1072	1161	1251	1340	1429
97	273	364	455	546	637	728	819	910	1001	1092	1183	1274	1365	1456
98	278	371	464	557	650	743	835	928	1021	1114	1207	1300	1392	1485
99	284	379	473	568	663	757	852	947	1041	1136	1231	1325	1420	1515
100	289	386	482	579	675	772	869	965	1062	1158	1255	1351	1448	1544
101	295	393	492	590	688	787	885	983	1082	1180	1278	1377	1475	1573
102	301	401	502	602	702	803	903	1003	1104	1204	1304	1405	1505	1605
103	307	409	512	614	716	819	921	1023	1126	1228	1330	1433	1535	1637
104	313	417	522	626	730	835	939	1043	1148	1252	1356	1461	1565	1669
105	319	425	532	638	744	851	957	1063	1170	1276	1382	1489	1595	1701
106	325	433	542	650	758	867	975	1083	1192	1300	1408	1517	1625	1733
107	331	442	553	663	773	884	995	1105	1216	1326	1437	1547	1658	1768
108	337	450	563	675	788	900	1013	1125	1238	1350	1463	1575	1688	1800
109	344	459	573	688	803	917	1032	1147	1261	1376	1491	1605	1720	1835
110	350	467	583	700	817	933	1050	1167	1283	1400	1517	1633	1750	1867
111	356	475	594	713	832	951	1069	1188	1307	1426	1545	1664	1782	1901
112	362	483	604	725	846	967	1087	1208	1329	1450	1571	1692	1812	1933
113	369	492	615	738	861	984	1107	1230	1353	1476	1599	1722	1845	1968
114	375	501	626	751	876	1001	1126	1252	1377	1502	1627	1752	1877	2003
115	382	509	637	764	891	1019	1146	1273	1401	1528	1655	1783	1910	2037
116	389	519	648	778	908	1037	1167	1297	1426	1556	1686	1815	1945	2075
117	396	528	660	792	924	1056	1188	1320	1452	1584	1716	1848	1980	2112
118	403	537	672	806	940	1075	1209	1343	1478	1612	1746	1881	2015	2149
119	410	547	683	820	957	1093	1230	1367	1503	1640	1777	1913	2050	2187
120	417	556	695	834	973	1112	1251	1390	1529	1668	1807	1946	2085	2224

Source: U.S. Forest Service.

APPENDIX TABLE 4
INTERNATIONAL LOG RULE, ¼-IN. SAW KERF, FOR LOGS 8 TO 20 FT IN LENGTH

Diameter (small end of log inside bark), in.	Length of log, ft							Diameter, in.
	8	10	12	14	16	18	20	
	Volume, bd ft							
4	....	5	5	5	5	5	10	4
5	5	5	10	10	10	15	15	5
6	10	10	15	15	20	25	25	6
7	10	15	20	25	30	35	40	7
8	15	20	25	35	40	45	50	8
9	20	30	35	45	50	60	70	9
10	30	35	45	55	65	75	85	10
11	35	45	55	70	80	95	105	11
12	45	55	70	85	95	110	125	12
13	55	70	85	100	115	135	150	13
14	65	80	100	115	135	155	175	14
15	75	95	115	135	160	180	205	15
16	85	110	130	155	180	205	235	16
17	95	125	150	180	205	235	265	17
18	110	140	170	200	230	265	300	18
19	125	155	190	225	260	300	335	19
20	135	175	210	250	290	330	370	20
21	155	195	235	280	320	365	410	21
22	170	215	260	305	355	405	455	22
23	185	235	285	335	390	445	495	23
24	205	255	310	370	425	485	545	24
25	220	280	340	400	460	525	590	25
26	240	305	370	435	500	570	640	26
27	260	330	400	470	540	615	690	27
28	280	355	430	510	585	665	745	28
29	305	385	465	545	630	715	800	29
30	325	410	495	585	675	765	860	30
31	350	440	530	625	720	820	915	31
32	375	470	570	670	770	875	980	32
33	400	500	605	715	820	930	1,045	33
34	425	535	645	760	875	990	1,110	34
35	450	565	685	805	925	1,050	1,175	35
36	475	600	725	855	980	1,115	1,245	36
37	505	635	770	905	1,040	1,175	1,315	37
38	535	670	810	955	1,095	1,245	1,390	38
39	565	710	855	1,005	1,155	1,310	1,465	39
40	595	750	900	1,060	1,220	1,380	1,540	40

APPENDIX TABLE 4 (*Continued*)

Diameter (small end of log inside bark), in.	Length of log, ft							Diameter, in.
	8	10	12	14	16	18	20	
	Volume, bd ft							
41	625	785	950	1,115	1,280	1,450	1,620	41
42	655	825	995	1,170	1,345	1,525	1,705	42
43	690	870	1,045	1,230	1,410	1,600	1,785	43
44	725	910	1,095	1,290	1,480	1,675	1,870	44
45	755	955	1,150	1,350	1,550	1,755	1,960	45
46	795	995	1,200	1,410	1,620	1,835	2,050	46
47	830	1,040	1,255	1,475	1,695	1,915	2,140	47
48	865	1,090	1,310	1,540	1,770	2,000	2,235	48
49	905	1,135	1,370	1,605	1,845	2,085	2,330	49
50	940	1,185	1,425	1,675	1,920	2,175	2,425	50
51	980	1,235	1,485	1,745	2,000	2,265	2,525	51
52	1,020	1,285	1,545	1,815	2,080	2,355	2,625	52
53	1,060	1,335	1,605	1,885	2,165	2,445	2,730	53
54	1,100	1,385	1,670	1,960	2,245	2,540	2,835	54
55	1,145	1,440	1,735	2,035	2,330	2,640	2,945	55
56	1,190	1,495	1,800	2,110	2,420	2,735	3,050	56
57	1,230	1,550	1,865	2,185	2,510	2,835	3,165	57
58	1,275	1,605	1,930	2,265	2,600	2,935	3,275	58
59	1,320	1,660	2,000	2,345	2,690	3,040	3,390	59
60	1,370	1,720	2,070	2,425	2,785	3,145	3,510	60

Source: U.S. Forest Service

APPENDIX TABLE 5
DOYLE LOG RULE FOR LOGS 6 TO 20 FT IN LENGTH

Diameter (small end of log inside bark), in.	Length of log, ft							
	6	8	10	12	14	16	18	20
	Volume, bd ft							
8	6	8	10	12	14	16	18	20
9	9	13	16	19	22	25	28	31
10	14	18	23	27	32	36	41	45
11	18	25	31	37	43	49	55	61
12	24	32	40	48	56	64	72	80
13	30	41	51	61	71	81	91	101
14	38	50	63	75	88	100	113	125
15	45	61	76	91	106	121	136	151
16	54	72	90	108	126	144	162	180
17	63	85	106	127	148	169	190	211
18	74	98	123	147	172	196	221	245
19	84	113	141	169	197	225	253	281
20	96	128	160	192	224	256	288	320
21	108	145	181	217	253	289	325	361
22	122	162	203	243	284	324	365	405
23	135	181	226	271	316	361	406	451
24	150	200	250	300	350	400	450	500
25	165	221	276	331	386	441	496	551
26	182	242	303	363	424	484	545	605
27	198	265	331	397	463	529	595	661
28	216	288	360	432	504	576	648	720
29	234	313	391	469	547	625	703	781
30	254	338	423	507	592	676	761	845
31	273	365	456	547	638	729	820	911
32	294	392	490	588	686	784	882	980
33	315	421	526	631	736	841	946	1,051
34	338	450	563	675	788	900	1,013	1,125
35	360	481	601	721	841	961	1,081	1,201
36	384	512	640	768	896	1,024	1,152	1,280
37	408	545	681	817	953	1,089	1,225	1,361
38	434	578	723	867	1,012	1,156	1,301	1,445
39	459	613	766	919	1,072	1,225	1,378	1,531
40	486	648	810	972	1,134	1,296	1,458	1,620

Source: U.S. Forest Service.

APPENDIX TABLE 6
THE DISTRIBUTION OF *t*

95% (handwritten)

					Probability				
df	0.5	0.4	0.3	0.2	0.1	0.05	0.02	0.01	0.001
1	1.000	1.376	1.963	3.078	6.314	12.706	31.821	63.657	636.619
2	0.816	1.061	1.386	1.886	2.920	4.303	6.965	9.925	31.598
3	0.765	0.978	1.250	1.638	2.353	3.182	4.541	5.841	12.941
4	0.741	0.941	1.190	1.533	2.132	2.776	3.747	4.604	8.610
5	0.727	0.920	1.156	1.476	2.015	2.571	3.365	4.032	6.859
6	0.718	0.906	1.134	1.440	1.943	2.447	3.143	3.707	5.959
7	0.711	0.896	1.119	1.415	1.895	2.365	2.998	3.499	5.405
8	0.706	0.889	1.108	1.397	1.860	2.306	2.896	3.355	5.041
9	0.703	0.883	1.100	1.383	1.833	2.262	2.821	3.250	4.781
10	0.700	0.879	1.093	1.372	1.812	2.228	2.764	3.169	4.587
11	0.697	0.876	1.088	1.363	1.796	2.201	2.718	3.106	4.437
12	0.695	0.873	1.083	1.356	1.782	2.179	2.681	3.055	4.318
13	0.694	0.870	1.079	1.350	1.771	2.160	2.650	3.012	4.221
14	0.692	0.868	1.076	1.345	1.761	2.145	2.624	2.977	4.140
15	0.691	0.866	1.074	1.341	1.753	2.131	2.602	2.947	4.073
16	0.690	0.865	1.071	1.337	1.746	2.120	2.583	2.921	4.015
17	0.689	0.863	1.069	1.333	1.740	2.110	2.567	2.898	3.965
18	0.688	0.862	1.067	1.330	1.734	2.101	2.552	2.878	3.922
19	0.688	0.861	1.066	1.328	1.729	2.093	2.539	2.861	3.883
20	0.687	0.860	1.064	1.325	1.725	2.086	2.528	2.845	3.850
21	0.686	0.859	1.063	1.323	1.721	2.080	2.518	2.831	3.819
22	0.686	0.858	1.061	1.321	1.717	2.074	2.508	2.819	3.792
23	0.685	0.858	1.060	1.319	1.714	2.069	2.500	2.807	3.767
24	0.685	0.857	1.059	1.318	1.711	2.064	2.492	2.797	3.745
25	0.684	0.856	1.058	1.316	1.708	2.060	2.485	2.787	3.725
26	0.684	0.856	1.058	1.315	1.706	2.056	2.479	2.779	3.707
27	0.684	0.855	1.057	1.314	1.703	2.052	2.473	2.771	3.690
28	0.683	0.855	1.056	1.313	1.701	2.048	2.467	2.763	3.674
29	0.683	0.854	1.055	1.311	1.699	2.045	2.462	2.756	3.659
30	0.683	0.854	1.055	1.310	1.697	2.042	2.457	2.750	3.646
40	0.681	0.851	1.050	1.303	1.684	2.021	2.423	2.704	3.551
60	0.679	0.848	1.046	1.296	1.671	2.000	2.390	2.660	3.460
120	0.677	0.845	1.041	1.289	1.658	1.980	2.358	2.617	3.373
∞	0.674	0.842	1.036	1.282	1.645	1.960	2.326	2.576	3.291

Source: This table is abridged from Table III of Fisher and Yates, "Statistical Tables for Biological, Agricultural, and Medical Research," published by Longman Group, Ltd., London (previously published by Oliver and Boyd, Edinburgh), and by permission of the authors and publishers.

5106	6521	7330	0064	0342	8376	3794	5226	5630	4639
5109	4365	1175	8043	0552	0109	1963	8058	4664	1046
6605	5140	6162	5301	2878	6123	9808	3114	8167	9569
7954	3300	0736	0116	7223	9616	0479	2600	0808	4399
0635	0975	6244	3885	3885	7600	4476	1079	4487	2894
7693	8898	5379	6291	9871	0856	9494	8963	1920	8409
0697	5711	8789	4862	2490	5880	8299	0900	7134	1145
3812	0447	0228	9987	3252	1178	3490	2145	7901	8146
3835	9073	1416	0746	0381	5707	7991	5730	1208	5624
6744	6923	8911	1915	8923	3686	6807	7062	8758	0445
9847	2209	8957	2180	2718	7975	2672	5854	7507	9686
7565	6940	2656	9836	1520	4019	8151	8997	5940	9373
2831	6736	3660	2894	8267	3581	3897	6492	0377	5172
2138	6670	3029	0203	4730	5756	2737	2678	1904	1243
3537	7843	2669	2790	9140	6871	4266	8227	7108	3481
2358	8095	1590	7508	8018	5833	7145	8927	1030	2419
0248	4620	9040	4894	5296	3774	6663	6856	9592	0205
6477	7864	6718	1057	7731	4015	2935	9428	8053	8680
3842	6597	4525	9405	5806	4568	7407	0158	3074	3564
7831	6156	4111	1943	4702	2627	3781	8013	2083	3608
8603	8276	4601	9016	3751	9715	0181	7216	6340	3096
9491	0826	2511	5656	7989	8832	9087	5137	8365	6479
2485	5821	0730	6042	8405	4543	5735	6409	4080	2659
7391	2742	7400	4011	1905	4348	3884	7257	7438	0911
9787	6104	5818	3266	3826	6367	0050	4895	1854	6814
7914	7925	1675	1327	6146	8498	0549	0758	9357	4528
4545	1012	1026	1568	2402	3238	6898	9141	1507	3283
6902	9340	8873	5173	1635	2207	1076	5048	8249	1615
9436	5049	1951	3116	7285	8381	2588	0932	5852	4177
3277	1524	0864	3500	8687	5594	0018	3146	4709	1853
4349	3943	3022	0011	9748	7430	6379	5381	4619	1021
8083	9890	7431	8239	8627	8572	7524	3313	5375	7964
1834	4041	2626	6707	9167	4754	6494	2060	9768	1187
3139	7084	9539	4824	5398	1345	0192	8587	5342	8815
8036	9923	8882	3870	5180	3796	3530	1500	7001	2732
1028	8813	5914	7967	5022	5706	8064	2971	2086	9686
5871	4857	9326	3953	4678	3468	0238	8851	7463	5951
6162	9300	1876	8044	5966	0054	8002	5159	4900	7532
5156	2167	8966	9018	4612	0267	8986	4829	3992	2683
3662	3661	4856	8588	0879	7279	5184	6173	3172	9959
7952	5863	5069	4285	9248	2924	4044	7962	3596	2736
9447	5086	8388	1739	2617	6069	4357	0458	8825	1598
3253	3208	5951	0534	4223	0799	2219	5173	8497	2379
2430	0986	7198	5921	0651	0879	9852	0971	3345	1923
7291	1147	2490	1461	1027	4246	0482	8286	3458	3061
2961	3230	3474	0266	6274	7704	2401	7898	2764	5172
2916	5584	4621	8444	4599	3566	7259	4792	8947	9416
0489	9446	8317	9611	3348	1207	3781	0837	0899	9107
7415	6205	7831	6455	0987	1066	2939	2702	5900	7205
7835	2788	2060	5129	5165	0851	4486	4097	3381	3342

4232	2291	6096	9588	4909	3970	0321	2927	0976	9060
0232	0553	6118	8473	5382	5604	5122	7643	2131	6916
0060	1455	6104	7205	2618	6538	9682	9839	7524	4804
2012	9153	6880	9953	1439	2034	3422	1883	2930	6426
0809	1349	7582	8816	2065	6693	8315	1879	1144	5355
0939	3409	1378	5741	1204	0691	6088	0756	1892	4240
0893	4650	4487	2871	5422	2293	6481	9696	6889	3049
4331	5921	6071	0083	9452	7508	1201	0653	4931	8162
2760	4919	7046	8364	6239	7545	3807	3338	5417	6593
7222	4156	4160	4492	9264	1872	8476	8749	8563	8986
0609	6968	2035	7842	6911	3410	2152	4860	0054	8954
2669	4013	6498	9935	9993	8010	0262	3276	1311	0665
5312	9279	2201	6086	8965	1938	4853	2685	2354	3812
5195	6222	3291	8079	4361	8608	6138	9132	3200	7390
7044	5124	7929	7821	9769	7732	3009	0068	0787	2245
1367	1081	4195	8323	3339	6349	7596	1300	4683	2621
3472	5020	2675	8024	1094	5783	6727	5717	3942	3376
9227	1518	8581	9474	7239	2016	9191	5839	7873	0756
4104	8180	0154	1917	8294	9445	8945	1465	9493	1125
8163	9217	0352	1693	7553	3710	0409	3327	5754	5501
4871	2763	8332	4654	0194	6332	8900	7126	2884	0646
7745	6673	8951	1201	4221	8890	5219	8764	4459	9852
0076	8803	1824	9954	8313	7131	6228	7901	9944	1502
9750	4687	6246	3601	5254	3336	0192	5318	8450	9542
5577	9502	0134	4160	0549	5028	1977	8012	1489	3709
7479	6841	8844	9864	2183	9192	5083	1082	7210	6825
2048	4657	4266	1286	6675	3410	7183	5954	4005	0931
3219	4613	5579	9916	3549	2634	5405	8978	8932	8100
4323	0080	3236	9041	9137	0527	6341	6228	7662	9647
6859	9073	9002	0971	6580	1906	0469	9730	2678	6857
1332	0411	1886	9938	8290	6498	3200	6607	2931	8453
1415	5595	9341	7806	1248	5860	0663	0401	0048	6774
2890	8534	1642	1993	4465	5194	4142	2188	6066	8750
7117	3646	6312	1277	7501	1846	3170	8378	6926	3597
6675	0991	7946	6696	5952	9717	7035	0640	8542	4826
2681	8167	5877	4173	7010	9985	6735	6535	3451	0230
0866	7119	8042	4648	6555	0938	9883	8758	5617	2163
9846	9723	5827	2812	3943	8811	7505	5101	7907	2998
0303	0491	9592	6618	5220	6793	2373	1488	2222	1513
0937	2662	9498	2127	1770	4264	9379	4904	4232	6071
1232	3295	1451	3715	6115	1625	6827	7782	2481	1409
9492	8024	3546	4445	6299	9946	7470	9994	3159	3091
7577	5748	3109	6903	1271	0302	5822	2033	1206	0514
8086	1336	9462	0062	6208	0972	4760	3750	4197	1526
9317	4149	9900	4568	4501	3389	2101	4461	9836	1014
7868	2035	9977	7349	9396	7563	1526	9165	6607	3374
0743	4378	0856	1042	5419	9104	3688	8834	9280	6515
1191	7844	2435	1522	5861	8829	9896	4540	3638	6301
8938	8655	1729	6938	4972	2946	3008	3770	7861	5844
1544	6587	1031	4704	4950	3438	5901	6820	7031	3800

APPENDIX TABLE 8
BASIC MATHEMATICAL OPERATIONS

Some constants

$\pi = 3.14159\ 26536$
$e = 2.71828\ 18285$

Common notation

$x = y$	x is equal to y	$Pr\{z\} = r$	the probability of the event z is r
$x \neq y$	x is not equal to y		
$x \doteq y$	x is approximately equal to y	$E[X]$	the expected value of the random variable X
$x \approx y$	x is approximately equal to y	$C(n,r)$	the number of combinations of n objects taken r at a time
$x \simeq y$	x is approximately equal to y	$P(n,r)$	the number of permutations of n objects taken r at a time
$x < y$	x is less than y	$A\|B$	A given B
$x \leq y$	x is less than or equal to y	$+\infty$	positive infinity
		$-\infty$	negative infinity
$x > y$	x is greater than y	$\lim_{x \to a^+} f(x)$	limit of $f(x)$ as x approaches a from the right
$x \geq y$	x is greater than or equal to y	$\lim_{x \to a^-} f(x)$	limit of $f(x)$ as x approaches a from the left
$y = f(x_1, x_2)$	y is a function of the variables x_1 and x_2	$\int_a^b f(x)\,dx$	the integral of $f(x)$ from a to b
$\|-3\| = 3$	the absolute value of -3 is 3		

$$\sqrt[n]{x} = x^{1/n} \qquad e^x = \exp\{x\}$$

$$\log_e x = \ln x \qquad f'(x) = \frac{df(x)}{dx}$$

$$\log_{10} x = \log x \qquad x^{-1} = 1/x$$

Powers and roots

$$a^x \times a^y = a^{(x+y)} \qquad a^0 = 1\,[\text{if } a \neq 0] \qquad (ab)^x = a^x b^x.$$

$$\frac{a^x}{a^y} = a^{(x-y)} \qquad a^{-x} = \frac{1}{a^x} \qquad \left(\frac{a}{b}\right)^x = \frac{a^x}{b^x}$$

$$\left(a^x\right)^y = a^{xy} \qquad a^{1/x} = \sqrt[x]{a} \qquad \sqrt[x]{ab} = \sqrt[x]{a}\sqrt[x]{b}$$

$$\sqrt[x]{\sqrt[y]{a}} = \sqrt[xy]{a} \qquad a^{x/y} = \sqrt[y]{a^x} \qquad \sqrt[x]{\frac{a}{b}} = \frac{\sqrt[x]{a}}{\sqrt[x]{b}}$$

Logarithms

$\log_a a^x = x,\ a^x > 0$

$\log_a xy = \log_a x + \log_a y$

$\log_a b^x = x \log_a b$

$\log_a(x/y) = \log_a x - \log_a y$

$\log_a 1 = 0$

$\log_a a = 1$

Note that logarithms are not defined for negative quantities

$\log_{10} \pi = 0.497\ 149\ 873$

$\log_e \pi = 1.144\ 729\ 886$

Change of base

$\log_a x = \log_b x / \log_b a$

$\log_{10} x = \log_e x / \log_e 10$

$\log_e x = \log_{10} x / \log_{10} e$

$\log_e x = 2.302\ 585\ 093\ \log_{10} x$

$\log_{10} x = 0.434\ 294\ 482\ \log_e x$

Summation relationships

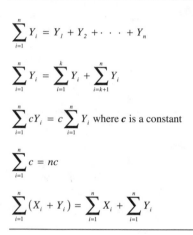

$$\sum_{i=1}^{n} Y_i = Y_1 + Y_2 + \cdots + Y_n$$

$$\sum_{i=1}^{n} Y_i = \sum_{i=1}^{k} Y_i + \sum_{i=k+1}^{n} Y_i$$

$$\sum_{i=1}^{n} cY_i = c\sum_{i=1}^{n} Y_i \text{ where } c \text{ is a constant}$$

$$\sum_{i=1}^{n} c = nc$$

$$\sum_{i=1}^{n} (X_i + Y_i) = \sum_{i=1}^{n} X_i + \sum_{i=1}^{n} Y_i$$

Solution of polynomial equations

Linear or first degree

$$ax + b = 0$$

$$x = -\frac{b}{a}$$

Quadratic or second degree.   Any quadratic equation may be reduced to the form,

$$ax^2 + bx + c = 0$$

$$x = \frac{-b \pm \sqrt{b^2 - 4ac}}{2a}$$

If a, b, and c are real then:

If $b^2 - 4ac$ is positive, the roots are real and unequal.

If $b^2 - 4ac$ is zero, the roots are real and equal.

If $b^2 - 4ac$ is negative, the roots are imaginary and unequal.

APPENDIX TABLE 8 (*Continued*)

Calculus

Derivatives. In the following formulas, u and v represent functions of x, while a, c, n represent fixed real constants and $f(u)$ is a function of x.

$$\frac{d}{dx}(a) = 0$$

$$\frac{d}{dx}(x) = 1$$

$$\frac{d}{dx}(au) = a\frac{du}{dx}$$

$$\frac{d}{dx}(u + v) = \frac{du}{dx} + \frac{dv}{dx}$$

$$\frac{d}{dx}(uv) = u\frac{dv}{dx} + v\frac{du}{dx}$$

$$\frac{d}{dx}\left(\frac{u}{v}\right) = \frac{v\frac{du}{dx} - u\frac{dv}{dx}}{v^2} = \frac{1}{v}\frac{du}{dx} - \frac{u}{v^2}\frac{dv}{dx}$$

$$\frac{d}{dx}(u^n) = nu^{n-1}\frac{du}{dx}$$

$$\frac{d}{dx}[f(u)] = \frac{d}{du}[f(u)] \cdot \frac{du}{dx}$$

$$\frac{d}{dx}(\log_a u) = (\log_a e)\frac{1}{u}\frac{du}{dx}$$

$$\frac{d}{dx}(\log_e u) = \frac{1}{u}\frac{du}{dx}$$

$$\frac{d}{dx}(a^u) = a^u(\log_e a)\frac{du}{dx}$$

$$\frac{d}{dx}(e^u) = e^u\frac{du}{dx}$$

$$\frac{d}{dx}(u^v) = vu^{v-1}\frac{du}{dx} + (\log_e u)u^v\frac{dv}{dx}$$

Integrals. In the following expressions $\ln x$ is the logarithm to the base e; u and v are variables that depend on x; a and n are constants; and the arbitrary integration constants are omitted for simplicity.

$$\int a\, dx = ax$$

$$\int au\, dx = a\int u\, dx$$

$$\int (u + v)dx = \int u\, dx + \int v\, dx$$

$$\int x^n\, dx = \frac{x^{n+1}}{n + 1} \qquad n \neq -1$$

$$\int \frac{dx}{x} = \ln x$$

$$\int e^x\, dx = e^x$$

$$\int e^{ax}\, dx = e^{ax}/a$$

$$\int b^{ax}\, dx = \frac{b^{ax}}{a\ln b}$$

$$\int \ln x\, dx = x\ln x - x$$

Source: Abridged and adapted from Section 25, "Mathematics and Statistics," of the *Forestry Handbook,* 2d ed. (K. F. Wenger, ed.), 1984, published by John Wiley & Sons, New York.

TRIGONOMETRIC AND AREA FORMULAS

Solution of right triangles

Sine $A = \dfrac{a}{c} = \dfrac{\text{opposite side}}{\text{hypotenuse}}$

Cosine $A = \dfrac{b}{c} = \dfrac{\text{adjacent side}}{\text{hypotenuse}}$

Tangent $A = \dfrac{a}{b} = \dfrac{\text{opposite side}}{\text{adjacent side}}$

Cotangent $A = \dfrac{b}{a} = \dfrac{\text{adjacent side}}{\text{opposite side}}$

Secant $A = \dfrac{c}{b} = \dfrac{\text{hypotenuse}}{\text{adjacent side}}$

Cosecant $A = \dfrac{c}{a} = \dfrac{\text{hypotenuse}}{\text{opposite side}}$

Areas of some plane figures

Figure	Formula for Area (A)	Diagram
Rectangle	$A = a\,b$	
Parallelogram	$A = a\,h$	
Triangle	$A = \dfrac{a\,h}{2}$	
Trapezoid	$A = \frac{1}{2}(a + b)\,h$	
Circle	$A = \pi\,r^2$	
Ellipse	$A = \pi\,a\,b$	
Parabola	$A = \left(\frac{2}{3}\right) a\,b$	

APPENDIX TABLE 10
EQUATIONS, LINEAR MODELS, AND GRAPHICAL REPRESENTATIONS OF SELECTED
FUNCTIONS AND CURVE FORMS

I. $Y = a + bX$ —— Straight line
Linear model: $Y = b_0 + b_1 X$

II. $(Y - a) = k(X - b)^2$ —— Second-degree parabola
Linear model: $Y = b_0 + b_1 X + b_2 X^2$

III. $(Y - a) = k/X$ —— Hyperbola
Linear model: $Y = b_0 + b_1\left(\dfrac{1}{X}\right)$

IV. $(Y - a) = k/X + bX$
Linear model: $Y = b_0 + b_1 \, 1/X + b_2 X$

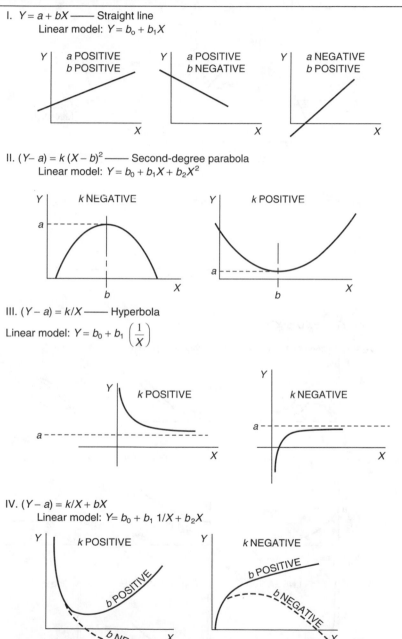

APPENDIX TABLE 10 (*Continued*)

V. $(Y - a) = k (X - b) (X - c) [X - (b + c) /2]$ ——— Cubic
   Linear model: $Y = b_0 + b_1 X + b_2 X^2 + b_3 X^3$

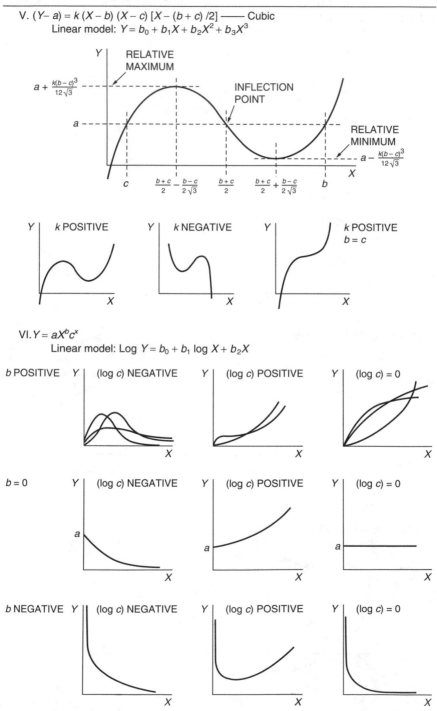

APPENDIX TABLE 10 (*Continued*)

VII. $Y = ab^{(x-c)^2}$
   Linear model: $\log Y = b_0 + b_1X + b_2X^2$

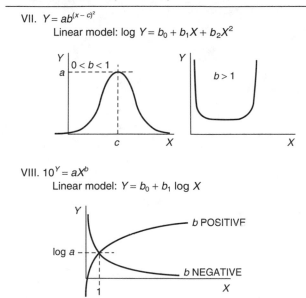

VIII. $10^Y = aX^b$
   Linear model: $Y = b_0 + b_1 \log X$

Source: Freese, Frank. 1964. Linear regression methods for forest research. *U.S. Forest Serv., Forest Products Lab., Res. Paper* FPL 17. 136 pp.

INDEX